Semiconductor Wafer Bonding: Science, Technology and Applications 14

Editors:

T. Suga

H. Baumgart

F. Fournel

M. S. Goorsky

K. D. Hobart

R. Knechtel

C. Seng Tan

Sponsoring Divisions:

Electronics and Photonics

Battery

Published by
The Electrochemical Society
65 South Main Street, Building D
Pennington, NJ 08534-2839, USA
tel   609 737 1902
fax   609 737 2743
www.electrochem.org

ecstransactions ™

Vol. 75, No. 9

Copyright 2016 by The Electrochemical Society.
All rights reserved.

This book has been registered with Copyright Clearance Center.
For further information, please contact the Copyright Clearance Center,
Salem, Massachusetts.

Published by:

The Electrochemical Society
65 South Main Street
Pennington, New Jersey 08534-2839, USA

Telephone 609.737.1902
Fax 609.737.2743
e-mail: ecs@electrochem.org
Web: www.electrochem.org

ISSN 1938-6737 (online)
ISSN 1938-5862 (print)
ISSN 2151-2051 (cd-rom)

ISBN 978-1-62332-368-4 (CD-ROM)
ISBN 978-1-60768-726-9 (PDF)

Printed in the United States of America.

Preface

The papers included in this issue of *ECS Transactions* were originally presented in the symposium "Semiconductor Wafer Bonding: Science, Technology and Applications 14", held during the PRiME 2016 joint international meeting of The Electrochemical Society, The Electrochemical Society of Japan, and the Korean Electrochemical Society, with the technical co-sponsorship of the Chinese Society of Electrochemistry, the Electrochemistry Division of the Royal Australian Chemical Institute, the Japan Society of Applied Physics, the Chinese Physical Society Semiconductor Division, and the Semiconductor Physics Division of the Chinese Physics Society. This meeting was held in Honolulu, Hawaii, from October 2-7, 2016.

 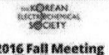

2016 Fall Meeting of
The Electrochemical
Society of Japan

230th Meeting of
The Electrochemical
Society

2016 Fall Meeting
The Korean
Electrochemical Society

technical co-sponsors:

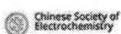

Chinese Society of
Electrochemistry

Electrochemistry Division of the
Royal Australian Chemical Institute

The Japan Society of
Applied Physics

Korean Physical Society Semiconductor Division   Semiconductor Physics Division of Chinese Physics Society

ECS Transactions, Volume 75, Issue 9
Semiconductor Wafer Bonding: Science, Technology and Applications 14

Table of Contents

Preface   *iii*

Chapter 1
Low Temperature Bonding by Surface Activation

(Invited) Surface Activated Wafer Bonding; Principle and Current Status   3
  H. Takagi, Y. Kurashima, T. Suga

Surface Activation and Planarization with Gas Cluster Ion Beam for Wafer Bonding   9
  N. Toyoda, T. Sasaki, I. Yamada, T. Suga

Low-Temperature Aluminum-Aluminum Wafer Bonding   15
  B. Rebhan, A. Hinterreiter, N. Malik, K. Schjølberg-Henriksen, V. Dragoi,
  K. Hingerl

Ultra-Thick Metal Ohmic Contact Fabrication Using Surface Activated Bonding   25
  J. Liang, K. Furuna, M. Matsubara, M. Dhamrin, Y. Nishio, N. Shigekawa

(Invited) Analysis of Defect Levels at GaAs/GaAs Surface-Activated Bonding   33
Interface for Multi-Junction Solar Cells
  M. Sugiyama, D. Yamashita, K. Watanabe, M. Fujino, T. Suga, Y. Nakano

The Roles of Band Bending, Surface Misorientation, and Passivation on Electrical   39
Transport Across III-V Bonded Structures
  M. Yee, M. Liao, M. Seal, M. S. Goorsky

Conductive Semiconductor Interfaces Fabricated by Room Temperature Covalent   45
Wafer Bonding
  C. Flötgen, N. Razek, V. Dragoi, M. Wimplinger

(Invited) High Output Power Deep Ultraviolet Light-Emitting Diodes with   53
Hemispherical Lenses Fabricated Using Room Temperature Bonding
M. Ichikawa, S. Endo, H. Sagawa, A. Fujioka, T. Kosugi, T. Mukai, M. Uomoto,
T. Shimatsu

Necessary Thickness of Au Capping Layers for Room Temperature Bonding of   67
Wafers in Air Using Thin Metal Films with Au Capping Layers
M. Uomoto, T. Shimatsu

Direct Wafer Bonding of SiC-SiC at Room Temperature by SAB Method   77
F. Mu, K. Iguchi, H. Nakazawa, Y. Takahashi, M. Fujino, T. Suga

Chapter 2
Heterogeneous & Photonic Integration

(Invited) Diverse Accessible Heterogeneous Integration (DAHI) Foundry at Northrop   87
Grumman Aerospace Systems (NGAS)
A. Gutierrez-Aitken, D. Scott, K. Sato, B. Poust, E. Nakamura, K. Thai, W. Chan,
E. Kaneshiro, C. Monier, I. Smorchkova, N. Lin, D. Ferizovic, X. Zeng, A. Oki,
R. Kagiwada

Suppressed Self-Heating in Multi-Finger InP-Based DHBTs with Au Subcollector   97
Fabricated on SiC Substrate by Surface-Activated Bonding
Y. Shiratori, T. Hoshi, N. Kashio, K. Kurishima, E. Higurashi, H. Matsuzaki

Au/SiO$_2$ Hybrid Bonding with 6-µm-Pitch Au Electrodes for 3D Structured Image   103
Sensors
Y. Honda, K. Hagiwara, M. Goto, T. Watabe, M. Nanba, Y. Iguchi, T. Saraya,
M. Kobayashi, H. Toshiyoshi, E. Higurashi, T. Hiramoto

Chapter 3
Hybrid Bonding by Surface Activation

Cu-Cu Die to Die Surface Activated Bonding in Atmospheric Environment Using Ar   109
and Ar/N$_2$ Plasma
S. L. Chua, C. S. Tan

Combined Surface Activated Bonding Technique for Hydrophilic SiO$_2$-SiO$_2$ and   117
Cu-Cu Bonding
R. He, M. Fujino, A. Yamauchi, T. Suga

Impact of Water Edge Absorption on Silicon Oxide Direct Bonding Energy 129
F. Fournel, M. Tedjini, V. Larrey, F. Rieutord, C. Morales, C. Bridoux,
H. Moriceau

Chapter 4
Bonding Mechanism & Photonic Integration

Control of Direct Bonding Behavior by Interlayers 137
M. Eichler, H. Dillmann, K. Nagel, C. P. Klages

Adhesion Energy and Bonding Wave Velocity Measurements 145
V. Larrey, G. Mauguen, F. Fournel, D. Radisson, F. Rieutord, C. Morales,
C. Bridoux, H. Moriceau

A Study of Void Formation in Fluorine Containing Plasma Activated Wafer Bonding 153
C. Wang, Y. Liu, T. Suga

Edge Water Penetration in Direct Bonding Interface 163
F. Rieutord, S. Tardif, D. Landru, O. Kononchuk, V. Larrey, H. Moriceau,
M. Tedjini, F. Fournel

(Invited) Locally Measuring the Adhesion of InP Membranes Directly Bonded on 169
Silicon
G. Patriarche, K. Pantzas, E. Le Bourhis, G. Beaudoin, A. Talneau

Chapter 5
Photonic Integration & Layer Transfer Technologies

(Invited) Heterogeneous Photonic Integration by Direct Wafer Bonding 179
M. L. Davenport, L. Chang, D. Huang, N. Volet, J. E. Bowers

Modified Surface Activated Bonding Using Si Intermediate Layer for Bonding and 185
Debonding of Glass Substrates
K. Takeuchi, M. Fujino, T. Suga

Nanomechanical Analysis of Polydimethylglutarimide Based Lift Off Resist Used for 191
Temporary Bonding and Film Transfers
Y. Mohammed, T. Matsumae, A. D. Koehler, T. Suga, H. Baumgart, K. D. Hobart,
A. A. Elmustafa

vii

Room Temperature Bonding with Lift-Off Resist Using the Surface Activated Bonding Method for a Layer Transfer Platform   197
  T. Matsumae, T. Suga

Thin Layer Transfer Using Room Temperature Wafer-Level Bonding Process   203
  K. Abadie, F. Fournel, C. Morales, H. Moriceau, M. Wimplinger

Chapter 6
Poster Session

Optical Isolator with Si Guiding Layer Fabricated by Photosensitive Adhesive Bonding   215
  H. Yokoi, S. Choowitsakunlert, K. Kobayashi, K. Takagiwa

Determination of Band Structure at GaAs/4H-SiC Heterojunctions   221
  J. Liang, S. Shimizu, M. Arai, N. Shigekawa

Surface Preparation and Eutectic Wafer Bonding   229
  M. Heller, M. Zoberbier, T. Fujita, M. Eichler

Decreased Surface Porosity and Roughness of InP for Epitaxially Grown Thin-Film Devices: A Path to Integration of High Performance Electronics   241
  M. Gervasoni, A. Machness, M. Goorsky

Transfer of Ultra-Thin Semi-Conductor Films onto Flexible Substrates   247
  P. Montméat, I. De Nigris Brandolisi, S. Tardif, T. Enot, G. Enyedi, R. Kachtouli, P. Besson, F. Rieutord, F. Fournel

Chapter 7
MEMS Integration

Glass Frit Wafer Bonding - Sealed Cavity Pressure in Relation to Bonding Process Parameters   255
  R. Knechtel, S. Dempwolf, H. Klingner

Wafer-Level Hermetic Seal Bonding at Low-Temperature with Sub-Micron Gold Particle Using Stencil Printing   265
  H. Ishida, T. Ogashiwa

Aluminum-Germanium Eutectic Bonding for MEMS: Behaviour and Solidification of   273
Liquid Al-Ge on Different Substrates
  V. Lumineau, F. Fournel, B. Imbert, F. Hodaj

Chapter 8
Bonding for 3D-Integration

(Invited) Self-Assembly Based Multichip-to-Wafer Bonding Technologies for   285
3D/Hetero Integration
  T. Fukushima, K. Lee, T. Tanaka, M. Koyanagi

Wafer-Level Vacuum Packaging by Thermocompression Bonding Using Silver after   291
Fly-Cut Planarization
  C. Liu, H. Hirano, J. Froemel, S. Tanaka

Low Temperature Thermo Compression Bonding with Printed Intermediate Bonding   299
Layers
  M. Wiemer, F. Roscher, T. Seifert, K. Vogel, T. Ogashiwa, T. Gessner

Plastic Deformation of Thin Si Membranes in Si-Si Direct Bonding   311
  E. Poppe, G. U. Jensen, S. T. Moe, D. Wang

Surface Protection for Semiconductor Direct Bonding   321
  R. Knechtel, H. Klingner

Chapter 9
Equipment & Applications

Direct Bonding of Multiple Curved, Wedged and Structured Silicon Wafers as X-Ray   331
Mirrors
  B. Landgraf Sr., R. Gunther Sr., G. Vacanti, N. Barriere, M. Vervest, D. Girou,
  A. Yanson, M. Collon Sr.

High Efficiency Cleaning Processes for Direct Wafer Bonding   339
  D. Dussault, J. Rothballer, F. Kurz, M. Reichardt, V. Dragoi

High Precision Low Temperature Direct Wafer Bonding Technology for Wafer-Level   345
3D ICs Manufacturing
  F. Kurz, T. Plach, J. Süss, T. Wagenleitner, D. Zinner, B. Rebhan, V. Dragoi

ix

Bonding of SiO_2 and SiO_2 at Room Temperature Using Si Ultrathin Film 355
J. Utsumi, K. Ide, Y. Ichiyanagi

Author Index 363

Facts about ECS

The Electrochemical Society (ECS) is an international, nonprofit, scientific, educational organization founded for the advancement of the theory and practice of electrochemistry, electronics, and allied subjects. The Society was founded in Philadelphia in 1902 and incorporated in 1930. There are currently over 7,000 scientists and engineers from more than 70 countries who hold individual membership; the Society is also supported by more than 100 corporations through Corporate Memberships.

The technical activities of the Society are carried on by Divisions. Sections of the Society have been organized in a number of cities and regions. Major international meetings of the Society are held in the spring and fall of each year. At these meetings, the Divisions and Groups hold general sessions and sponsor symposia on specialized subjects.

The Society has an active publication program that includes the following:

Journal of The Electrochemical Society — (JES) is the leader in the field of electrochemical science and technology. This peer-reviewed journal publishes an average of 550 pages of 85 articles each month. Articles are published online as soon as possible after undergoing the peer-review process. The online version is considered the final version and is fully citable with articles assigned specific page numbers within specific issues. The date of online publication is the official publication date of record.

Journal of Solid State Science and Technology — (JSS) is one of the newest peer-reviewed journals from ECS launched in 2012. JSS covers fundamental and applied areas of solid state science and technology including experimental and theoretical aspects of the chemistry and physics of materials and devices. Articles are published online as soon as possible after undergoing the peer-review process. The online version is considered the final version and is fully citable with articles assigned specific page numbers within specific issues. The date of online publication is the official publication date of record.

Electrochemistry Letters — (EEL) is one of the newest journals from ECS launched in 2012. It is dedicated to the rapid dissemination of peer-reviewed and concise research reports in fundamental and applied areas of electrochemical science and technology. Articles are published online as soon as possible after undergoing the peer-review process. The online version is considered the final version and is fully citable with articles assigned specific page numbers within specific issues. The date of online publication is the official publication date of record.

Solid State Letters — (SSL) is one of the newest journals from ECS launched in 2012. It is dedicated to the rapid dissemination of peer-reviewed and concise research reports in fundamental and applied areas of solid state science and technology. Articles are published online as soon as possible after undergoing the peer-review process. The online version is considered the final version and is fully citable with articles assigned specific page numbers within specific issues. The date of online publication is the official publication date of record.

Electrochemical and Solid-State Letters — (ESL) was the first rapid-publication electronic journal dedicated to covering the leading edge of research and development in the field of solid-state and electrochemical science and technology. ESL was a joint publication of ECS and IEEE Electron Devices Society. Volume 1 began July 1998 and contained six issues, thereafter new volumes began with the January issue and contained 12 issues. The final issue of ESL was Volume 16, Number 6, 2012. Preserved as an archive, ESL has since been replaced by SSL and EEL.

Interface— *Interface* is an authoritative yet accessible publication for those in the field of solid-state and electrochemical science and technology. Published quarterly, this four-color magazine contains technical articles about the latest developments in the field, and presents news and information about and for members of ECS.

ECS Meeting Abstracts— *ECS Meeting Abstracts* contain extended abstracts of the technical papers presented at the ECS biannual meetings and ECS-sponsored meetings. This publication offers a first look into the current research in the field. ECS Meeting Abstracts are freely available to all visitors to the ECS Digital Library.

ECS Transactions— (ECST) is the online database containing full-text content of proceedings from ECS meetings and ECS-sponsored meetings. ECST is a high-quality venue for authors and an excellent resource for researchers. The papers appearing in ECST are reviewed to ensure that submissions meet generally-accepted scientific standards. Each meeting is represented by a volume and each symposium by an issue.

Monograph Volumes — The Society sponsors the publication of hardbound monograph volumes, which provide authoritative accounts of specific topics in electrochemistry, solid-state science, and related disciplines.

For more information on these and other Society activities, visit the ECS website:

www.electrochem.org

Chapter 1

Low Temperature Bonding by Surface Activation

2

Surface Activated Wafer Bonding; Principle and Current Status

H. Takagi[a], Y. Kurashima[a], and T. Suga[b]

[a] Research Center for Ubiquitous MEMS and Micro Engineering, National Institute of Advanced Industrial Science and Technology (AIST), Tsukuba, 305-8564, Japan
[b] School of Engineering, The University of Tokyo, Tokyo, 113-8656, Japan

The surface activated bonding (SAB) is a method which achieves strong bonding at room temperature. It is assumed that at the bonding interface prepared by SAB atoms bind to each other by the same interatomic bonds with those in bulk materials, such as covalent bond, metallic bond, etc. SAB well matches the wafer direct bonding, because intimate contact at the bonding interface is spontaneously achieved by the attractive force between very smooth wafer surfaces. In SAB, the surfaces of bonded materials are sputter-etched, in other word "cleaned", before bonding. Various materials, such as semiconductor wafers, single crystal oxide wafers, metal films on wafers, etc. have been successfully bonded. Recently, other surface treatments have been proposed to apply SAB to wide variety of materials. SAB has been used to fabricate integrated substrates for RF-filters for more than 10 years, and has been already applied in MEMS packaging in industry. Other applications, such as power semiconductors, solar cells, etc. are now extensively developed.

Introduction

The surface activated bonding (SAB) originally based on a simple idea that atoms on two clean surfaces can make strong interatomic bonds even at room-temperature when they are mated. Actually it was defined in a very general expression as "the bonding at atomically clean material surfaces obtained by energetic particles bombardment" in the original paper (1). The idea is inspired from the studies on adhesion phenomena in UHV, especially in developments of space technologies in -1970s. SAB originally uses cleaning of material surfaces by sputter etching using high energy ion/atom beam of inert gases, typically Ar, as shown in figure 1. The cleaning process removes adsorbed atoms and compound layers, typically oxides, which stabilize the surface. Therefore, after the cleaning process the surfaces become unstable and/or "active" states. Mating two such activated surfaces in vacuum enables strong bond formation at room temperature.

SAB was first applied to the bonding between soft metals. Deformation of the soft metal by load application was necessary to achieve atomic-level intimate contact at room temperature between relatively rough surfaces without CMP (2,3). Figure 2 shows a high resolution TEM image of Al-Al bonding interface prepared by SAB at room temperature. Direct bonding between metal atoms achieved at the bonding interface (4). This method was successfully applied to bonding between other soft metals such as Cu, Au, Sn, etc., metal/ceramics bonding such as Al/Si_3N_4, and metal/semiconductor bonding such as

Al/Si. This method was also applied to bonding of 0.5 μm thick Al films deposited on a wafer (5).

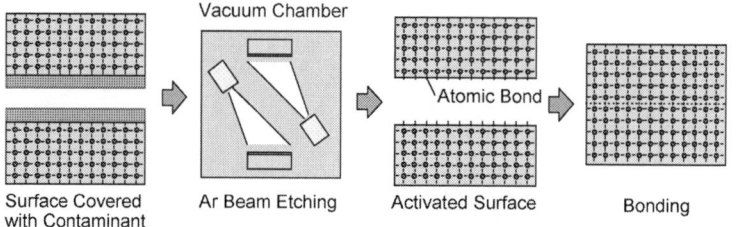

Figure 1. SAB using sputter etching by high-energy Ar ion/atom beam

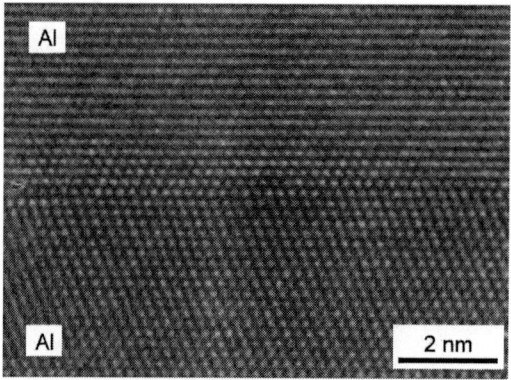

Figure 2. High resolution TEM image of Al-Al interface prepared by SAB.

Wafer Direct Bonding by SAB

In wafer direct bonding process, intimate contact between two wafer surfaces is achieved spontaneously in case that the wafer surfaces are well polished in atomic-level (6). The spontaneous bonding is assumed to be achieved by the attractive force between two surfaces. This property is quite advantageous for intimate contact formation at room temperature. Therefore, wafer direct bonding fits well to SAB. Si wafers were successfully bonded by SAB without applying any heat treatment and mechanical load (7-10). Strong bonding equivalent to bulk strength is achieved by the room-temperature process as shown in figure 3.

This method has been successfully applied to various semiconductor wafers such as GaAs, SiC, etc (11-13). Atomically smooth surfaces of metals can be also bonded by SAB. Cu layers finished by CMP was bonded by SAB, and applied to high density inter-connection of ICs (14,15). Thin metal films deposited on well-polished wafers are also bonded by SAB. Figure 4 shows bonding interface of 10 nm thick Ta films by SAB. Although the bonding interface looks slightly brighter, no voids are observed.

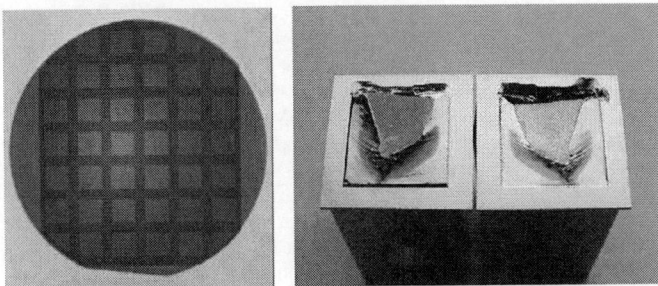

Figure 3. IR image of bonded patterned Si wafers and fracture surface after tensile test.

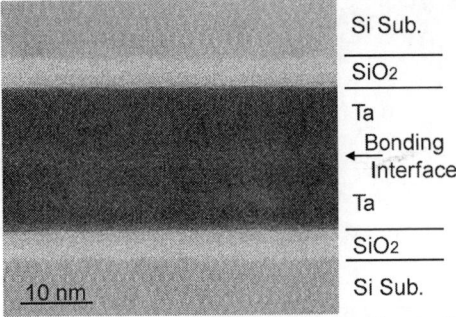

Figure 4. TEM image of Ta/Ta bonding interface by SAB

As is well known, conventional wafer direct bonding is achieved by the hydrogen bonds between -OH groups on the wafer surfaces and adsorbed water molecules (6). Even the plasma activated wafer bonding method bases on similar principle (16). Therefore, annealing process is necessary to stabilize the bonding, and at least a thin oxide layer remains at the bonding interface. Contrary, in SAB, bonding is assumed to be achieved by interatomic bonds, which are the same as those in bulk material. For example, Si atoms bind to etch other by covalent bonds, and metal atoms are bound by metallic bonds based on shearing of free electrons.

Therefore, in SAB, bonding properties are largely influenced by interatomic bonds in materials to be bonded. Ionic compounds are also successfully bonded to metals and semiconductors. For example, Si_3N_4 ceramics strongly bonds to Al (3), and $LiNbO_3$, $LiTaO_3$, Al_2O_3, and $Gd_3Ga_5O_{12}$ single crystal wafers were successfully bonded to Si wafers (17.18). On the other hand, bonding between oxide materials is relatively weck, and in some cases annealing process is effective to improve the bonding strength (18).

Expansions of SAB concept and Applications

The concept of SAB has been expanded in order to bond wide range of materials. Materials such as SiO_2 and polymers cannot be directly bonded by SAB. To bond such materials, deposition of very thin intermediate layers of metals or Si has been proposed (19, 20). The deposited metal atoms firmly adhere to the surface of SiO_2 and/or polymers and simultaneously form active metal surface. The thin film deposition is regarded as a new process for the surface activation. Figure 5 shows TEM images of bonding interface between PEN (poly-ethylene 2,6-naphthalate) sheets using Si intermediate layers. Concerning metal thin layers, especially in case of metals with relatively large atomic diffusion coefficient, atom diffusion and recrystallization plays important role in intimate contact formation at the bonding interface when two wafers are mated (21-23). This bonding method was named as atomic diffusion bonding.

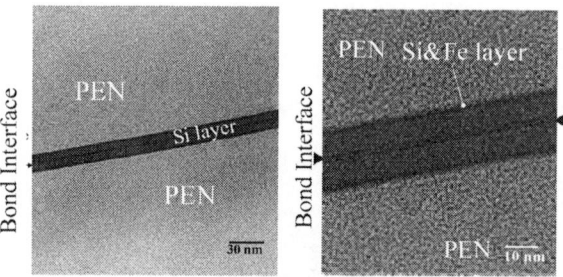

Figure 5. TEM image of the PEN sheets using Si intermediate layers.

In the bonding processes using metal film deposition, Au film has a special advantage. Wafers with Au films are successfully bonded at room temperature in atmospheric air (22-26). Because Au is not oxidized, it is assumed that Au surface remains in an active state even in air. Room-temperature bonding in air is quite advantageous in various industrial applications. In addition, it is assumed that Au films can be bonded in various atmospheres such as low vacuum, gas ambient, etc.

In SAB, influence of the surface sputter etching is also an important issue. High energy ion/atom beam used for sputter etching introduce crystal defects near the surface. In case of metals, the damage can recover even at room temperature and a directly bonded interface between two crystals is formed (3,4). On the other hand, Ar beam irradiation create a damaged layer on Si surface and an amorphous layer remains at the bonding interface (9). Surface sputter etching influences also to the surface roughness. Ar beam irradiation roughen the Si surface (8), whereas Ne beam irradiation smoothen the surface (27, 28).

Various applications of the surface activated bonding have been developed in the field of wafer-level packaging and engineered substrates. In wafer-level packaging field, various MEMS devices have been already commercialized and 3D-integration of heterogeneous devices are under investigation. In engineered substrate application, RF filters have been already commercialized (29,30) and still succeeding developments continue. Engineered substrates for micro-electronics, power-electronics (13, 31, 32), solar cells (33-35), etc. are now extensively developed by the method.

Summary

SAB was firstly based on the surface cleaning by sputter etching using high energy ion/atom beam of Ar. It achieves strong bonding at room temperature, and well matches wafer direct bonding. Various materials are successfully bonded by SAB, although some materials such as SiO_2 and polymers do not form strong direct bonding. Some modified SAB methods have been proposed to improve bonding properties and to apply SAB to wide range of materials. We believe SAB will be used in wide range of technological fields in the near future.

References

1. F. S. Ohuchi, T. Suga, Trans. Mat. Res. Soc. Jpn., M. Sakai et al., editors, **16B**, p1195, Elsevier Science B.V., Amsterdam (1994)
2. T. Suga, K. Miyazawa, Y. Yamagata, in Metal-Ceramic Joints, Proc. MRS Int. Meeting Adv. Mater., **8**, p. 257(1989).
3. T. Suga, Y. Takahashi, H. Takagi, B. Gibbesch and G. Elssner, Acta Metall. Mater., **40**, S113 (1992).
4. T.Akatsu, N.Hosoda, T.Suga, M.Rühle, J. Mater. Sci., **34**, 4133 (1999).
5. T. Yamada, M. Takahashi, Proc. 13th IEEE Micro Electromech. Syst. (MEMS), Miyazaki, 574 (2000)
6. R. Stengl, T. Tan, U. Gösele, Jpn. J. Appl. Phys., **28**, 1735 (1989).
7. H. Takagi, K. Kikuchi, R. Maeda, T.R. Chung, T. Suga, Appl. Phys. Lett. **68**, 2222 (1996).
8. H. Takagi, R. Maeda, T.R. Chung, N. Hosoda, T. Suga, Jpn. J. Appl. Phys., **37**, 4197 (1998)
9. H. Takagi, R. Maeda, N. Hosoda, T. Suga, Jpn. J. Appl. Phys., **38**, 1589 (1999).
10. H. Takagi, R. Maeda, T. Suga, Sens. Actuat. A, **105**, 98 (2003).
11. T.R. Chung, N. Hosoda, T. Suga, H. Takagi, Appl. Phys. Lett. **72**, 1565 (1998).
12. M. M. R. Howlader, T. Watanabe, T. Suga., J. Vac. Sci. Technol. B, **19**, 2114 (2001).
13. J. Suda, T. Okuda, H. Uchida, A. Minami, N Hatta, T. Sakata, T. Kawahara, K. Yagi, K.Imaoka, Y. Kurashima, H. Takag, Proc. Int. Conf. Silicon Carbide Related Mater. 2013, Miyazaki, 358 (2013).
14. A. Shigetou, T. Itoh, T. Suga, J. Mater. Sci., **40**, 3149 (2005).
15. A. Shigetou, T. Itoh, T. Suga, Proc. IEEE 56th Electron. Components Technol. Conf. (ECTC), San Diego, 1223 (2006) .
16. G. Kräuter, A. Schumacher, U. Gösele, Sens. Actuat. A, 70, 271 (1998).
17. H. Takagi, R. Maeda, N. Hosoda, T. Suga, Appl. Phys. Lett., **74**, 2387 (1999).
18. H. Takagi, R. Maeda, J. Cryst. Growth., **292**, 429 (2006).
19. R. Kondou, T. Suga, Scripta Materialia **65**, 320 (2011).
20. T. Matsumae, M. Fujino, T. Suga, Jpn. J. Appl. Phys., 54, 101602 (2015).
21. T. Shimatsu, M. Uomoto, J. Vac. Sci. Technol. B, **28**, 706 (2010).

22. T. Shimatsu, M. Uomoto, ECS Trans., **PV33-4**, 61, (2010).
23. T. Shimatsu, M. Uomoto, H. Kon, ECS Trans., **PV64-5**, 317, (2014).
24. K. Okumura, E. Higurashi, T. Suga, K. Hagiwara, Proc. 2014 Int. Conf. Electron. Packaging (ICEP), Toyama, Japan, 716 (2014).
25. K. Okumura, E. Higurashi, T. Suga, K. Hagiwara, Proc. 4th IEEE Int. Workshop Low Temperature Bonding 3D Integration (LTB-3D), Tokyo, Japan, 26 (2014).
26. Y. Kurashima, A. Maeda, H. Takagi, Microelectron. Eng., **129**, 1 (2014).
27. Y. Kurashima, A. Maeda, H. Takagi, Appl. Phys. Lett., **102**, 251605 (2013).
28. H. Takagi, Y. Kurashima, A. Maeda, ECS Trans., **PV64-5**, 69, (2014).
29. M. Miura, T. Matsuda, Y. Satoh, M. Ueda, O. Ikata, Y. Ebata, H. Takagi, Proc. IEEE Ultrasonic Symp. 2004, Montreal, 1322 (2004).
30. M. Miura, T. Matsuda, M. Ueda, Y. Satoh, O. Ikata, Y. Ebata, H. Takagi, Proc. IEEE Ultrasonic Symp. 2005, Rotterdam, 573 (2005).
31. K. Yagi, N. Hatta, T. Sakata, A. Minami, T. Kawahara, H. Uchida, K. Imaoka, T. Okuda, J. Suda, Y. Kurashima, H. Takagi, Proc. 4th IEEE Int. Workshop Low Temperature Bonding 3D Integration (LTB-3D), Tokyo, Japan, 56 (2014).
32. J. Liang, S. Nishida, M. Arai, N. Shigekawa, Appl. Phys. Lett., **104**, 161604 (2014).
33. J. Liang, S. Nishida, M. Morimoto, N. Shigekawa, Electorn. Lett., **49**, 830 (2013).
34. K. Derendorf, S. Essig, E. Oliva, V. Klinger, T. Roesener, S. P. Philipps, J. Benick, M. Hermle, M. Schachtner, G. Siefer, W. Jäger, F. Dimroth, IEEE J. Photovoltaics, **3**, 1423 (2013).
35. S. Uchida, T. Watanabe, H. Yoshida, T. Tange, M. Arimochi, M. Ikeda, P. Dai, W. He, L. Ji, S. Lu, H. Yang, Appl. Phys. Exp., 7, 112301 (2014)

Surface activation and planarization with gas cluster ion beam for wafer bonding

N. Toyoda[a], T. Sasaki[a], I. Yamada[a], T. Suga[b]

[a] Graduate school of engineering, University of Hyogo, Himeji, Hyogo, 671-2280, Japan
[b] School of Engineering, The University of Tokyo, Bunkyo-ku, Tokyo, 113-0033, Japan

In this paper, preliminary irradiation effects of glancing incidence gas cluster ion beam (GCIB) for surface activated bonding are reported. Unique irradiation effects of GCIB, such as low-damage irradiation and surface smoothing effects, might be beneficial for surface activated bonding. From XPS analysis, Ar-GCIB irradiation at 70° removed native oxide on Si(100) substrate efficiently. On the contrary, mixing of native oxide by normal incidence GCIB occurred, which caused residual oxide layer formation. The surface roughness at oblique incidence showed lowest value (Ra ~ 0.5 nm) without ripple formations. In addition, Cu-Cu bonding were demonstrated as preliminary results of wafer bonding with GCIB.

Introduction

Surface activated bonding (SAB)[1,2] has been widely used for various materials and devices. In the SAB process, atomic Ar ions or fast atom beam bombard wafer surface to form activated surface. In order to form activated surface on materials which are susceptible to irradiation damage, we have been investigating the gas cluster ion beam (GCIB)[3] as an alternative to atomic beams. Gas clusters are aggregates of several thousands of gaseous atoms. Therefore, energy per atom of a gas cluster ion can be easily reduced to several eV/atom. Although energy/atom of a gas cluster ion is low, thousands of low-energy atoms bombard a target surface at the same location simultaneously. As a result, the energy density on the target surface became extremely high, and non-linear collisions of atoms are induced [4]. They induces unique irradiation effects of GCIB such as high-yield sputtering [5], low-damage analysis of organic materials [6], thin film deposition at low-temperature [7], and surface smoothing [8]. These effects of GCIB might be beneficial for SAB. In this paper, preliminary irradiation effects of surface activation and planarization with glancing incidence GCIBs are reported.

Experiments

Figure 1 shows a configuration of GCIB system [9] for wafer bonding. Neutral cluster beams were formed by supersonic expansion of high-pressure gas (~ 1 MPa) through a nozzle. In this study, Ar gas was used as a source gas. Subsequently, neutral clusters were ionized by electron bombardments. Gas cluster ions were accelerated by an electrostatic electrode (< 30 kV). Atomic Ar ions were removed by a permanent magnet, as a result, only gas cluster ions with heavy mass were transported to the target chamber. At the target chamber, a pair of sample holder facing each other was mounted on linear motion stages. After irradiation of GCIB at oblique incidence, the samples were put together by linear motion stages inside the vacuum chamber. Additional pressing and heating of the bonded sample was carried out outside of the vacuum chamber. For XPS characterization of irradiated surface, XPS system with GCIB source was used. In this study, mainly two experiments were carried out. First experiment was native oxide removal with glancing incidence GCIB on Si(100). Irradiated surfaces were characterized with in-vacuum XPS. In addition, surface morphologies were observed with an atomic force microscope (AFM). Subsequently, preliminary bonding experiments were carried out for Cu bonding. The acceleration voltage and the ion fluence were 10 kV and 1×10^{15} ions/cm^2, respectively. The incident angle was 70° from the surface normal of Cu samples.

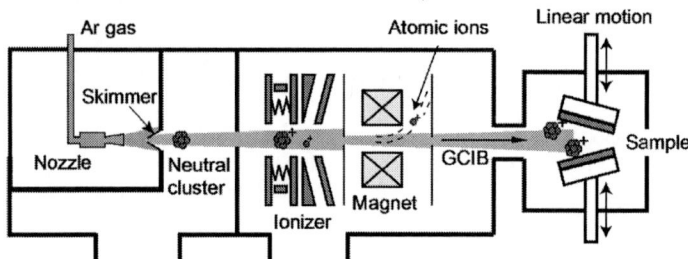

Figure 1: Schematic diagram of GCIB system for wafer bonding

Results and Discussions

At first, native oxide removal on Si(100) substrates with Ar-GCIB at various incident angle was studied using in-vacuum XPS analysis. The acceleration voltage was 10 kV. The incident angle of Ar-GCIB were varied between 0° and 70° from the surface normal. The ion fluence was varied between 1×10^{14} and 9×10^{14} ions/cm^2. After irradiation, the samples were transferred to the XPS analysis chamber by a transfer rod, and subsequently, XPS measurements were carried out. The vacuum pressure in the target chamber was 1×10^{-6} Pa. Figure 2 shows Si2p XPS after Ar-GCIB irradiation with various ion fluence at incident angle of 70°. On the virgin Si(100) substrate, there was a native oxide peak around 104 eV. The peak intensity of this peak gradually decreased with increasing the ion fluence, which indicated that Ar-GCIB irradiation at oblique incidence was effective for oxide removal on Si.

Next, the ion fluence dependence of oxide peak at different incident angle was studied. Figure 3 shows the relationship between the ion fluence and the normalized oxygen peak intensity of XPS on Si(100) irradiated with Ar-GCIB at 0°, 45° and 70°. In the case of 0°, the oxygen peak dropped quickly at low-fluence. However, it did not attain complete removal of oxide, which might be explained by mixing effect of surface oxide. At 45°, it also dropped dramatically at low-fluence, however, oxygen intensity increased at high-fluence. Figure 4 shows AFM images of Si after Ar-GCIB irradiation at incidence angle of (a) 0°, (b) 45° and (c) 70°, respectively. The ion fluence was 3×10^{14} ions/cm^2. From surface morphology measurements (Figure 4(b)), surface roughening ($R_a = 5.1$ nm, $R_{rms} = 6.4$ nm) occurred due to ripple formation at 45°. This will be the reason for the phenomenon mentioned above.

From figure 3, the oxygen intensity at incident angle of 70° gradually decreased with ion fluence, and it saturated at lowest value. It takes 20 min to reach the saturation value with the GCIB system used in this experiment. From AFM observation (Figure 4 (c)), the surface after oxide removal was quite smooth (R_a = 0.5 nm, R_{rms} = 0.7nm) compared to other incident angle. These results indicated the possibility of oblique GCIB for wafer bonding. However, there was still residual oxide on Si surface. Since the vacuum conditions in the irradiation chamber was 1×10^{-6} Pa, adsorbed residual water might form the oxide layer. We would like to improve the vacuum condition, and discuss the complete removal of oxide in the future.

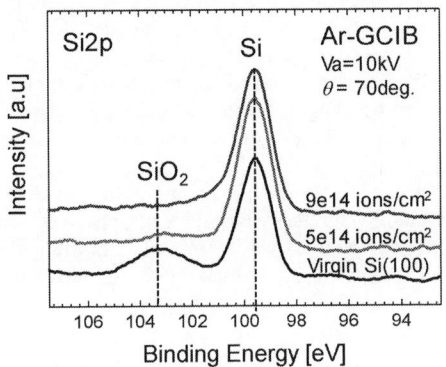

Figure 2. Si2p XPS after Ar-GCIB irradiation with various ion fluence at incident angle of 70°.

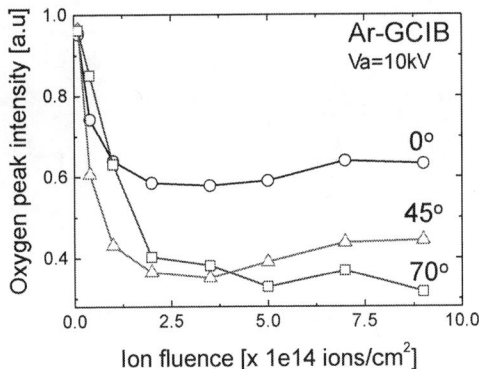

Figure 3: Relationship between the ion fluence and the normalized oxygen peak intensity of XPS on Si(100) irradiated with Ar-GCIB at 0°, 45° and 70°.

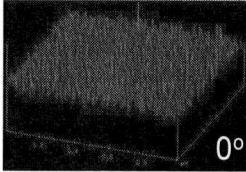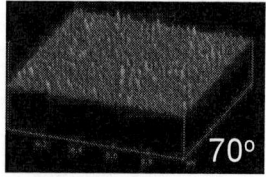

(a)0°(R$_a$:2.1nm, R$_{rms}$:2.6nm) (b)45°(R$_a$:5.1nm, R$_{rms}$:6.4nm) (c)70° (R$_a$:0.5nm, R$_{rms}$:0.7nm)
Figure 4: AFM images of surface morphology, and roughness values (R$_a$ and R$_{rms}$) of
Si(100) irradiated with Ar-GCIB at incident angle of 0°, 45°, and 70°.
(Accel. voltage: 10 kV, Ion fluence: 3×10^{14} ions/cm^2)

As a demonstration, irradiation of oblique incidence Ar-GCIB on Cu and subsequent wafer bonding was carried out. Electroplated Cu films on Si substrates were used as specimens. Prior to bonding experiments, surface states of Cu was characterized with XPS. Figure 5 (a) shows Cu2p XPS of pristine Cu and that irradiated with Ar-GCIB. The acceleration voltage and the ion fluence were 10 kV and 1×10^{15} ions/cm^2, respectively. The incident angle was 70°. From figure 5(a), it was shown that CuO peak, which existed around 934 eV in the virgin Cu, disappeared after oblique Ar-GCIB irradiation. Subsequently, Cu samples with/without GCIB irradiation were put together by linear motion stages. Pressing and heating of the bonded sample was carried out outside of the vacuum chamber. In the case of the virgin Cu samples, they did not bond together. On the contrary, Cu samples with oblique Ar-GCIB irradiation bonded after pressing and heating process owing to the removal of oxide and contaminant on Cu surface.

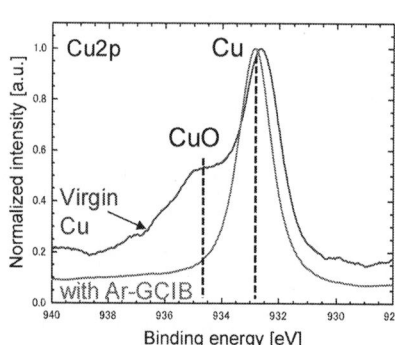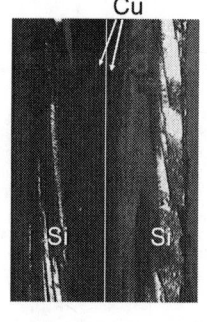

(a) Cu2p XPS of virgin Cu and that with Ar-GCIB   (b) Sample after wafer bonding
(Accel. voltage: 10 kV, Ion fluence: 1×10^{15} ions/cm^2, θ : 70°)
Figure 5: (a) XPS of virgin Cu and that irradiated with Ar-GCIB irradiation, (b) Optical image of sample after wafer bonding

Summary

In this study, preliminary study of irradiation effects of glancing incidence GCIB for wafer bonding were reported. Oblique incidence GCIB realized removal of surface oxide without mixing of oxide on the target surface together with the smooth surface. Preliminary bonding experiment of Cu-Cu showed promising results. Further investigation of optimum irradiation conditions of GCIB will be carried out together with the investigation of bonding interface with GCIB irradiation.

References

1. T. Suga, K. Miyazawa, Y. Yamagata, Material Research Soc., **8**, 257 (1989).
2. H. Takagi, K. Kikuchi, R. Maeda, T. R. Chung, T. Suga, Appl. Phys. Lett., 68016, 2222 (1996).
3. I. Yamada, J. Matsuo, N. Toyoda, A. Kirkpatrick, Mat. Sci. and Eng. R, **34**, 231 (2001).
4. Z. Insepov, I. Yamada, Nucl. Instr. and Meth. B, **112**, 16 (1996).
5. J. Matsuo, N.Toyoda, M. Akizuki, I. Yamada, Nucl. Instr. and Meth. B, **121**, 459 (1997).
6. S. Ninomiya, Y. Nakata, K. Ichiki, T. Seki, T. Aoki, J. Matsuo, Nucl. Instr. and Meth. B, **256**, 493, (2007).
7. N. Toyoda, Y. Fujiwara, I. Yamada, Nucl. Instr. and Meth. B, **206**, 875, (2003).
8. N. Toyoda, T. Hirota, I. Yamada, H. Yakushiji, T. Ono, H. Matsumoto, IEEE Trans. on Magn., **46**, 1599 (2010).
9. N. Toyoda, I. Yamada, IEEE Trans. on Plasma Sci., **36**, 1471 (2008).

14

Low-Temperature Aluminum-Aluminum Wafer Bonding

B. Rebhan[a], A. Hinterreiter[b,c], N. Malik[d], K. Schjølberg-Henriksen[e], V. Dragoi[a] and K. Hingerl[b]

[a] EV Group, DI E. Thallner Straße 1, St. Florian/Inn 4782, Austria
[b] Center for Surface- and Nanoanalytics, Johannes Kepler University, Linz 4040, Austria
[c] Christian Doppler Laboratory for Microscopic and Spectroscopic Material Characterization
[d] Centre for Materials Science and Nanotechnology, University of Oslo, PO Box 1032 Blindern, N-0315 Oslo, Norway
[e] SINTEF, Department of Microsystems and Nanotechnology, PO Box 124 Blindern, N-0314 Oslo, Norway

Aluminum-aluminum thermo-compression wafer bonding is becoming increasingly important in the production of microelectromechanical systems (MEMS) devices. As the chemically highly stable aluminum oxide layer acts as a diffusion barrier between the two aluminum metallization layers, up to now the process has required bonding temperatures of 300°C or more. By using the EVG®580 ComBond® system, in which a surface treatment and subsequent wafer bonding are both performed in a high vacuum cluster, for the first time successful Al-Al wafer bonding was possible at a temperature of 100°C. The bonded interfaces of blank Al wafers and Al wafers with patterned frames were characterized using C-mode scanning acoustic microscopy (C-SAM) and transmission electron microscopy (TEM) as well as dicing yield and pull tests representative for the bonding strength. The investigations revealed areas of oxide-free, atomic contact at the Al-Al bonded interface.

Introduction

Thermo-compression wafer bonding is a key technology for the wafer-level production of hermetically sealed cavities, which are essential for the functioning of many microelectromechanical systems (MEMS). Aluminum, with its low material price, high thermal and electrical conductivities and its complementary metal oxide semiconductor (CMOS) compatibility, is a promising candidate for the fabrication of CMOS-MEMS, in which the sensor/actuator part is bonded to the electrical circuit.

The highly chemically stable native oxide layer on the Al surface cannot be removed by conventional methods. The thin oxide acts as a diffusion barrier layer between the two aluminum metallization layers, and therefore inhibits successful low temperature Al-Al wafer bonding. So far, effective Al-Al wafer bonding has required processing temperatures of >300°C and high contact pressures. TABLE I summarizes the experimental parameters extracted from a number of reports on Al-Al wafer bonding and the current work. In the referenced processes, a high contact pressure (usually several tens of MPa) (1-3) is used to break the oxide in order to establish diffusion channels for Al atoms. As a calculation shows (4), the elastic energy is too low to influence the bonding between atoms directly, but the applied stress and the resulting strain breaks up the surface layer. The wafers are bonded at high temperatures, usually in the range of 400°C to 550°C (1-3, 5-7). In recent experiments, Malik et al. were able to reduce the required bonding temperature to about 300°C by depositing the Al metallization layer onto an intermediate SiO_2 layer (3).

TABLE I Comparison of experimental parameters from different reports in literature and the present work on successful Al-Al wafer bonding.

First Author	Al Thickn. (μm)	Cu Content (%)	Bond Area (cm^2)	Force (kN)	Pressure (MPa)	Temp. (°C)
Martin (5)	1-2	1	n. a.	n. a.	30-117	450
Yun (8)	2	0-4	6-12[a]	60[a]	50-100	450
Yun (9)	2	2	n. a.	9-18	n. a.	450
Cakmak (6)	0.5	0	175	60	3.4	400-550
Froemel (7)	1	n. a.	n. a.	n. a.	4.5	450
Malik (2)	1	0	5.25	18-36	34-69	400-550
Malik (3)	1	0	5.25	36-60	69-114	300-550
Rebhan	0.3-1	0-0.5	5.25-314	60	1.9-114	100-550

n. a. = not available   [a]Values estimated from the description of frame structure

In the references listed in TABLE I, typically temperature of 400°C-550°C led to bonded Al-Al wafers with Al_2O_3 precipitates present at the bonded interface and still with reasonably good bonding quality. In the EVG®580 ComBond® system, the aluminum oxide is first removed physically, followed by bonding of the two metal layers, both performed in a high vacuum cluster. The first Al-Al wafers were successfully bonded at 100°C using EVG®580 ComBond® equipment, which allows for preparation of oxide-free surfaces enhancing atomic contact. Notably, Akatsu et al. used a surface activated bonding set-up (10) to bond cubes of single crystalline aluminum at room temperature (with a bonding pressure of 40 MPa) (11). However, to our knowledge, there is no account for surface activated Al-Al bonding on wafer-level.

In the present work, Al-Al bonding of blank layers bonded in the EVG®580 ComBond® equipment was compared to Al-Al wafers bonded conventionally in an EVG®520IS equipment. The microstructure of the aluminum films on the silicon substrates was characterized by atomic force microscopy (AFM) and transmission electron microscopy (TEM). C-mode scanning acoustic microscopy (C-SAM), TEM and energy dispersive x-ray spectroscopy (EDXS) interface studies of the Al-Al interfaces bonded with this novel method revealed oxide-free, atomic contact and grain growth across the original interface. Further, the bond strength was characterized based on dicing yield and pull tests of bonded Al-Al wafers with patterned frame structures.

Experimental

An overview of the samples processed in this work is shown in TABLE II. All wafers were bonded with a bonding force of 60 kN at 100-550°C for 1 h, if not otherwise mentioned.

TABLE II Experimental parameters for each used bonding equipment.

Sample type	Metal	Sputtering	Equipment	Temperature
1	Blank Al+0,5% Cu	Standard	EVG®520	150-550°C
2	Blank Al+0,5% Cu	Standard	ComBond®	150-550°C
3	Blank Al+0,5% Cu	ALPS	EVG®520	150-550°C
4	Blank Al+0,5% Cu	ALPS	ComBond®	150-550°C
5	Patterned pure Al	Standard	ComBond®	100-150°C

The first types of substrates used for the bonding experiments were non-patterned 200 mm diameter silicon wafers. Within this work, they are referred to as "blank wafers". First, a

20 nm Ti adhesion layer/diffusion barrier was deposited on the Si wafers, followed by full-sheet metallization layers of 99.5% Al with 0.5% Cu concentration and a thickness of 300 nm. Two different techniques – standard sputtering deposition and aluminum low pressure seed (ALPS) – were used. ALPS wafers differ from the standard deposition mainly in terms of processing pressure and temperature. While the standard deposition was performed at 215°C with an argon pressure of 3.3×10^{-3} mbar, the ALPS process was carried out at only 30°C with an argon pressure of 5.33×10^{-5} mbar. The surface roughness of the wafers was determined from atomic force microscopy (AFM) scans recorded in tapping mode on areas of 2×2 μm^2.

Figure 1 shows AFM measurements of the surfaces of the ALPS and standard sputtered Al films. The standard sputtered Al layer developed grains with lateral size ranging from 300 nm to 700 nm, while the grains of the ALPS wafers were significantly smaller, ranging from 200 nm to 300 nm grain size. The root mean square (RMS) surface roughness was found to be about 1.2 nm for both films.

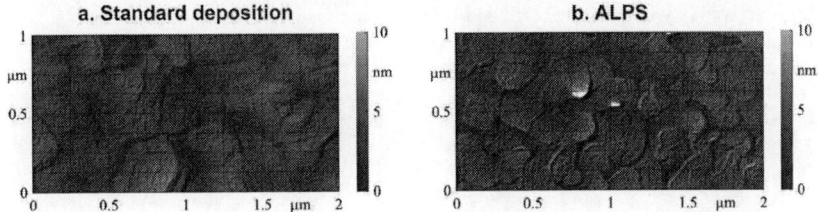

Figure 1. AFM measurements of the blank Al wafer surfaces. The different processing conditions result in larger grains for the standard deposition and smaller grains for the ALPS wafers.

The cross-section TEM investigation revealed that the grains extended throughout the entire Al layer thickness for both deposition types. As an example, the cross-section of a standard sputtered Al layer is shown in Fig. 2. In XTEM high-resolution mode, the thickness of the native aluminum oxide was determined as ~3.5 nm, which is in agreement with typical values reported in literature (12, 13).

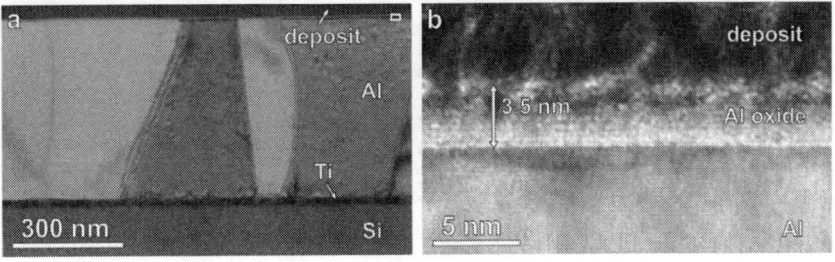

Figure 2. XTEM measurements after "standard deposition" of the Al layer. In (b) the marked inset of (a) at the top right demonstrates the Al surface and its native oxide layer with high magnification.

While the conventional bonds were performed in a standard thermo-compression wafer bonding system (EVG®520IS), the low-temperature bonds were performed in the EVG®580 ComBond®, a fully automated, high-vacuum wafer bonding system. In the latter system, the wafers are transferred from a central chamber to several modules. This way, a surface preparation step can be performed prior to bonding without exposing the wafers to an oxidizing atmosphere.

The proprietary ComBond® surface preparation process can be tuned to perform an oxide removal while only negligibly changing the sample's surface roughness. Prior to the bonding experiments, aluminum oxide removal rates of up to 15 nm/min were confirmed by thickness measurements of Al_2O_3 films on Si substrates. The bonding chamber in the EVG®580 ComBond® equipment is identical in functionality to that of the EVG®520IS used for standard bonds, the major difference is that in this equipment the wafers are handled between the ComBond® surface preparation chamber and the bond chamber under high vacuum environment and not at ambient conditions. In both setups, a bonding force of 60 kN was used. Since these bonded wafer pairs are full area bonds of non-patterned wafers (bonding area: 314 cm²), the applied force corresponds to a bonding pressure of 1.9 MPa.

MEMS devices often consist of a cavity produced by etching in one or both substrates and closed by a wafer bonding process. Therefore, besides bonding substrates with blank Al layers, two wafers with patterned Al bond frames were bonded to flat wafers in the EVG®580 ComBond® system. Within this paper these wafers are referred to as "frame wafers". These wafers are patterned with MEMS-relevant size dummy dies with bonding frames having a line width of 100 µm, 200 µm, or 400 µm, with straight corners, named F100, F200 and F400, as well as 200 µm wide frames with rounded corners, named F200R, respectively (see Fig. 3 and (3)). The outer dimension of all frames was 3x3 mm². Two 400 µm thick Si wafers (150 mm diameter) with (100) orientation were patterned by deep reactive ion etching (AMS 200 I-Prod, Alcatel) to realize 6 µm high frames. All four wafers were thermally wet oxidized to a nominal thickness of 150 nm SiO_2. Subsequently an adhesion layer of 100 nm Ti was sputtered on the oxide surfaces. Without breaking the vacuum between the deposition steps, an approximately 1.2 µm thick layer of pure Al (99.999%) was deposited in the same sputtering chamber. In order to allow for optimum bonding conditions, chemical mechanical polishing (CMP) of the wafers was performed, resulting in a final Al layer thickness of 1 µm.

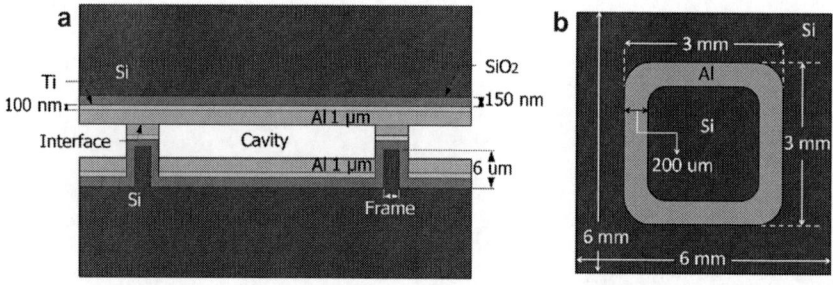

Figure 3. (a) Schematic cross-section and (b) top view of a single F200R bond frame.

A wafer with frames was bonded to a flat wafer in the ComBond® system applying a bonding force of 60 kN and a bonding temperature of 100°C or 150°C. The applied bonding force corresponded to a bonding pressure of 114.3 MPa. The bonded wafer pairs were diced into individual dies using a DAD321 (Disco) saw. The dicing yield, defined as the percentage of dies that were not delaminated after the dicing process, was recorded. A random selection of 12 non-delaminated dies of frame type F200R was made from each bonded pair, glued to flat headed bolts and pull tested using a MiniMat2000 equipment (Rheometric Inc.). During pull testing, displacement versus applied force was recorded and the maximum force, at which the fracture occurred, designated as the fracture force, was determined. The bond strength was calculated by dividing the fracture force by the bonding area.

Results and Discussion

The discussion of the experimental results is split into three sections, each corresponding to a bonding type (materials and processes used):
- conventional Al-Al wafer bonding with blank wafers
- low-temperature surface pre-treated Al-Al wafer bonding with blank wafers
- low-temperature surface pre-treated Al-Al wafer bonding with frame wafers

Conventional Al-Al Wafer Bonding: Blank Wafers

The attempts to bond two wafers with blank Al films in an EVG®520IS resulted in a bonded interface of low quality at bonding temperatures between 400°C-550°C. In Fig. 4 a typical C-SAM result of a wafer pair (standard Al deposition) bonded at 550°C. The different tones of gray in the C-SAM image represent areas which are weakly bonded or even unbonded areas. The bond quality was found to be highly sensitive to local pressure variations. The relatively low bonding pressure of 1.9 MPa (this was the maximum available for this setup) was not enough to reproduce the results of Cakmak et al., who were able to bond non-patterned 150 mm diameter wafers (6) with a higher pressure of 3.4 MPa. The root cause for this difference might be explained by differences in the Al film properties (4,14) or by the bonding pressure difference.

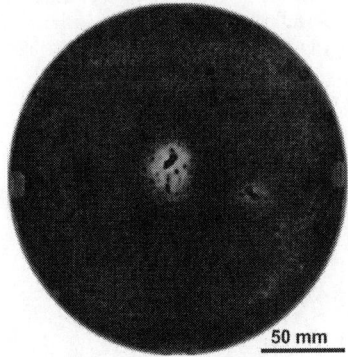

Figure 4. Typical C-SAM result of 200 mm diameter (standard deposition) Al wafers bonded in an EVG®520IS at 550°C for 3 h with 60 kN.

The aluminum oxide properties (e.g. thickness, chemistry) seem to be crucial for the bonding process. Between the two aluminum layers, non-damaged Al oxide was visible in the SEM cross section images, and Al oxide was also detectable by Auger electron spectroscopy (AES). Figure 5 shows the depth profile of the bonded interface for the wafers with ALPS deposited films at a position with relatively good bonding quality; this corresponds to one of the dark areas in Fig. 4. Close to the bonded interface the metallic Al concentration drops. At the same time, the signal for Al oxide increases. It was assumed that the oxide layer obstructs diffusion of Al atoms between the two Al metal layers, hence preventing significant grain growth across the initial wafer interface. The results of wafers with standard sputtered Al layers were showing similar aspect (not shown here).

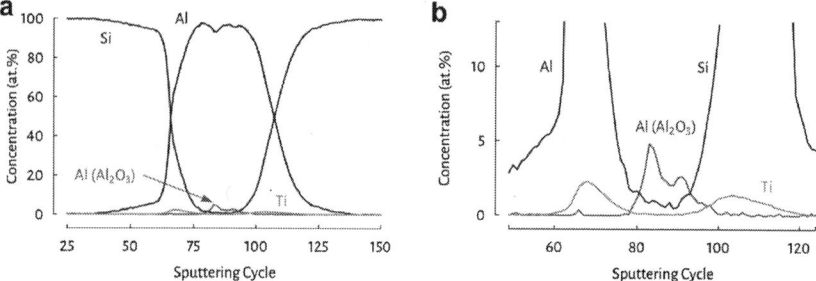

Figure 5. AES depth profile of bonded 300 nm thick Al ALPS wafer pair showing the wafer surface structure (Si-Ti-Al) as well as the presence of aluminum oxide at the interface. In (b) a closer view of the sputtering cycles around the bonding interface is presented.

Low-Temperature Surface Pre-treated Al-Al Wafer Bonding: Blank Wafers

Figure 6 shows a comparison of a typical C-SAM result of bonded wafer pairs with ALPS Al films, bonded in an EVG®520IS at 550°C (left) and at 150°C in an EVG®580 ComBond® with surface pre-treatment (right). Both wafers were bonded for 1.5 h using 60 kN piston force. Due to the surface pre-treatment, which removes the native Al oxide prior to the bonding process, the wafers can be bonded at temperatures significantly lower than any values reported in the literature (see TABLE I). Although the bonding pressure was relatively low (1.9 MPa), the C-SAM image shows a high quality bonding interface at almost any position of the wafer pair bonded in the EVG®580 ComBond®.

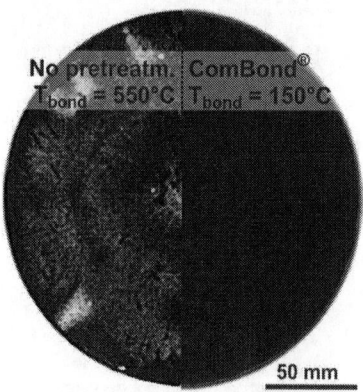

Figure 6. C-SAM images of bonded Al ALPS wafer pairs (left) without pretreatment and (right) with ComBond® surface treatment prior wafer bonding. Both wafer pairs were bonded at 60 kN (1.9 MPa) for 1.5 h.

Compared to bonded ALPS wafer pairs, the samples produced from standard Al-sputtered films had significantly more weakly bonded or larger unbonded areas upon C-SAM inspection. For a qualitative comparison of the bond energy, a razor blade was inserted at the bond interface (15) and the approximate length of the resulting crack was measured with C-SAM. The crack lengths in the samples with standard Al-sputtered films were 5% to 10% longer than the ones in samples with ALPS Al films. The C-SAM and crack length measurement results indicate that the ALPS Al films gave higher bond energy than standard sputtered Al films. The smaller grains observed in the ALPS Al films could be a likely explanation for the observed difference, since smaller grains result in a higher density of grain boundaries, which in turn promotes the diffusion of Al atoms, as the concentration of short-circuit diffusion paths is increased (14, 4).

The bonded interface of the wafer pair with ALPS Al films bonded in the EVG®580 ComBond® system at low temperature (150°C), was inspected by TEM and EDXS. The high resolution TEM image in Fig. 7 (left) shows that no amorphous layer separated the two Al metal films. Similar observations were made by Akatsu et al. (11) for single crystalline samples. The EDXS mapping shown in Fig. 7 (right) revealed no additional oxide near the bonded interface. The oxide signal stemmed exclusively from the native oxide on the surface of the TEM specimen (noise signal), which was exposed to air during the TEM preparation.

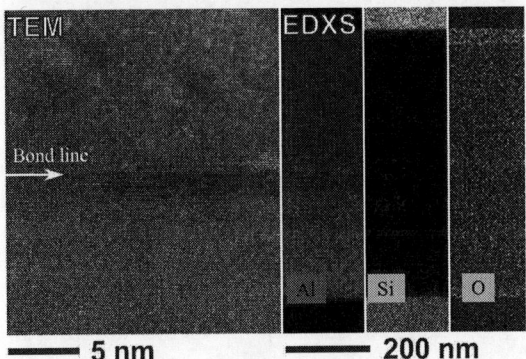

Figure 7. High-resolution TEM image and EDXS mapping of the interface of a ComBond® surface pre-treated, low-temperature bonded Al ALPS wafer pair, showing atomic, oxide-free contact. The wafer pair was bonded at 60 kN (1.9 MPa) and at a temperature of 150°C for 1 h.

Low-Temperature Surface Pre-treated Al-Al Wafer Bonding: Frame Wafers

Figure 8 shows the C-SAM result of a frame wafer pair, bonded after ComBond® surface pre-treatment at 100°C. The successful bonding of the wafers is shown by the absence of trapped gas, which would have given significant acoustic wave reflections. As the wafers were placed in water during C-SAM inspection and the edge was not sealed, water penetrated from the edge into the unbonded wafer areas, explaining the black irregular pattern close to the wafer edge. The frame wafer pair bonded at 150°C with the same bonding time (1 h) and force showed no difference to the C-SAM result of the 100°C bonded wafer pair.

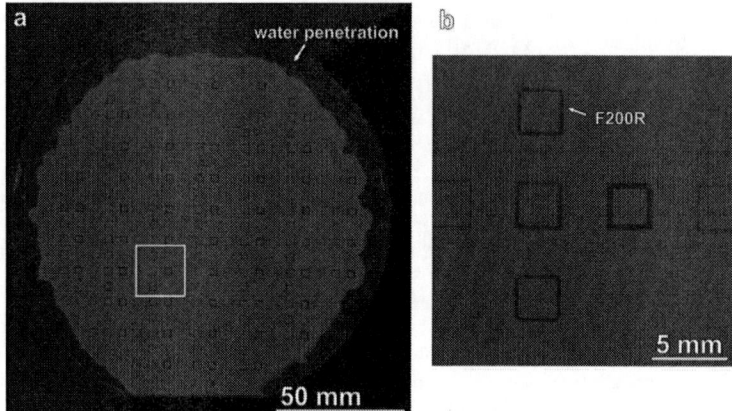

Figure 8. C-SAM measurement of frame wafers after ComBond® surface treatment and subsequent bonding at 100°C for 1 h with 60 kN (114 MPa). In (b) a detailed scan of the bond frames is shown. The dark irregular pattern at the wafer edge in (a) is generated by the water which penetrated into the unbonded wafer areas during the C-SAM measurement.

The dicing yield results of wafers bonded at temperatures of 100°C and 150°C are shown in Fig. 9a. A dicing yield of 100% was obtained for wafers bonded at both temperatures for all frame types. The dicing yield results showed that the bonds formed with both types of bonds were strong enough to survive the force exerted on them by the dicing saw. The pull test results are shown in Fig. 9b. The average tensile bond strength of chips from the wafer pair bonded at 100°C was 23 MPa, while for the wafer pair bonded at 150°C it was 37 MPa. This shows that the average bond strength increased with increasing bonding temperature from 100°C to 150°C. However, the standard deviation was overlapping.

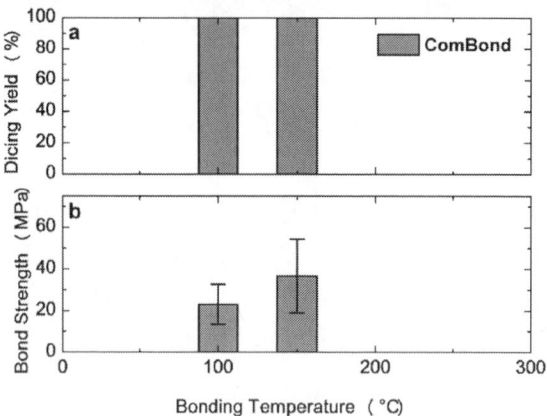

Figure 9. Comparison of (a) dicing yield and (b) tensile bond strength results. Al wafers were bonded with the ComBond® surface activation at 100°C and 150°C for 1 h at 60 kN (114 MPa).

Figure 9 shows that a bond strength which is sufficient for most MEMS applications was obtained with the EVG®580 ComBond® system at only 100°C bonding temperature. The bond strength values obtained at 100°C and 150°C are comparable to bond strength values obtained at temperatures ranging from 300-400°C using an EVG®520IS standard bonder (3). The high tensile bond strength and dicing yield obtained at bonding temperatures of only 100°C indicate that the oxide removal procedure performed using the ComBond® surface preparation has a high impact on the bonding ability of Al films. It is reasonable that the Al_2O_3 removal enables direct contact between the two Al metal surfaces and subsequent bonding by metal diffusion.

Conclusion

Conventional Al-Al wafer bonding requires extremely high processing temperatures and pressures, mainly due to the native chemically-stable Al oxide layer, which obstructs diffusion of Al atoms between the two metal layers. A dry surface pre-treatment process, which removes the native oxide, is crucial to enable Al-Al wafer bonding at low temperatures. It was shown that the bonding temperature could be reduced to as low as 150°C for wafers with blank Al films, and 100°C for wafers with frame Al pattern. The bonding interface of bonded blank wafers was inspected by C-SAM and TEM, and featured areas of oxide-free, atomic contact. The bonding quality was better for Al films deposited by ALPS sputtering than for Al films deposited by standard sputtering, probably due to the smaller grain size of the former films. The more application-relevant frame wafers showed 100% dicing yield for 150°C and 100°C, and high tensile bond strength of 37 MPa and 23 MPa, respectively. In both cases the measured bond strength is sufficient for most MEMS applications. High-quality Al-Al bonded wafer pairs were produced with high, but also with low bonding pressures of 114 MPa and 1.9 MPa, respectively. Compared to Al-Al thermo-compression wafer bonding without *in situ* removal of native oxide, when using the ComBond® pre-treatment, the bonding pressure is no longer a key parameter for successful Al-Al bonding.

Acknowledgment

Financial support by the Austrian Federal Ministry of Science, Research and Economy and by the Austrian National Foundation for Research, Technology and Development is gratefully acknowledged. The work was also partially supported by the Norwegian Research Council through the NBRIX project (contract 247781).

References

1. C.H. Yun, J. Martin, L. Chen, T.J. Frey, *ECS Trans.*, **16**, 117 (2008).
2. N. Malik, K. Schjølberg-Henriksen, E. Poppe, M.M.V. Taklo, T.G. Finstad, *Sens. Actuat.*, **A 211**, 115 (2014).
3. N. Malik, K. Schjølberg-Henriksen, E. Poppe, M.M.V. Taklo, T.G. Finstad, *J. Micromech. Microen.*, **25**, 035025 (2015).
4. B. Rebhan, K. Hingerl, *J. Appl. Phys.*, **118**, 135301 (2015).
5. J. Martin, SPIE Proceedings series, vol. 6463, p. 64630M, SPIE, (2007).
6. E. Cakmak, V. Dragoi, E. Pabo, T. Matthias, T.L. Alford, Materials Research Society Proceedings Series, Vol. 1222, p. DD04, Materials Research Society (2009).

7. J. Froemel, M. Baum, M. Wiemer, F. Roscher, M. Haubold, C. Jia, T. Gessner, in *16th International Solid-State Sensors, Actuators and Microsystems Conference*, IEEE Proceedings Series, p. 990, (2011).
8. J.R. Lloyd, *J. of Physics D*, **32**(17), 109 (1999).
9. C. Yun, J. Martin, E. Tarvin, J. Winbigler, in *IEEE 21st International Conference on Micro Electro Mechanical Systems*, IEEE Proceedings Series, p. 810, (2008).
10. T. Akatsu, G. Sasaki, N. Hosoda, T. Suga, *J. Mater. Res.*, **12**, 852 (1997).
11. T. Akatsu, N. Hosoda, T. Suga, *J. Mater. Sci.*, **34**, 4133 (1999).
12. R.K. Hart, *Proc. Roy. Soc.*, **236**(1204), 68 (1956).
13. J. Evertsson, F. Bertram, F. Zhang, L. Rullik, L.R. Merte, M. Shipilin, M. Soldemo, S. Ahmadi, N. Vinogradov, F. Carl`a, J. Weissenrieder, M. Göthelid, J. Pan, A. Mikkelsen, J.O. Nilsson, E. Lundgren, *Appl. Surf. Sci.*, **349**, 826 (2015).
14. B. Rebhan, M. Wimplinger, K. Hingerl, *ECS Trans.*, **64**, 369 (2014).
15. W.P. Maszara, G. Goetz, A. Caviglia, J.B. McKitterick, *J. Appl. Phys.*, **64**, 4943 (1988).

Ultra-thick metal ohmic contact fabrication using surface activated bonding

J. Liang[a], K. Furuna[a], M. Matsubara[b], M. Dhamrin[b], Y. Nishio[b], and N. Shigekawa[1]

[a] Graduate School of Engineering, Osaka City University, Sumiyoshi, Osaka 5588585, Japan
[b] Core Technology Center, Toyo Aluminum K. K., Chuo-ku, Osaka 5410056, Japan

We successfully bonded aluminum foils to Si substrates to fabricate p-Si/Al, n-Si/Al, p^+-Si/Al, and n^+-Si/Al junctions by surface activated bonding (SAB). The effects of the annealing temperature process on the electrical properties of the junctions were investigated by measuring their current voltage (I-V) characteristics. It was found that the leakage current of the reverse bias of n-Si/Al junctions was improved and the I-V characteristics of p-Si/Al revealed excellent linearity properties after the junctions annealing at 400 °C. The interface resistance of p^+-Si/Al, and n^+-Si/Al junctions decreased with increasing annealing temperature and decreased to 0.021 and 0.032 $\Omega\cdot cm^2$ after the junction annealing at 300 and 400 °C, respectively. These results demonstrated that thick metal Ohmic contact in devices could be realized by SAB.

Introduction

Ohmic contact to semiconductor substantially degrades the overall performance of the high-power devices and circuits with large-current and/or high-voltage capability (1). The major loss of performance is usually due to high ohmic contact resistance between metal electrode and semiconductor. Thus, the realization of excellent ohmic contact on semiconductor is absolutely necessary to obtain optimum device performance. The primary factors determining contact resistance are carrier concentration, semiconductor surface preparation and cleaning, and contact metal work functions hence Schottky barrier height (2, 3). In addition, it was reported that the ohmic contact resistance decreased with increasing the metal thickness.[4] However, the deposition of a thick metal layer is extremely difficult to obtain because it would take a large quantity of time and the production cost by the conventional coating method such as Electron-Beam-Evaporation and Sputter-Deposition. On way to circumvent these difficulties is surface-activated bonding (SAB) (5, 6), in which different substrate materials could be directly bonded to each other without heating. We previously successfully fabricated p-Si/Al junctions and found that the current-voltage characteristics of the junctions revealed Schottky properties (7). The formation of Ohmic contact frequently requires a high temperature step so that the evaporated metals could alloy with the semiconductor to reduce the barrier height of the interface. Thus, the investigation of the annealing temperature-dependent electrical properties of Si/Al junctions is very important to realize the thick film metal contacts.

In this study, we demonstrated the fabrication of p-Si/Al, n-Si/Al, and p$^+$-Si/Al, and n$^+$-Si/Al junctions by SAB method and examined the annealing temperature dependence of the electrical properties of all the junctions. The electrical properties of the junctions without and with annealing were investigated by I-V measurements and the feasibility of the thick film metal contact was discussed. Furthermore, their interfaces were investigated by field emission-scanning electron microscopy (FE-SEM).

Experimental Process

Four types (p-, n-, p$^+$-, and n$^+$-) of (100) Si substrates and aluminum foils (Al) with a thickness of 38 μm (it is commercially available) are used for our bonding experiment. The carrier concentrations of Si substrates are shown in Table I . The Al/Ni/Au and Ti/Au multilayers were evaporated on the back surfaces of p-Si and n-Si substrates prior to the bonding, respectively, and the Ohmic contacts of p-Si substrates were achieved by rapid thermal annealing at 400 °C for 1 min in N$_2$ gas ambient. Al foils were bonded to each Si substrates by using SAB (5, 6), so that p-Si/Al, n-Si/Al, p$^+$-Si/Al, and n$^+$-Si/Al foil junctions were obtained. In addition, p$^+$-Si and n$^+$-Si substrates with Al contacts on both surfaces were fabricated by deposition of 100 nm Al film on both surfaces of the substrates. The samples were not heated during the bonding process. After the bonding, a 4-by-6 mesa array was fabricated on the Al foils by using Al wet etching for 12 h. The wide and length of the mesa were all 1.4 mm. A top view photograph of the fabricated Si/Al junctions is shown in Figure 1a. An Agilent B2902A Precision Measurement Unit was used for measuring the I-V measurements of the junctions at different annealing temperature. The cross-sectional structures of Al/Si junctions were analyzed using an FE-SEM facility (JEOL JSM6500F).

TABLE I. The carrier concentration and the thickness of substrates.

Type	Carrier concentration (cm^{-3})	Thickness (μm)
p-Si (100)	2.4×10^{17}	525
n-Si (100)	8.5×10^{15}	525
p$^+$-Si (100)	2.6×10^{19}	525
n$^+$-Si (100)	2.6×10^{19}	525

Results and Discussion

We observed the microscopy images of the cross-section of the fabricated Si/Al junctions by using FE-SEM equipment. The observation results are shown in Figure. 1b. A straight line can be clearly recognized at the center of the junction, which corresponds to the bonded interfaces. Furthermore, no cracks or hollow spaces were observed at the bonded interface. This indicate that Al foils were firmly bonded to the Si substrates. The I-V characteristics of n-Si/Al junctions as a Schottky diode measured at room temperature are shown in Figure 2(a). Their I-V characteristics showed rectification properties. The ideality factor for the forward bias voltages between 0.1 and 0.3 V was calculated to be 2.08, 2.15, 2.14, 2.07, and 2.24 for the junction without and with annealing at 100, 200, 300, and 400 °C, respectively. The value is insensitive to the annealing temperature, which is larger than 2 suggesting that the recombination current mechanism mainly dominates the transport properties of carriers across the junction interface. We observed that the magnitude of the forward bias current decreased as the annealing temperature increased, which should be attributed to the oxidation of the Al foils for the highr annealing process. In addition, the magnitude of the current at -3 V significantly decreased from 2.7×10^{-2} to 5.0×10^{-4} A/cm^2 as the annealing temperature increased to 400 °C. The obtained values of parameter for the respective junctions are summarized in Table I. Note that the reveres bias current continuously increased as the reverse bias voltage increased.

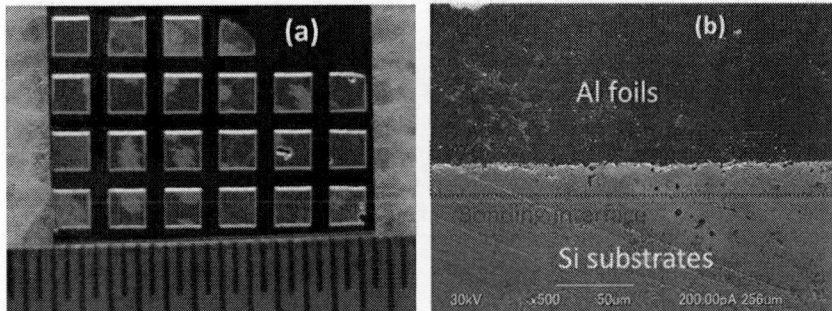

Figure 1. (a) A top view photograph of and (b) An FE-SEM cross-sectional image of the bonded Si/Al junctions.

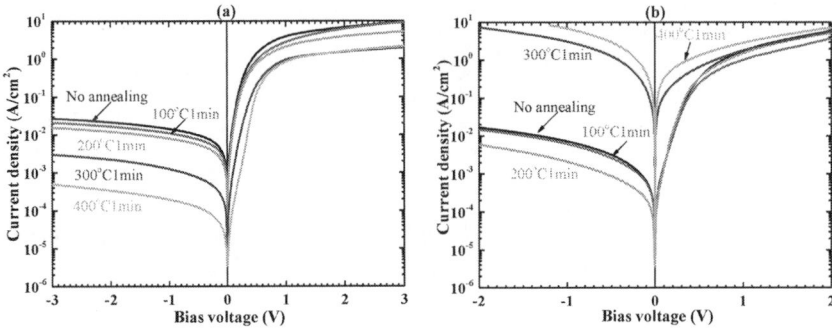

Figure 2. *I-V* characteristics of (a) n-Si/Al foil and (b) p-Si/Al junctions without and with annealing at 100, 200, 300, and 400 °C measured at room temperature.

Table II. The reverse-bias current and ideality factor of n-Si/Al junctions and the interface resistance of p-Si/Al junctions

Annealing temperature	n-Si/Al junction		p-Si/Al junction
	Reverse-bias current (A/cm^2)	Ideality factor	Resistance (Ω·cm^2)
Without annealing	2.7×10^{-2}	2.08	102.04
100 °C	2.1×10^{-2}	2.15	109.89
200 °C	1.6×10^{-2}	2.14	178.57
300 °C	3.0×10^{-3}	2.07	0.73
400 °C	5.0×10^{-4}	2.24	0.26

The influence of the annealing temperature on the *I-V* characteristics of p-Si/Al junctions measured at room temperature is shown in Figure 2(b). We found that the *I-V* characteristics of the junction without annealing shown in this figure revealed Schottky rectification properties. The current density of the revere bias voltage decreased as the annealing temperature increased up to 200 °C and then increased after the annealing temperature increased from 300 to 400 °C. Moreover, the *I-V* characteristics of the junctions with annealing at 300 and 400 °C show linear property. The interface resistances were extracted to be 0.73 and 0.26 $\Omega \cdot cm^2$ for the junction annealed at 300 and 400 °C, respectively, by least-square fitting around 0 V. Similarly, the interface resistances of the junctions without and with annealing at 100 and 200 °C were calculated and the results are shown in Table I . It was found that the interface resistance decreased with increasing the annealing temperature. Furthermore, the junction after annealing at 400 °C brought about the smallest value of the interface resistance in all the samples. However, this value is two orders of magnitude larger than the resistance of p-Si substrate (not shown in this paper) Ohmic contacts.

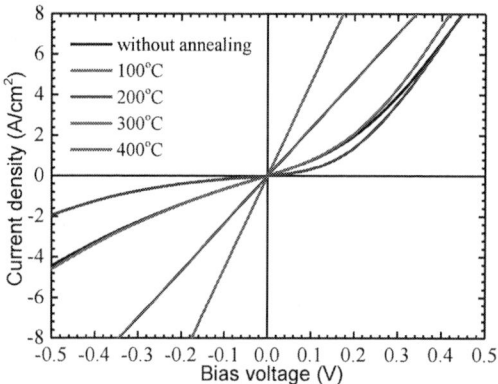

Figure 3. *I-V* characteristics of p^+-*Si/Al* foil junctions without and with annealing at 100, 200, 300, and 400 °C measured at room temperature.

The *I-V* characteristics of p^+-*Si/Al* foil junctions without and with annealing from 100 to 400 °C measured between - 0.5 and 0.5 V at room temperature are shown in Figure 3. We found that, in the bias voltage for the measurements, the *I-V* characteristics of p^+-*Si/Al* foil junctions without and with annealing at 100 and 200 °C showed nonlinear properties. However, after the junctions annealing at 300 and 400 °C their *I-V* characteristics show excellent linearity. Furthermore, their interface resistances were found to be 0.021 and 0.042 $\Omega \cdot cm^2$, respectively, by least-square fitting at approximately 0 V. In the same way, the resistances of the junctions with other various annealing temperature are obtained and shown in Figure 5. It was found that the interface resistance increased slightly as the annealing temperature increased up to 200 °C, and then decreased sharply as the annealing temperature increased to 300 °C, finally increased again when the temperature increased to 400 °C.

Figure 4 shows the *I-V* characteristics of n⁺-Si/Al foil junctions without and with annealing from 100 to 600 °C measured between - 0.5 and 0.5 V at room temperature. It was found that the *I-V* characteristics in this figure showed excellent linear property. By least-squares fitting at approximately 0 V, the interface resistance of the junctions without and with annealing at 100, 200, 300, 400, 500, 600 °C were determined to be 0.04, 0.042, 0.038, 0.036, 0.032, 0.060, 0.167 $\Omega \cdot cm^2$, respectively, and are shown in Figure 6. It is evident that the interface resistance decreases little by little with the annealing temperature increasing to 400 °C. Moreover, the junctions after annealing at 400 °C showed the smallest interface resistance among all n^+-Si/Al foil junctions. However, when the annealing temperature exceeded 400 °C, the interface resistance substantially increased.

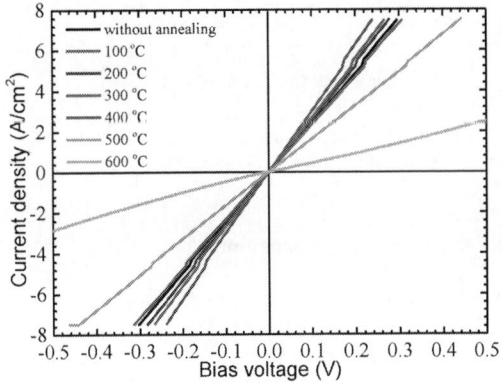

Figure 4. *I-V* characteristics of (a) n^+-Si/Al foil junctions and (b) n^+-Si substrates with Al contacts on both surfaces without and with annealing at 100, 200, 300, and 400 °C measured at room temperature.

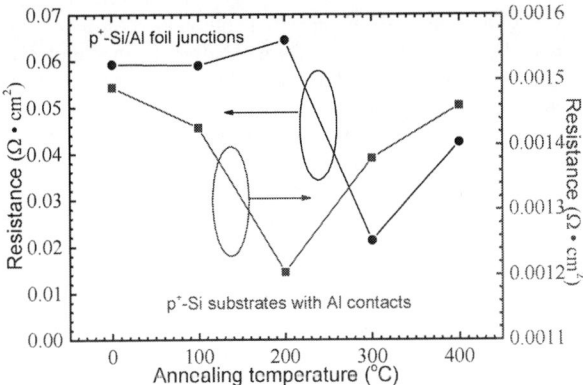

Figure 5. Resistances of the respective samples at various annealing temperatures for p^+-Si/Al foil junctions and p^+-Si substrates with Al contacts on both surfaces.

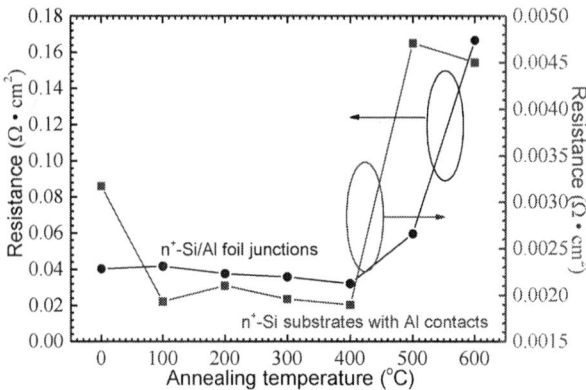

Figure 6. Resistances of the respective samples at various annealing temperatures for n$^+$-Si/Al foil junctions and n$^+$-Si substrates with Al contacts on both surfaces.

Discussion

It was found that the *I-V* characteristics of n-Si/Al, p-Si/Al, p$^+$-Si/Al, and n$^+$-Si/Al junctions largely depended on the annealing temperature. The reverse bias current of n-Si/Al junctions decreased as the annealing temperature increased. In contrast the current of p-Si/Al junctions markedly increased when the temperature is higher than 200 °C. With increasing the annealing temperature, the interface resistances of p$^+$-Si/Al and n$^+$-Si/Al junctions became more and more small. An upward trend in the resistance of p$^+$-Si/Al junctions with annealing at 300 °C was observed, which is due to the oxidation of Al foil surfaces for the high temperature annealing. Similar results were also observed in the contact resistance of p$^+$-Si substrates with Al film on both surfaces with annealing at various temperatures. The annealing temperature dependence of the ohmic contacts of p$^+$-Si substrates with Al film is shown in Figure 5. In addition, the interface resistance of n$^+$-Si/Al junctions continually decreased as the annealing temperature increased up to 400 °C and then increased as the annealing temperature increased to 600 °C. A similar change was observed in the contact resistance of n$^+$-Si substrates with Al film on both surfaces. The contact resistances of n$^+$-Si substrates with annealing at various temperatures are shown in Figure 6. In addition, the electrical conductivity of n$^+$-Si/Al junctions degraded at higher than 400 °C, which may be due to aluminum as acceptor impurities diffuses into n$^+$-Si substrates and causes the neutralization of acceptor and donor impurities near the bonded interfaces. A similar behavior has also been reported in multi-crystalline silicon (8).

The abovementioned the electrical conductivity of all the samples were improved after the junctions annealing at high temperature, which should be attributed to the recovery of the interface damaged layers due to the annealing process. In the bonding process, the Ar atom fast beam irradiation are used to the activation of substrate surfaces so that the amorphous layer with a thickness of several nanometers are formed on the bonding surface (9, 10). A large number of interface states should be existed at the bonding interface and distributed in the amorphous layer according to our previous report

(11, 12), which could directly act as traps and recombination centers. Such interface states can introduce a large charge at the interface so that a potential barrier and depletion layers form in the conduction band (valence band) of Si/Al bonded interface. It was reported that the thickness of the amorphous layers and the density of interface states depended on the annealing temperature in Si/Si junctions fabricated by SAB, which decreased as the annealing temperature increased (12). Furthermore, the electrical conductivity of Si/Si junctions was significantly improved after annealing at high temperature. Consequently, the high annealing process larger than 400 °C is essential for obtaining low interface resistance of p-Si/Al junctions. These results suggest that the bonding processes of thick metal foil has the potential of providing low resistance Ohmic contacts and reducing the fabrication cost of devices.

Conclusion

We fabricated the n-Si/Al, p-Si/Al, n^+-Si/Al, and p^+-Si/Al junctions using SAB method and demonstrated the influence of thermal annealing process on their electrical properties. The *I-V* characteristics of the n-Si/Al junctions after annealing at 400 °C showed good rectification properties. In contrast, the *I-V* characteristics of p-Si/Al junction with annealing at 400 °C revealed linear properties. Moreover, the interface resistance of 0.26 $\Omega \cdot cm^2$ was obtained. The interface resistance of n^+-Si/Al and p^+-Si/Al junctions decreased with a rise in the annealing temperature. They were reduced to 0.032 and 0.021 $\Omega \cdot cm^2$ after the junction annealing at 400 and 300 °C, respectively. These results indicate that SAB of metal foils is an effective way for fabricating thick film electrode with a thickness of several-ten micrometers.

Acknowledgements

This work was partly supported by "Creative Research for Clean Energy Generation Using Solar Energy" project in Core Research for Evolutional Science and Technology (CREST) programs of the Japan Science and Technology Agency (JST).

References

1. T. C. Shen, G. B. Gao, and H. Morkog, *J. Vac. Sci. Technol. B*, **10**, 2113 (1992).
2. S. M. Sze and K. K. Ng, *Physics of Semiconductor Devices*, p. 188, John Wiley and Sons, New Jersey (2007).
3. E. F. Chor, D. Zhang, H. Gong, W. K. Chong, and S. Y. Ong, *J. Appl. Phys.*, **87**, 2437 (2000).
4. R. S. Chen, C. C. Tang, W. C. Shen, and Y. S. Huang, *Nanotechnology*, **25**, 415706 (2014).
5. H. Takagi, K. Kikuchi, R. Maeda, T. R. Chung, and T. Suga, *Appl. Phys. Lett.*, **68**, 2222 (1996).
6. J. Liang, T. Miyazaki, M. Morimoto, S. Nishida, N. Watanabe, and N. Shigekawa, *Appl. Phys. Express*, **6**, 021801 (2013).

7. K. Furuna, J. Liang, N. Shigekawa, M. Matsubara, M. Dhamrin, and Y. Nishio, 2016 IEEE International Meeting for Future of Electron Devices, Kansai, p. 86, Kyoto, Japan (2016).
8. D. Margadonna, F. Ferrazza, R. Peruzzi, S. Pizini, C. Acerboni, L. Tarchini, W. Xiwen, and A. De Lilo, in Tenth E. C. Photovoltaic Solar Energy Conference/1991, A. Luque, G. Sala, W. Palz, G. Dos Santos, P. Helm, Editors, PV 8-12, p.678, Proceedings of the International Conference, Lisbon, Portugal (1991).
9. J. Liang, T. Miyazaki, M. Morimoto, S. Nishida, and N. Shigekawa, *J. Appl. Phys.*, **114**, 183703 (2013).
10. J. Liang, S. Nishida, M. Arai, and N. Shigekawa, *Appl. Phys. Lett.*, **104**, 161604 (2014).
11. J. Liang, S. Nishida, T. Hayashi, M. Arai, and N. Shigekawa, *Appl. Phys. Lett.*, **105**, 151607 (2014).
12. M. Morimoto, J. Liang, S. Nishida, and N. Shigekawa, *Jpn. J. Appl. Phys.*, **54**, 030212 (2015).

Analysis of Defect Levels at GaAs/GaAs Surface-Activated Bonding Interface for Multi-Junction Solar Cells

M. Sugiyama [a], D. Yamashita [a], K. Watanabe [b], M. Fujino [a], T. Suga [a], and Y. Nakano [a, b]

[a] School of Engineering, The University of Tokyo, Bunkyo-ku, Tokyo, 113-8656, Japan
[b] RCAST, The University of Tokyo, Meguro-ku, Tokyo, 153-8904, Japan

Quantitative evaluation of crystal defects introduced by fast atom beam (FAB) treatment was developed for surface-activated wafer bonding. The surface of n-GaAs was treated with the FAB using Ne, Ar and Kr, and Au Schottky electrodes were formed on the surfaces. Capacitance of a Schottky diode as a function of both probe frequency and DC bias allowed us to characterize both energy depth of the defects and their density profile along the physical depth from the GaAs surface. The results indicated that atoms with the smaller diameter generate high-density defects to the deeper region from the surface. When the defect density exceeding the doping level of GaAs spreads to wider than 5 nm, significant Schottky characteristics appeared in the interfacial current-voltage characteristics, as suggested by simulations. Such a tendency was semi-quantitatively in good agreement with the measured current-voltage characteristics.

Introduction

Wafer-bonded III-V multi-junction solar cells are attractive because their bandgap combination can be engineered beyond the constraint of lattice mismatching (1). Most of such multi-junction structure need the bonding between III-V compound semiconductor surfaces.

For high efficiency, the electrical resistance at the bonding interface should be low enough to avoid power loss due to joule-heating. Surface activated bonding (SAB) is a promising technique to fabricate low resistance bonding interfaces. In SAB process, native oxide layers and contaminations at material surfaces are removed by fast-atom-beam (FAB) treatment in vacuum prior to the bonding process. FAB irradiation, however, introduces crystal defects on the surface of III-V semiconductors, resulting in the increase in interfacial electrical resistance when those damaged surfaces are bonded together (2). It is, therefore, essential to minimize such defects by optimizing the ion species and irradiation conditions, for which quantitative analysis on the defect profile in the surface vicinity is indispensable. We here will demonstrate the effectiveness of admittance spectroscopy (3) for such defect profiling by forming a Schottky electrode on the damaged surface. GaAs has been chosen as a test material since its electrical properties are quite susceptive to defects and it is one of the most commonly used material at the bonding interface of multi-junction cells.

Experimental

As a simple structure for clarifying the impact of SAB process conditions on the electrical conduction property of the bonded interface, two n-GaAs wafers were bonded by SAB process.

Prior to the bonding process, AuGe/Ni ohmic electrode was formed on the backside of n-GaAs wafers. The wafers were then diced into chips and their surfaces were treated with the FAB of Ne, Ar and Kr in a high-vacuum chamber, with an acceleration voltage of 1.4kV and the duration of 3 min as shown in Fig. 1.

For the admittance spectroscopy to analyze the defects introduced by the FAB treatment, the n-GaAs diced substrates were taken out of the vacuum chamber and Au Schottky electrode was evaporated on the surface.

For the conductivity measurement of the boded interface, two GaAs dices after the FAB treatment were put together and pressurized to complete bonding. Subsequently, the bonded dices were further cut into 2×2 mm^2 chips and current-voltage characteristics were measured by 4 terminal method.

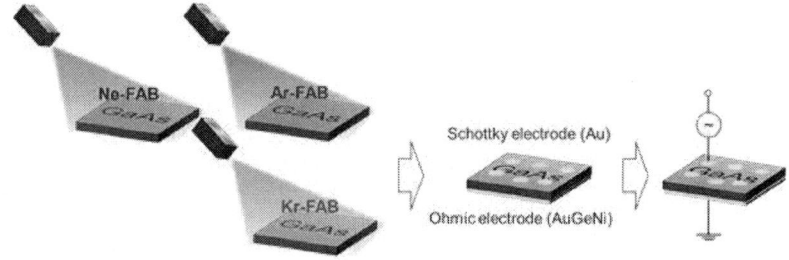

Figure 1. Process flow of the defect analysis on the GaAs surface treated with fast atom beam (FAB).

Results and Discussion

Admittance spectroscopy

Figure 2 shows typical capacitance-frequency curves for the Schottky junction using the GaAs substrate treated with Ar-FAB. When electrons trapped by the defects in the surface vicinity of the substrate respond to the small AC field for capacitance measurement, capacitance in the depletion region of the junction increases. This is why we observe the increase in capacitance in the low-frequency region for the substrate treated with Ar-FAB. DC bias changed the band bending near the junction and quasi Fermi level hits the defects existing in different depth-position according to the value of the DC bias. When a large reverse bias (a negative value in the figure) is applied to the junction, the position where defect respond to the AC field moves apart from the surface, and the reduction in the capacitance indicates that defects density decreases to the depth direction of the substrate. Too large a reverse bias induces junction leak and measurement became unsuccessful.

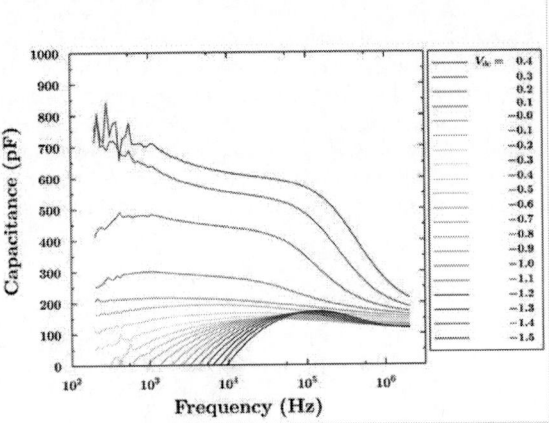

Figure 2. Typical Capacitance-Frequency characteristics for a Au/GaAs Schottky junction whose GaAs surface had been treated with FAB for varied DC bias voltages.

From such a dependence of the junction capacitance on both AC frequency and DC bias, with an assumption of a typical value of carrier-capture cross section, it is possible to estimate both the energy position and the defect density profile along the depth direction of the defects existing in the surface vicinity of the substrate, or in the depletion region of the Schottky junction, as shown in Fig. 3. Remarkably high defect density (around 10^{19} cm^{-3}) was obtained exceeding the doping level of the substrate and the depth profile indicates that the FAB of the smaller atom introduces defects in the deeper region from the surface. The energy level of the defects were estimated to be 0.8 eV below the conduction-band edge.

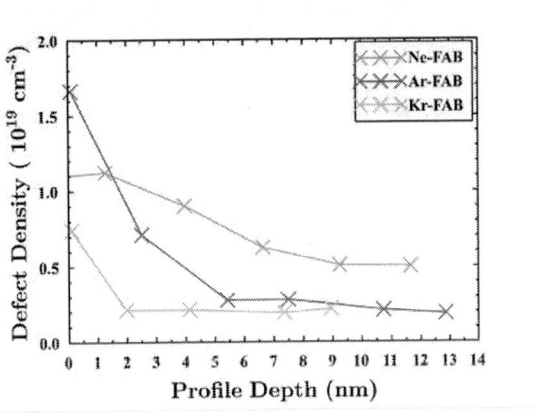

Figure 3. Depth profile of the defect density from the GaAs surface treated with FAB using different atomic species.

Impact of defects on the interfacial electrical conductance

Such high-density interfacial defects in the middle of the bandgap will certainly distort the band edges at the n-GaAs/n-GaAs junction and degrade electrical conductance. Figure 4 shows the simulated band lineup in the vicinity of the junction, where it is assumed that a fixed density of defects (2×10^{19} cm^{-3}) exist uniformly along the depth direction (1 to 7 nm) at the junction. The acceptor-like defect imposes negative potential and the band-edge sticks up, imposing a potential barrier for current conduction.

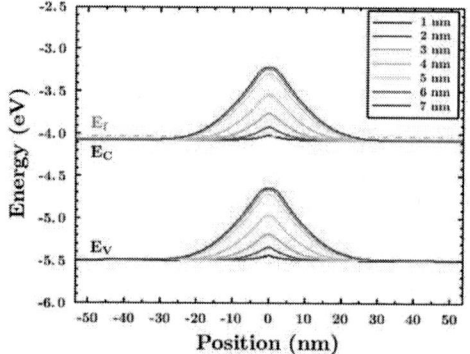

Figure 4. Simulated band-edge profiles at the n-GaAs/n-GaAs bonded interface with the thickness of the region containing 2×10^{19} cm^{-3} defects at an energy level 0.8 eV below the conduction band edge. The doping level of n-GaAs is assumed to be 2×10^{18} cm^{-3}.

Conduction behavior at n-GaAs/n-GaAs interface

The current-voltage characteristics at the junction was simulated as shown in Fig. 5. The defect region thicker than 4 nm results in nonlinear, Schottky-like conduction behavior, which we should avoid in high-efficiency cells.

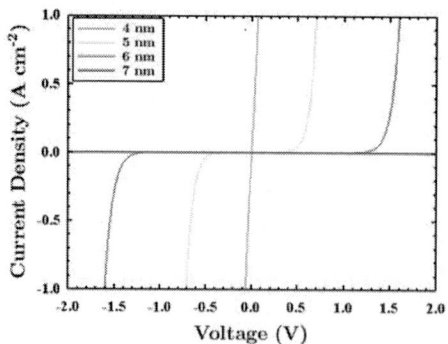

Figure 5. Simulated current-voltage characteristics for the n-GaAs/n-GaAs bonded interface in Fig. 4.

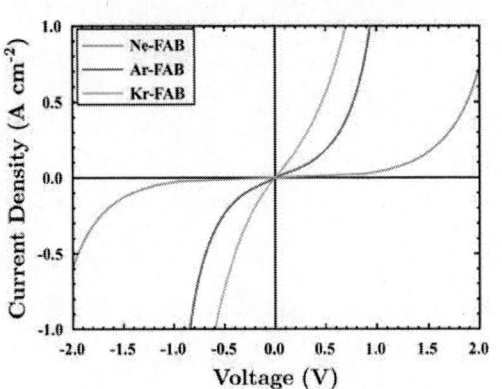

Figure 6. Measured current-voltage characteristics for the n-GaAs/n-GaAs interface by surface activation bonding using FAB treatment with Kr, Ar and Ne.

The measured current-voltage characteristics at the n-GaAs/n-GaAs junctions by the SAB using the FAB of Ne, Ar and Kr are shown in Fig. 6. The tendency of Schottky-like conduction becomes more apparent in the sequence of Ne, Ar and Kr. This is the sequence of smaller atomic radius. In the defect density profile in Fig. 3, if we focus on a certain defect density level such as 0.5×10^{19} cm^{-3}, the penetration depth of the defects is larger also in the sequence of Ne, Ar and Kr.

Summary

It is strongly suggested that the high-density defects at the interface by wafer bonding distorts band edges and makes potential barrier for electrical conduction. In SAB process, FAB treatment introduces the defects in 10^{19} cm^{-3} level and it has significant impact on the interfacial electrical conduction. A suggestion for reducing such introduction of defects is the use of larger atoms in FAB treatment, which reduces penetration depth of defects. It is clear that the condition of FAB such as acceleration voltage of primary ions and exposure time impacts the density and distribution of defects. The optimization of such process parameters is quite possible using the admittance spectroscopy proposed in this work.

Acknowledgement

A part of this study is supported by New Energy and Industrial Technology Development Organization (NEDO), Japan (P15003).

References

1. F. Dimroth et al., *Prog. Photovoltaics Res. Appl.*, **22**, 277 (2014).
2. L. Chai, J. Liang, S. Nishida, M. Morimoto, and N. Shigekawa, *Proc. 13th International Meeting for Future of Electron Devices*, 44 (2015).

3. T. Walter, R. Herberholz, C. Müller, and H. W. Schock, *J. Appl. Phys.*, **80**, 4411 (1996).

The roles of band bending, surface misorientation, and passivation on electrical transport across III-V bonded structures

M. Yee, M. Liao, M. Seal, and M. S. Goorsky

Department of Materials Science and Engineering, University of California, Los Angeles, California, 90095, USA

With results from bonding of several different materials combinations, surface orientations, and passivation treatments, as well as an assessment of barrier heights reported in the literature for transport in polycrystalline III-V materials as well as for other III-V bonded materials combinations, a general model has emerged. Lattice mismatch does not play a major role in limiting conductivity across an interface; however, misalignment (through twist or tilt) produces interfaces with increased resistance. Passivation techniques are most important for these systems which possess significant interface band-bending. The electrical characteristics of a grain boundary, whether in polcrystalline material or at a bonded interface, are determined by the presence of defect states and interfacial charge which create band bending and Fermi level pinning. The barrier heights and widths from zero bias conductance vs bias measurements over a wide range of temperatures and for different materials can be matched to band structure simulations of bonded semiconductor heterojunctions with resultant I-V curves showing good agreement between experiment and theory. Based on this study, which combines experimental results and transport modelling, it is expected that narrow bandgap III-V semiconductors should show lower barrier heights; but more importantly, modelling indicates that high doping at mismatched interfaces helps to significantly reduce the interface barrier height.

Introduction

Over the past few years, high electrical quality III-V interfaces have been demonstrated (1,2) at bonding temperatures and times that were more suitable for device structures that require low thermal budgets, such as solar cells. Indeed, record performance one-sun (3) and concentrated (4) multi-junction solar cells were based on bonded III-V interfaces. A more detailed understanding of the impact of different factors on the electrical quality of the bonded interface would better guide new applications for III-V bonding.

Jackson, et al., (1) demonstrated that sulfur-based passivation can improve the electrical conductivity across n-GaAs/ n-GaAs interfaces for wafers bonded in air. Low temperature, short time annealing led to interfaces with a resistance of ~ 0.01 $\Omega \cdot cm^2$, which was significantly lower than that reported under conditions which did not utilize interface passivation. In subsequent publications, the role of relative miscut was shown to be important as well, with an increased difference in miscut between the two bonding faces leading to increased interface resistance. More recent results (5) showed that bonded n-InP/ n-InP, albeit at a higher doping concentration, led to reduced interface resistance (0.005 $\Omega \cdot cm^2$) compared to n-GaAs / n-GaAs. Interestingly, n-GaAs / n-InP interfaces showed interface resistances that were very similar to n-InP / n-InP and, therefore, much lower than n-GaAs / n-GaAs.

A better understanding of the impact of different factors (e.g., doping concentration and III-V material) was pursued through modelling of the interface by one-dimensional band structure simulations. This work describes the insight gained from such a study.

Experimental

Experimental results from previous studies (1,2,6) are used here to provide experimental data for analysis through modelling. The band structure simulation program utilized here is based on the AFORS-HET program (7) in which an interface trap layer is introduced to mimic the dangling bonds present at the interface between the two materials. In all cases, the interface defects are considered to be mid-gap defects. While there are known defect states – and even a distribution of defects states – due to different impurities or even structural defect configurations, single energy, mid-gap states are employed here to show the general trends. Figure 1 shows the schematic of the structure used. For the modelling, substrate thicknesses of 150 μm were used. Doping levels were in the range of 10^{18} cm^{-3}.

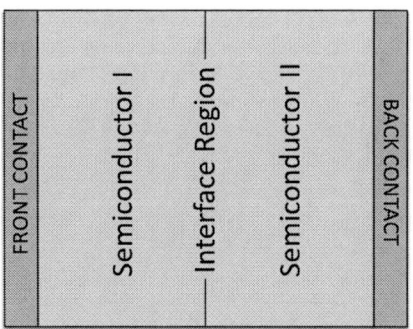

Figure 1. Schematic of structure used for modelling studies.

Results and Discussion

Modelling results using an n-GaAs / n-GaAs structure with interface defects is shown below. In this example, both GaAs layers are n-type doped at 3×10^{18} cm^{-3}. Figure 2 shows the presence of the interface barrier that is due to the addition of interface charge. Figure 3 plots the barrier height as a function of the interface charge density. It can be seen that interface charge values below about 10^{11} cm^{-2} do not introduce a measureable increase to the barrier height. The barrier height matches the Fermi level (for 3×10^{18} cm^{-3}

doping) when the interface charge density is ~ 2×10^{12} cm^{-2}. The barrier height increases substantially with interface charge when the interface charge is greater than about 10^{12} cm^{-2}. Experimentally, GaAs – GaAs interfaces without sulfur passivation exhibited a barrier height of approximately 0.8 eV. For bi-crystal specimens of GaAs, the energy barrier is reported to vary between 0.5–1.1 eV, although there is no information given about the orientation or chemistry at these grain boundaries from those earlier studies. A barrier height of 0.8 eV corresponds to an interface charge density of approximately 2×10^{13} cm^{-2}. The addition of sulfur to passivate the interface reduces the barrier height to 0.55-0.60 eV. This corresponds to an interface charge of approximately 6×10^{12} cm^{-2} or a reduction of about a factor of 3-4. On the other hand, increasing the relative miscut increases the barrier height to approximately 1.0 eV. This corresponds to an interface charge increase of about an order of magnitude. The full width at half maximum of the barriers is about 15 nm. This value compares well with the barrier width deduced from zero bias conductance vs temperature measurements (1). Work is continuing to address the role of both tilt and twist orientations - both experimentally and through modelling such as this.

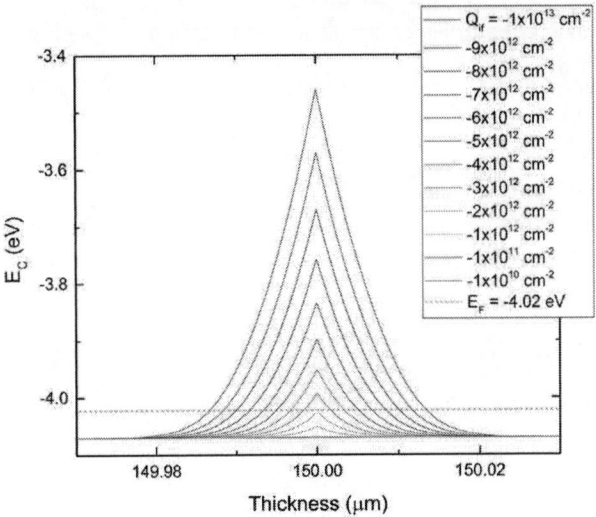

Figure 2. The barrier height across a GaAs-GaAs junction for different interfacial charge densities up to 10^{13} cm^{-2}. The dotted red line represents the Fermi level.

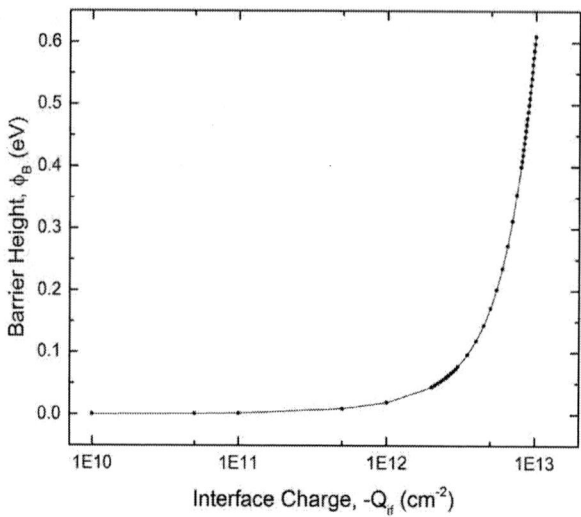

Figure 3. The dependence of the barrier height on the interface charge density.

The I-V curves from the structures shown in figure 1 are calculated based on the barrier heights and interface charge. Figure 4 shows the I-V curves for the different interfacial charge densities. There is negligible change in the I-V curves for defect densities below $2x10^{12}$ cm^{-2}, which corresponds to the case where the barrier height becomes comparable to the Fermi level (Figure 2). Qualitatively, these simulated curves match the experimental results which show that the interface resistance increases and the I-V curves become non-ohmic with increasing misorientation.

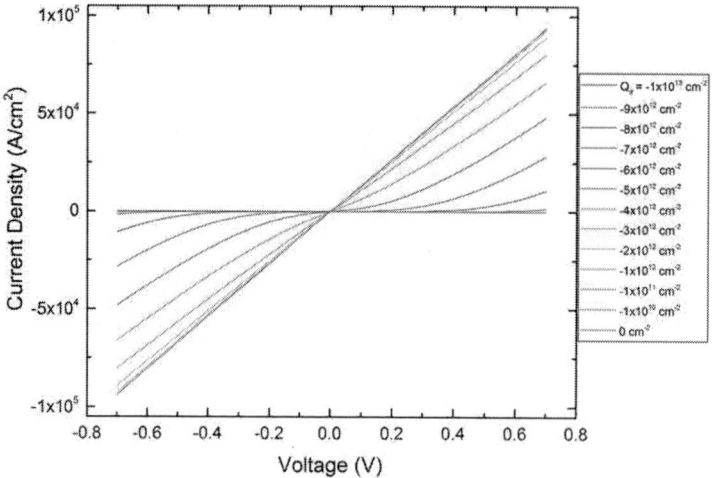

Figure 4. Simulated current density vs voltage curves for structures with increasing interface charge densities.

The understanding gained from the GaAs interface simulations indicate that similar understanding should apply to other materials as well. Indeed, the InP barrier heights appear to be lower than GaAs barrier heights. However, the InP wafers used are doped higher than the GaAs wafers. The reduced barrier heights for InP (~0.25 eV), therefore, are due primarily to the increased doping and not substantially to differences in materials properties.

The similarity between GaAs and InP results lead to a question about whether one would expect different barrier heights for different materials. Simulations based on homojunctions of InAs, InSb, and GaSb, do demonstrate that barrier heights are reduced for those materials for a given interface charge density. For example, with a doing concentration of $1x10^{18}$ cm^{-3} and an interface charge density of $2x10^{12}$ cm^{-2}, the barrier height for InAs is less than 0.07 eV, GaSb is 0.10 eV, and GaAs and InP are about 0.11 eV. The effect is more dramatic at higher charge density levels but this comparison indicates that lower barrier heights – and hence lower contact resistances are expected for materials with generally lower bandgaps.

Conclusion

A better understanding of the impact of different factors (e.g., doping concentration and III-V material) on the barrier height and I-V curves for III-V/III-V bonded structures was demonstrated through modelling of the interface by one-dimensional band structure simulations. Interface charge densities in the range of 10^{12} cm^{-2} have a significant impact on the interfacial barrier height. These results suggest that sulfur doping reduces the interface charge by about a factor of 3-4, and hence plays a major role in reducing barrier heights at III-V interfaces. The barrier widths also closely match the results from fitting zero bias conductance vs temperature measurements, further supporting the use of these models for understanding how to control barrier heights in III-V bonded structures. Initial results show that narrow bandgap semiconductors – under otherwise identical conditions – are expected to possess lower barrier heights.

References

1. M. Jackson, B. Jackson, M. Goorsky, *J. Appl. Phys.*, **110**, 104903 (2011).
2. K. Yeung, J. Mc Kay, C. Roberts, M.S. Goorsky, *ECS Trans.*, **50**(7), 99 (2012).
3. P. Chiu, D. Law, R. Woo, S. Singer, D. Bhusari, W. Hong, A. Zakaria, J. Boisvert, S. Mesropian, R. King, N. Karam, *IEEE J. Photovoltaics*, **4**, 1, (2014).
4. F. Dimroth, M. Grave, P. Beutel, U. Fiedeler, C. Karcher, T. Tibbits, E. Oliva, G. Siefer, M. Schachtner, A. Wekkeli, et al., *Prog. Photovolt: Res. Appl.* **22**, (2014).
5. M. Seal, J. Mc Kay, M. Liao, M.S. Goorsky, "Electrical Conductivity Across InP and GaAs Wafer-Bonded Structures with Miscut Substrates", in *42nd IEEE Photovoltaic Specialist Conference*, 2015, 1-6.
6. J. Mc Kay, M. Seal, K. Yeung, M. Jackson, M.S. Goorsky, *ECS Trans.*, **64**(5), 225 (2014).
7. R. Varache, C. Leendertz, M. E. Gueunier-Farret, J. Haschke, D. Muñoz, and L. Korte. "Investigation of Selective Junctions Using a Newly Developed Tunnel

Current Model for Solar Cell Applications." *Solar Energy Materials and Solar Cells* **141** (2015) 14–23.

**Conductive Semiconductor Interfaces Fabricated by
Room Temperature Covalent Wafer Bonding**

C. Flötgen, N. Razek, V. Dragoi, and M. Wimplinger

EV Group, DI E. Thallner 1, 4782 – St. Florian am Inn, Austria

Low temperature direct bonding in the range of maximum 400°C
opened an entirely new applications field as it allowed for the
direct bonding of wafers with metallization (e.g CMOS) while
maintaining a high alignment accuracy as well as bonding of
thermally mismatched materials. Despite the clear benefits of the
low temperature direct bonding process, the requirements of some
categories of applications in terms of allowing for room
temperature covalent bonding or oxide-free, electrically conductive
interfaces fabrication cannot be fulfilled. A new technology was
developed to address these types of applications: EVG®580
Combond®. This paper reports on fabrication of oxide-free,
electrically conductive Si-Si and Si-GaAs bonded interfaces based
on the newly developed technology. Experimental results on
bonding quality and electrical performance of the bonded
interfaces are presented.

Introduction

Direct wafer bonding is both an attractive and challenging bonding process. The plasma
activated wafer bonding process allows for low temperature processes (<400°C) (1)
providing the best conditions for extremely high wafer-to-wafer alignment accuracy,
clean process (CMOS-compatible) (2) and heterogeneous integration (3). However,
despite the benefits for numerous applications, low temperature plasma activated bonding
cannot cover a specific applications areas requiring formation of covalent bonds at room
temperature (no thermal annealing) or fabrication of oxide-free, conductive bonded
interfaces.

A new technology was developed to provide a solution for such applications:
EVG®580 Combond® (4). This wafer bonding technique is based on an *in situ* removal of
native oxides and bonding under high vacuum ambient in order to prevent surface re-
oxidation. Different materials were bonded using this new room temperature technology
with very encouraging results (e.g. SiC-SiC, Si-SiC, InP-GaAs, and even Al-Al).

This paper reports on fabrication of electrically conductive bonded interfaces using
Si-Si as well as Si-GaAs room temperature bonding. The Si-Si bonds were used as
baseline, silicon being a standard qualification material. The Si-GaAs materials
combination was selected for study due to its high application potential in high-efficiency
multi-junction solar cells (5, 6) and optoelectronic integrated circuits (7, 8).

Experimental

Two different types of bonds were performed in order to test the electrical performance of the bonded interfaces: Si-Si and Si-GaAs.

The Si-Si bonding experiments were performed on 200 mm diameter p-type Si wafers, (100) orientation, having only native oxide on their surfaces. The wafers were loaded to the EVG®580 Combond® equipment for surface preparation and bonding. Preliminary experiments were studying the influence of surface preparation method on surface microroughness by measuring with Atomic Force Microscope (AFM).

After bonding the wafers samples were diced for the electrical testing and metal contacts were deposited. Before the electrical measurements some of the samples were thermally annealed. The thermal annealing temperature was studied by measuring the amount of amorphous layer at the interface as a function of thermal annealing temperature and time.

The bonded interface quality was investigated using Scanning Acoustic Microscopy (SAM – not shown here) and high resolution Transmission Electron Microscopy (HR-TEM) cross section analysis was performed in order to verify the bonded interface structure and measure the amorphous layer thickness.

For the Si-GaAs experiments were used 100 mm diameter mirror polished n-GaAs:Si (2.5×10^{18} cm^{-3}), p-GaAs:Zn (2×10^{19} cm^{-3}) wafers and n-Si: Sb (1×10^{19} cm^{-3}) wafers.

Preliminary experiments were investigating the influence of the surface preparation on surface by AFM measurements of surface microroughnes.

After bonding the wafers were unloaded and 3×3 mm^2 samples were diced for the electrical testing and ohmic metal contacts (silicon: Ti/Pd/Ag,GaAs: Pd/Ge-based) were deposited by evaporation. The metal contacts were alloyed for 1 min at 290°C in forming gas environment (9). Before the electrical measurements some of the samples were thermally annealed using rapid thermal annealing (RTA) for1 minute at 290°C.

The bonded interface quality was investigated using Scanning Acoustic Microscopy (SAM) and high resolution Transmission Electron Microscopy (HR-TEM) cross section analysis was performed in order to verify the bonded interface structure and measure the amorphous layer thickness.

Results and Discussion

Si-Si Covalent Wafer Bonding

First experiments were performed for checking the various surface preparation process conditions influence on surface microroughness. The AFM measurements performed on 2×2 μm^2 areas for three different process recipes as well as for an as-received Si wafer (reference) are summarized in TABLE I.

TABLE I. Si-Si covalent wafer bonding: surface microroughness vs. surface preparation condition.

Process	Surface Microroughness (nm)
As received	0.18
Recipe 1	0.147
Recipe 2	0.092
Recipe 3	0.066

As all values are within the acceptable values for fusion wafer bonding all process conditions were compatible with bonding requirements.

In previous work was reported the presence of an amorphous layer at the interface as a result of the surface preparation. As far as not only oxide but also amorphous material has an impact on the electrical performance the selection of the proper surface activation condition will be based on the minimization of the amorphous layer thickness.

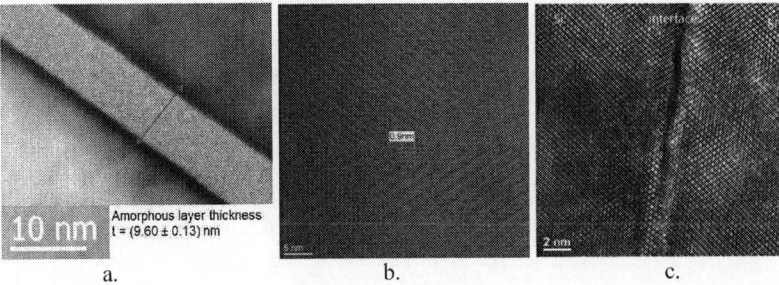

Figure 1. HR-TEM images showing the amorphous layer thickness for three different surface preparation processes using a. high power, b. medium power and c. low power.

It can be observed that the activation power is directly influencing the thickness of the amorphous layer so moderate process conditions were chosen.

Initial electrical characterization experiments showed no conductivity through the bonded interface for as-bonded samples (fig. 2).

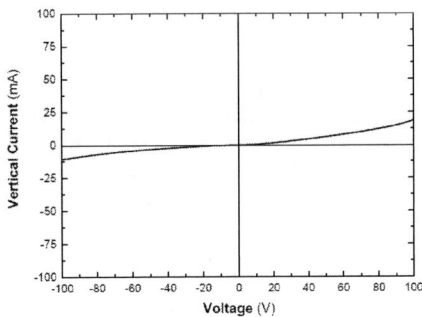

Figure 2. Current-Voltage measurement showing poor electrical performance of the bonded interface.

As prior work showed the absence of oxide from the bonded interface after the surface preparation and bonding the effect illustrate in fig. 2 is probably caused by the presence of amorphous layer.

Another possibility to further reduce the amorphous layer thickness is by thermal annealing. Performing a thermal annealing is not against the principle of room temperature bonding: the Combond® surface preparation is removing the oxide and provides full bond strength at room temperature, allowing for stress management (no stress due to thermal expansion) and even for maintaining highly accurate wafer-to-wafer alignment (not needed for this work but an interesting feature for other applications). The amorphous layer has a content of crystalline Si, which is also affected by thermal annealing conditions

Figure 3 shows an exemplary evolution of the amorphous layer thickness and Si crystalline phase content in the amorphous layer as function of the annealing temperature.

For both amorphous thickness and crystalline phase content can be observed that they depend strongly on the annealing temperature and just slightly on the annealing time.

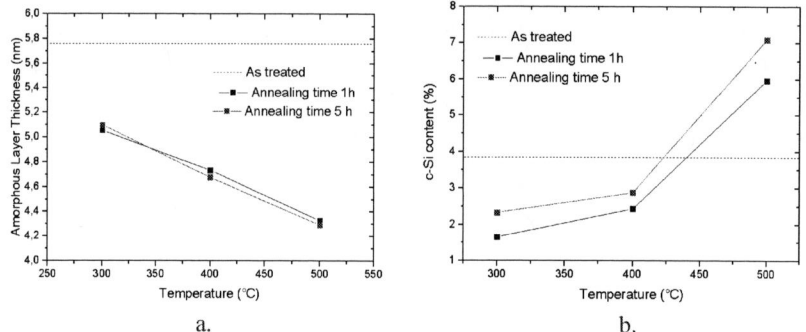

a.   b.

Figure 3. a. Amorphous layer thickness, and b. crystalline phase content in the amorphous layer vs. thermal annealing temperature for two different annealing times.

Based on the above results it was decided to perform a thermal annealing of 1 hour at 450°C in forming gas (to prevent potential oxidation) prior to the current-voltage measurements. The results are shown in fig. 4.

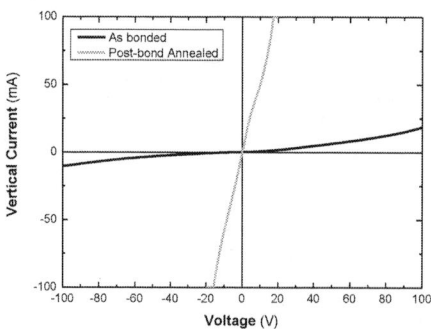

Figure 4. Current-Voltage measurement showing the difference between poor electrical performance of the as-bonded samples (black) and of the samples thermally annealed at 450°C for 1 hour.

It can be observed that after the thermal annealing the interface resistivity changes significantly and the bonded interface shows good electrical conductivity.

Si-GaAs Covalent Wafer Bonding

The microroughness measurements performed by AFM showed good quality surfaces after the surface activation (Table II).

TABLE II. Si-Si covalent wafer bonding: surface microroughness vs. surface preparation condition.

Process	Surface Microroughness (nm)
As received GaAs	0.16
Activated GaAs	0.10
As received Si	0.14
Activated Si	0.07

After bonding, the samples were inspected for bonding defects using SAM (fig. 5).

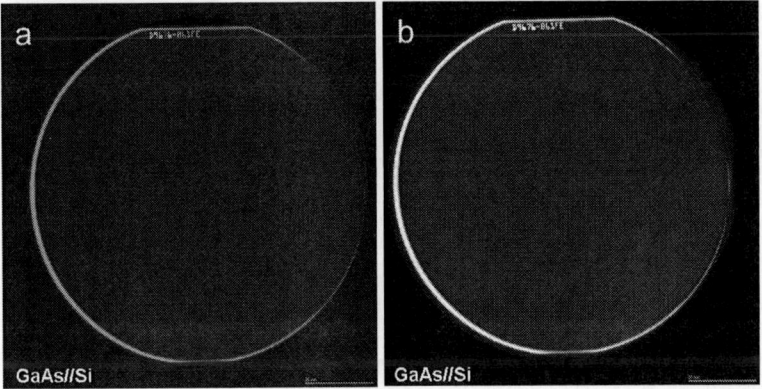

Figure 5. SAM image of a 100 mm wafer pairs (a) after room temperature bonding (no thermal annealing), and (b) after post-annealing at 250°C for 2 hour.

Some of the wafers were still thermally annealed for bond strength comparison as well as to check for potential interface damaging/wafer breakage during thermal annealing (thermal expansion coefficients ratio for GaAs/Si is 2). The bond strength was measured by crack opening method by inserting a blade at the edge: none of the samples (with or without thermal annealing) could be measured; the edge broke during attempting to insert the blade.

The quality of the Si/GaAs bonded interface was further studied by high-resolution cross-sectional TEM (fig. 6).

The interface is very thin and smooth, as well as there is no damage of the crystalline lattice. The presence of a Ga-amorphous layer is visible, most likely a side effect of the Combond® process or/and from surface misfit dislocation with twist angle close to 2°. Another possible root cause for these defects can be contamination, such as carbon at the bonded interface. At some positions, Si islands were observed within the interface layer, probably caused by the stress at the interface or micro-roughness of surfaces.

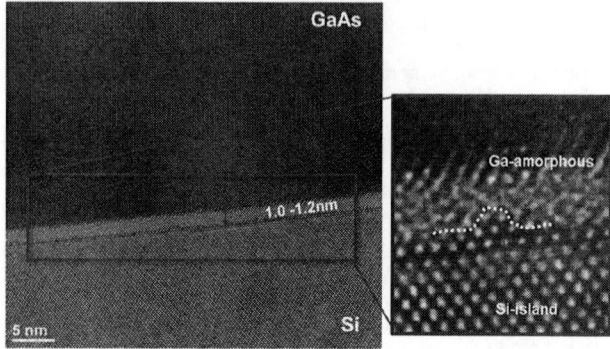

Figure 6. HR-TEM cross section image of a Si-GaAs room temperature bonded pair.

The current–voltage curves presented in fig. 7 were directly measured after ohmic contacts for Si and GaAs were evaporated on the GaAs/Si bond pairs for electrical conductivity evaluation.

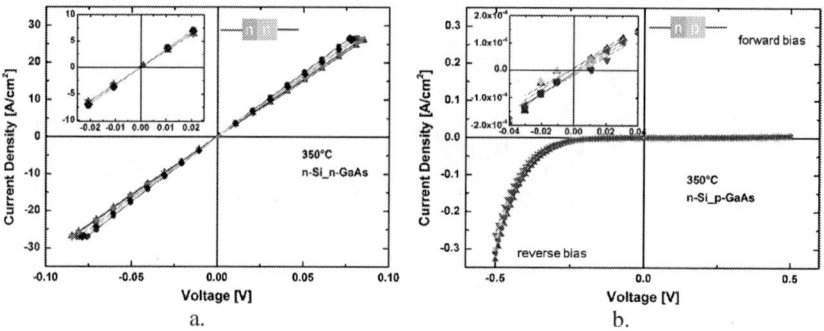

Figure 7. Electrical performance characterization of the room temperature bonded Si-GaAs interfaces and RTA processed 350°C/1 min: a. ohmic behavior of a nSi-nGaAs bonded interface, and b. diode behavior of a nSi-pGaAs bonded interface.

In fig. 7 the I-V characteristics of six n-GaAs/n-Si and p-GaAs/n-Si junctions show that the I–V characteristics of n-Si/n-GaAs in this figure (a) showed excellent linearity and for n-Si/p-GaAs with the interface resistivity was found to be about $0.003 \cdot \Omega cm^2$ (b.) showed standard diode-like behavior without tunneling, and the interface resistivity was found to be about $200 \cdot \Omega cm^2$.

Conclusion

The newly developed room temperature covalent wafer bonding was applied for Si-Si and Si-GaAs bonding in order to study the electrical performance of the bonded interfaces.

The impact of thermal annealing on amorphous layer thickness introduced during surface treatment for covalent wafer bonding was investigated on single p-type prime Si wafers by means of spectroscopic ellipsometry. Thermal annealing was performed directly after surface treatment in high vacuum equipment. Results suggest a growth of microcrystalline-Si grains in/through the amorphous layer during the first hour for temperatures above 250°C. Longer annealing times did neither show significant further increase in the microcrystalline-Si content, nor reduction in general a-Si / microcrystalline-Si layer thickness. Microcrystalline-Si grain growth seems to start at the a-Si/c-Si interface, growing in the direction to the vacuum/a-Si interface. At the vacuum/a-Si interface further recrystallization seems to be inhibited in the temperature range investigated. However, as TEM investigation of bonded samples suggest, further reduction of the a-Si content is possible, if the mixed layer is terminated on both sides by crystalline Si.

The Si-GaAs could be successfully demonstrated using Combond® technology. This process allows for removing all surface contaminations and native oxides from GaAs and Si surfaces, and leads to successful wafer bonding of GaAs/Si with high bond strength at room temperature.

The measured I-V curves of the n-Si /n-GaAs showed excellent linearity forward characteristics at room temperature and ohmic behavior with resistivity below $0.003 \cdot \Omega cm^2$ could be obtained at room temperature. In the case of n-Si/p-GaAs a standard diode-like behavior without tunneling was found, and the interface resistivity was found to be about $200 \cdot \Omega cm^2$.

Acknowledgments

The authors would like to thank Dr. F. Dimroth and Dr. R. Cariou from Fraunhofer Institute for Solar Energy Systems ISE, Germany, for performing the electrical conductivity measurements.

References

1. T. Plach, K. Hingerl, S. Tollabimazraehno, G. Hesser, V. Dragoi and M. Wimplinger, *J. Appl. Phys.*, **113**, 094905 (2013).
2. V. Dragoi, E. Pabo, J. Burggraf and G. Mittendorfer, *Microsys. Techn.*, **18**, 1065 (2012).
3. V. Dragoi, G. Mittendorfer, C. Thanner and P. Lindner, *Microsys. Techn.*, **14**, 509 (2008).
4. C. Flötgen, N. Razek, V. Dragoi and M. Wimplinger, *ECS Trans.*, **64**(5), 103 (2014).
5. E. Radziemska, *Progress in En. and Combust. Sci.*, **29**, 407 (2003).
6. J.Arokiaraj, H. Okui, H.Taguchi, T. Soga, T. Jimbo and M. Umeno, *Solar Energy Materials and Solar Cells*, **66**, 607 (2001).
7. T. Soga, K. Baskar, T. Kato, T. Jimbo and M. Umeno, *J. Cryst. Growth*, **174**, 579 (1997).
8. S. F. Fang, K. Adomi, S. Iyer, H. Morkoc, H. Zabel, C. Choi and M. Otsuka, *J.Appl. Phys.*, **68**, R31 (1990).
9. J. R. Bickford, D. Qiao, P. Yu, and S. Lau, *Appl. Phys. Lett.*, **89**, 012106 (2006).

52

ECS Transactions, 75 (9) 53-65 (2016)
10.1149/07509.0053ecst ©The Electrochemical Society

Invited: High Output Power Deep Ultraviolet Light-Emitting Diodes with Hemispherical Lenses Fabricated Using Room-Temperature Bonding

M. Ichikawa[a,b], S. Endo[a], H. Sagawa[a], A. Fujioka[a], T. Kosugi[a], T. Mukai[a], M. Uomoto[b], and T. Shimatsu[b,c]

[a] LED Development Division, Nichia Corporation, Anan, Tokushima 774-8601, Japan
[b] Frontier Research Institute for Interdisciplinary Sciences (FRIS), Tohoku University, Sendai, 980-8578, Japan
[c] Research Institute of Electrical Communication (RIEC), Tohoku University, Sendai, 980-8577, Japan

We reviewed the structural and optical properties of high output power 255 and 280 nm light-emitting diodes (LEDs) bonded with hemispherical lenses made of inorganic materials. The optimal LED structure with a lens to improve the output power for deep ultraviolet LEDs was designed using Monte Carlo simulation. The LED chips were bonded to sapphire lenses at room temperature using either atomic diffusion bonding or surface activated bonding. The bonding of the lenses to the LEDs is experimentally shown to improve the light extraction efficiency, and the light output power of the LEDs with lenses was significantly higher than that of conventional structure LEDs. The output power of a 255 nm LED with a lens was 2.8 times larger than that without a lens, with a maximum external quantum efficiency of 4.56%, and that of the 280 nm LED with a lens was larger by a factor of 2.3.

Introduction

There has been growing interest in AlGaN-based light-emitting diodes (LEDs) because they are expected as next-generation deep ultraviolet (DUV) light sources to replace mercury lamps (1,2). DUV LEDs have attracted interest because of their potential application in disinfection, sensing, and polymer curing. DUV LEDs are designed to emit different wavelengths, according to the specific application. Al in the epitaxial layer of DUV LEDs can change the spectral wavelength range from 210 to 365 nm. The outstanding characteristics of AlGaN-based LEDs make them potential candidates as mercury-free UV light sources. In particular, there are high expectations for the application of DUV LEDs in sterilization processes.

However, the present status of DUV LEDs is insufficient for practical applications; while visible LEDs have achieved significantly high external quantum efficiencies over 70% (3), those of most DUV LEDs remain lower than 5% (4-8). In the past, efforts were mainly focused on improvement of the epitaxial structure to increase output power. However, researchers have also recently attempted to improve the extraction efficiency of light from the LED chip. For instance, application of fluorocarbon resin molding over the chip (9) and processing of the growth substrate (10) have been reported in the literature. However, further improvement in the emission efficiency is necessary before mercury lamps can be replaced with LEDs.

53

We have bonded hemispherical lenses made of inorganic materials to LEDs using room temperature bonding methods, to improve the light extraction efficiency, and the light output power of the resultant LEDs was significantly higher than that of conventional structure LEDs (11).

In this paper, we examine the structural and optical properties of high output power 255 and 280 nm LEDs with hemispherical lenses in light of new experimental data. First, the DUV LED structure is explained briefly. We then explain the optimal LED structure with a lens to improve the output power of DUV LEDs, which was designed using Monte Carlo simulation. The LED lenses were bonded to LEDs at room temperature using two bonding methods, surface-activated bonding (12-14) and atomic-diffusion bonding (15-17). In the latter part of this paper, the bonded interface structures obtained using these bonding methods are presented, and the characteristics of the fabricated DUV LEDs are defined.

Structure of DUV LED

Figure 1 shows a cross-sectional schematic view of a DUV LED structure. All epitaxial layers are grown on a c-face sapphire substrate by low-pressure metal-organic chemical vapor deposition (LP-MOCVD). This is the same technique as that used for visible LEDs. First, an AlN layer is deposited onto the sapphire substrate, which acts as a low-dislocation density template to optimize the crystal growth conditions (7,8,18). This is very important for improvement of the quality of the active region (19). A strain-relieving AlN/AlGaN superlattice buffer is employed before undoped and Si-doped AlGaN n-type contact layers are deposited. The active layers consist of two pairs of wells and barriers. A thin undoped AlN layer and Mg-doped AlGaN layers are subsequently grown for carrier confinement and hole-injection, respectively. A p-type GaN contact layer is used for current distribution because AlGaN cannot become a p-type semiconductor.

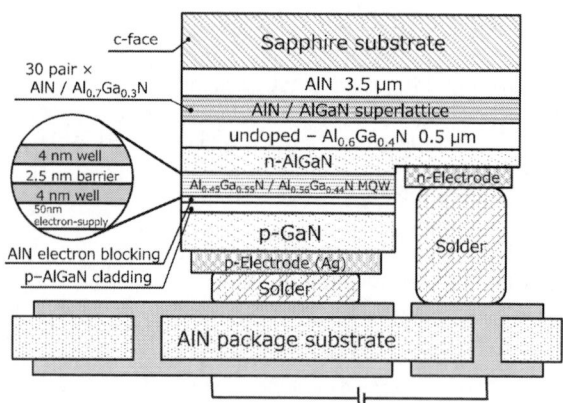

Figure 1.   Cross-sectional schematic view of a DUV die, which was flip-chip bonded to an AlN package by soldering.

The n- and p-type contact electrodes are Ti/Al and indium tin oxide (ITO), respectively. The LED die is flip-chip bonded to an AlN package substrate by soldering.

Attention must be directed toward GaN at the p-type contact in the DUV LED structure. The band gap of GaN is 3.4 eV, which corresponds to an emission wavelength of 365 nm. Thus, GaN absorbs all light that is less than 365 nm, as shown in Figure 2. These results are reflectance of 255 and 280 nm AlGaN-based LED wafers that were measured using spectrophotometry as shown in the inset. Light extraction from the p-layer side cannot be expected due to absorption. Therefore, a flip-chip type configuration is required to extract light through the substrate. In addition, the flip-chip structure has excellent heat dissipation.

Light extraction efficiency

Improvement of light extraction efficiency

In principle, the output power of LEDs is described by the following equations:

$$\phi_e = \eta_{ex} \times I_f \times h\nu, \qquad [1]$$

$$\eta_{ex} = \eta_{pkg} \times \eta_{int} \times \eta_{ext}, \qquad [2]$$

where ϕ_e is the output power of the LED, I_f is the operating current, $h\nu$ is the photon energy, η_{ex} and η_{int} are the external and internal quantum efficiencies, respectively, η_{pkg} is the extraction efficiency of light from the LED package to air, and η_{ext} is the extraction efficiency of light from the LED chip to air. Equation [1] shows that to enhance ϕ_e, it is necessary to improve η_{ex}, I_f, and $h\nu$. Here, I_f and $h\nu$ are fixed values; therefore, to enhance η_{ex}, it is necessary to improve η_{pkg}, η_{ext}, and η_{int} in the LED, as shown in Equation [2]. In this work, η_{pkg} is not considered, because the LED chip is the only form.

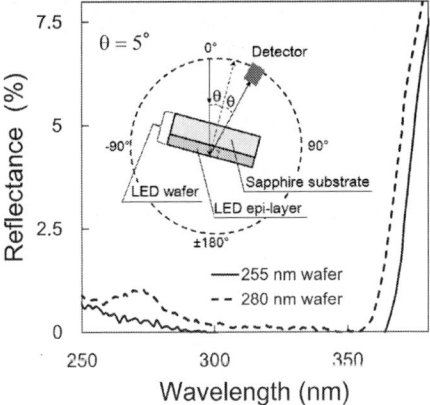

Figure 2.   Reflectance from 255 and 280 nm LED wafers measured with a spectrophotometer as a function of the measurement wavelength from 250 to 380 nm. The angle of incident light on the wafer is 5°, as shown in the inset.

Two major reasons for low external quantum efficiency are a high density of threading dislocations in the epilayers and low extraction efficiency of light. η_{int} can be improved by optimization of the crystal growth conditions (20,21).

In visible LEDs, η_{ext} has been successfully improved using three major techniques: reduction of the absorption loss of the electrode, light scattering inside the LED chip with patterned sapphire substrates (PSSs) (22), and covering the LED chip with a high refractive index material. All of these techniques share a basic principle, namely, to reduce the multiple reflections of light within the chip for enhancement of η_{ext}.

However, DUV LEDs have characteristic problems that visible LEDs do not have with respect to η_{ext}. The p-contact layer typically consists of Mg-doped GaN and absorbs the emitted UV light. This implies that a reflection-type p-electrode is not so effective for an improvement of η_{ext}. Moreover, conventional epoxy and silicone resin, which are used as graded refractive index media outside the chip in visible LEDs for efficient light extraction, exhibit large absorption in the DUV region. Recently, other research groups have proposed the use of a fluorine resin molding material (9). This strategy has significant value with respect to increasing η_{ext}; however, the long-term reliability has not yet been fully confirmed.

Simulation of light transport in the DUV region

Light extraction capability using PSS

First of all, the effect of PSSs on the light extraction efficiency in the DUV region was studied. Optical simulations were conducted and η_{ext} was calculated by the Monte Carlo method. Figure 3 shows the calculated values of η_{ext} with, as well as without (i.e., flat sapphire substrate), the PSS as a function of the reflectivity of the p-electrode, including the p-contact layer. A simulation model of the LED structure is shown as an inset. The chip size was assumed to be 1×1 mm^2 and the emission wavelength was set to 280 nm. The reflectivity of the p-electrode was varied from 0% to 90%. The variation of reflectivity of the p-electrode implies that both the p-electrode and p-contact layer have absorption losses; however, the contribution of each to the simulation results cannot be distinguished. All the light exiting from the chip was assumed to be collected by an omnidirectional detector.

When the reflectivity was 90%, η_{ext} with the PSS was 75%, which is significantly higher than that without the PSS (45%). However, η_{ext} decreases steadily with the reflectivity. When the reflectivity was 0%, η_{ext} was almost the same value of 15%, regardless of the substrate type, which indicates that the PSS does not effectively contribute to an enhancement of η_{ext} in the low reflectivity region.

The reduction in the optical absorption of the p-contact layer, such as with a mixed crystal of Al to GaN (6) for instance, is important to realize high reflectivity of the p-electrode including the p-contact layer. However, Ag used for the p-electrode has a reflectivity of only ca. 30% at 280 nm; therefore, the reflectivity of the p-electrode including a p-contact layer should be approximately 30%, even if the absorption of the p-contact layer is zero at 280 nm. The improvement of η_{ext} with the PSS at a reflectivity of 30% is small, as shown in Figure 3, which indicates that the PSS does not have a significant effect on η_{ext} in the DUV region.

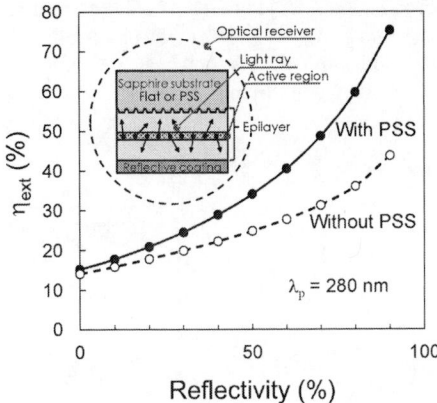

Figure 3. Calculated values of η_{ext} by Monte Carlo simulation with and without a PSS (flat sapphire substrate) as a function of the reflectivity of the p-electrode, including p-contact layer. A simulation model of the LED structure is also shown as an inset.

Hemispherical lens to enhance the extraction efficiency

A structure formed by combination with a hemispherical lens on the backside of the sapphire substrate, as shown in Figure 4, was proposed to enhance the extraction efficiency, and the values of η_{ext} were calculated by Monte Carlo simulation. The LED chip size was assumed to be 1×1 mm^2. The diameter of the hemispherical lens was varied between 1.5 mm and 5 mm, and the refractive index of the lens material was between 1.43 and 2.00. A flat sapphire substrate was assumed. The reflectivity of the p-electrode was set to 0%.

Figure 5 presents the values of η_{ext} calculated as a function of the refractive index of the lens material n, for various lens diameter, $\varnothing$. The calculated values of η_{ext} with the lens were much larger than that without (15%), which indicates that η_{ext} is enhanced by the lens. The enhancement of η_{ext} became almost saturated when $\varnothing$ exceeds 3 mm (11). For a lens with $\varnothing = 3$ mm, the value of η_{ext} was approximately 17.5% at n=1.43 (n value for general molding resin). η_{ext} increased gradually with n and has a broad maximum of approximately 24% at around n=1.83 (n value for sapphire).

Figure 6 presents the UV light beams obtained by Monte Carlo simulation for LEDs with lenses of n=1.83 (a-1) and n=1.43 (b-1). Figures 6(a-2) and (b-2) show expanded images around the LED chips of (a-1) and (b-1), respectively. A large numbers of light beams exit the lens with n=1.83 compared to that with n=1.43. Total reflection occurs with the n=1.43 lens at the interface between the lens and the LED chip due to the difference in the refractive index, and many DUV light beams are confined inside the LED, which results in multiple reflections of light within the chip. In contrast, total reflection does not occur in the LED chip with the n=1.83 lens, but instead DUV light exits the LED chip to the lens without any losses.

The simulation results revealed that the use of a hemispherical lens with n=1.83, which is equal to the refractive index of the sapphire substrate used for the LED chip, is the most effective way to enhance the light extraction efficiency. However, it should be noted that η_{ext} has a broad maximum at around n=1.83 in Figure 5, so that large values of η_{ext} are expected in the range of n = 1.75–2.0. In practical mass production, the use of sapphire lenses would result in high costs; however, the simulation results indicated that glass materials with n = 1.75–2.0 at 280 nm are also good candidates for use as lenses to enhance the extraction efficiency.

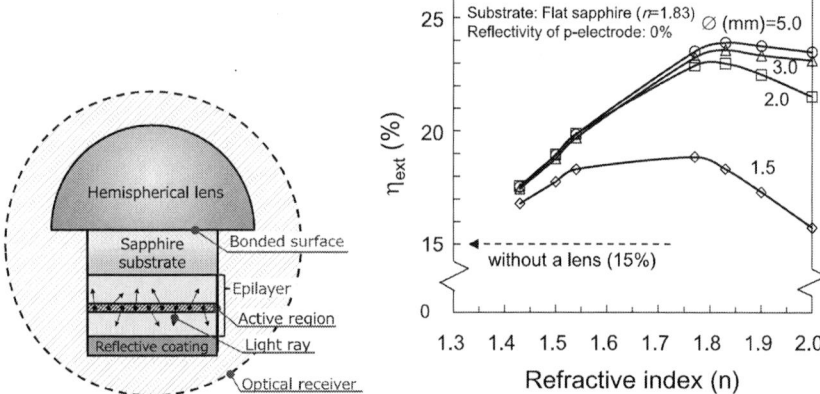

Figure 4. Schematic structure of a LED chip bonded to a hemispherical lens.

Figure 5. Values of η_{ext} calculated as a function of the refractive index of the lens material for various lens diameters.

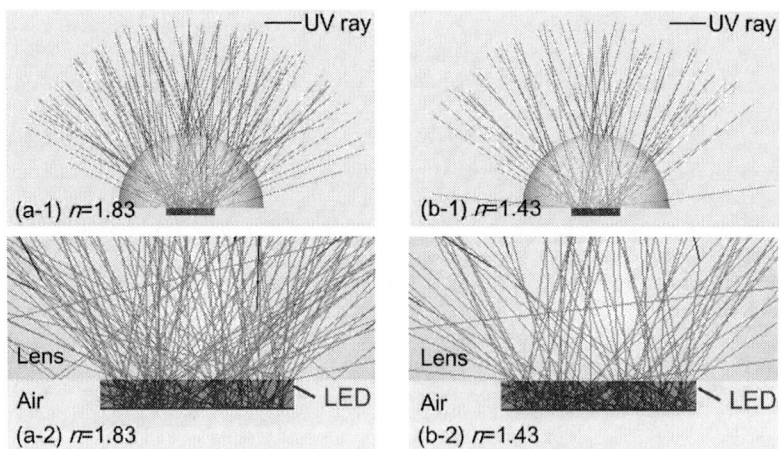

Figure 6. UV-rays obtained by Monte Carlo simulation for LEDs with lenses of n=1.83 (a-1) and n=1.43 (b-1). (a-2) and (b-2) show expanded images around LED chips shown in (a-1) and (b-1), respectively.

Room temperature bonding

Methods and principles of room temperature bonding

The room temperature bonding of lenses to LEDs is essential to realize the structure obtained by simulation because bonding at room temperature is necessary to maintain the LED characteristics. Two bonding methods were used here: surface-activated bonding (SAB) (12-14) and atomic diffusion bonding (ADB) (15-17).

Figure 7 shows schematic illustrations of the SAB and ADB processes. In SAB, the bonding surfaces are sputter-etched and activated by an Ar fast ion beam. The base pressure of the vacuum chamber was less than 6.3×10^{-6} Pa. It is reasonable to infer that there is no light absorption at the interface of wafers bonded using SAB because no adhesive material is used. However, the bonding strength of the SAB method is dependent on the materials involved. For example, sapphire can be bonded by SAB (23) but SiO_2 cannot. Moreover, SAB typically requires significant loading pressure for bonding.

In the ADB process, the samples are introduced into a vacuum chamber with a base pressure of less than 5×10^{-7} Pa. Thin metal films are fabricated on two flat wafer surfaces using sputter deposition with subsequent bonding of the two metal films on the wafers under vacuum. The high surface energies of metal films and the large atomic diffusion coefficient at the grain boundaries and film surfaces enable bonding at room temperature without unusually high loading pressures. ADB is performed using almost any metal films, even with film thickness of a few angstroms on each wafer. The ADB method typically involves optical absorption at the bonding interface due to the use of thin metal films. However, any mirror-polished wafers such as sapphire, SiO_2, metal, ceramics, or resin can be bonded by the ADB process; therefore, this process is potentially applicable for the bonding of lenses made of any material to LED chips.

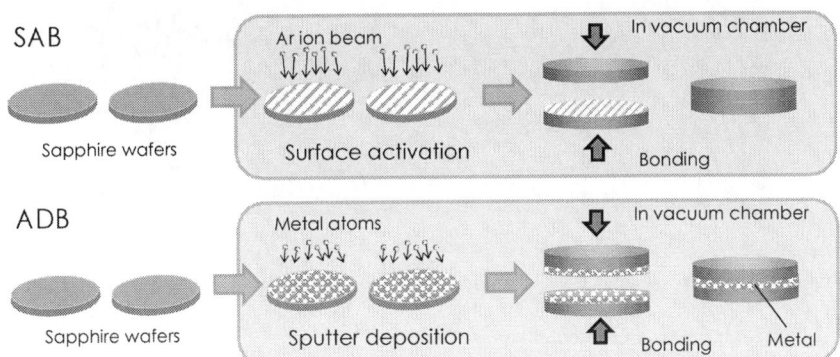

Figure 7. Schematic illustrations of the SAB and ADB processes for the room temperature bonding of sapphire wafers.

Interface structure bonded using the SAB method

Two c-face sapphire wafers were bonded using the SAB method and the resultant interface structure was examined. Figure 8(A) presents a cross-sectional transmission electron microscopy (TEM) image of the bonded wafer. Figure 8(B) shows selected-area electron diffraction (SAED) patterns measured at points (a) and (c) indicated in Figure 8(A), and a two-dimension fast Fourier transform (2D-FFT) for point (b) indicated in Figure 8(A). A thick amorphous layer exists at the interface of the bonded sapphire wafers. It is reasonable to infer that the sapphire surfaces were damaged by the Ar ion beam irradiation and bonding proceeded between the two amorphous layers on top of the sapphire wafers. A slight difference of contrast in the amorphous layer indicates that the thickness of the amorphous layer is 4 nm on one side and 7 nm on the other. The difference of the amorphous layer thickness is attributed to the Ar ion beam irradiation conditions. However, no unusual contrast attributable to internal stress was observed at the bonded interface. No elements other than the elements that make up the sample were observed at the bonded interface, according to energy-dispersive X-ray spectroscopy (EDS) measurements (data not shown). The diffraction patterns at points (a) and (c) are different, which is due to a slight misorientation of the crystals in the two bonded sapphire wafers, although this not important in the present discussion.

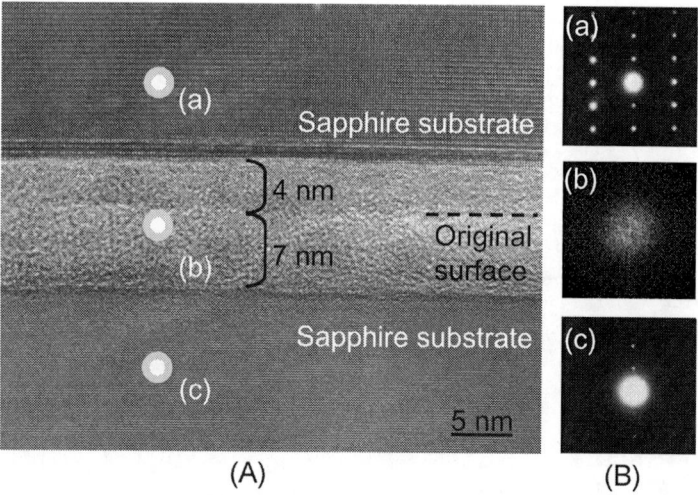

Figure 8. (A) Cross-sectional TEM image of c-face sapphire wafers bonded using SAB. (B) SAED patterns measured at points (a) and (c) indicated in (A), and two-dimension 2D-FFT for point (b) indicated in (A).

Interface structure bonded using the ADB method

The interface structure of c-face sapphire wafers bonded using the ADB method with 0.2 nm Al film on each wafer was examined. Figure 9(A) presents a cross-sectional TEM image of the bonded sapphire wafers. Figure 9(B) shows SAED patterns measured at points (a) and (c), and a nanobeam diffraction (NBD) pattern measured at point (b). No vacancies are observed at the bonded interface. Although the lattice images at the interface are modulated with a thickness of a few nanometers, it is likely that the thickness of the bonded interface is very thin. The thickness of the TEM observation sample (thickness along the direction normal to the image of Figure 9(A)) was approximately 100 nm in the present study, which is much thicker than the interface thickness and results in the lattice image modulation. Streaks of diffraction spots are observed along the direction normal to the bonded interface in the NBD pattern of point (b); however, no halo pattern was observed in the image, which supports the existence of a thin bonded interface without amorphous structure.

Although only Al and O peaks are observed in the EDS spectra for points (a), (b), and (c) (data not shown), forepart from carbon contaminants, the composition ration of Al to O at the bonding interface (point (b)) was higher than that in the sapphire substrates (points (a) and (c)). Thus, it is reasonable to infer that the thin Al film at the interface reacts with oxygen adsorbed on the sapphire surface or dissociated from the sapphire substrate, as observed for thin Ti films on quartz crystal wafers (17).

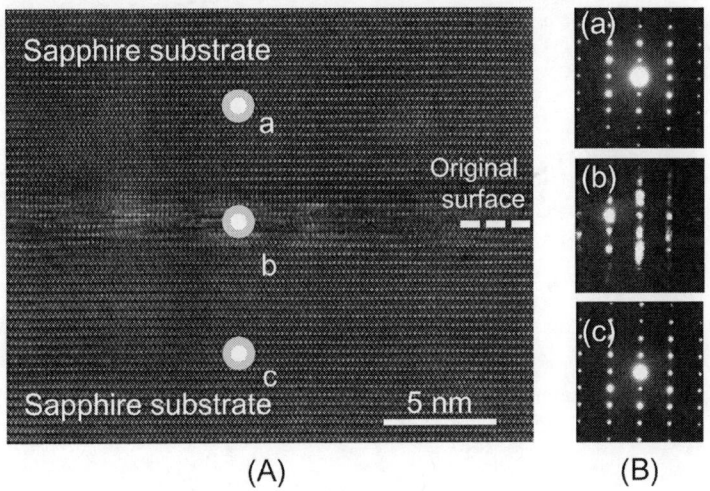

(A) (B)

Figure 9. (A) Cross-sectional TEM image of bonded sapphire wafers. (B) SAED patterns measured at points (a) and (c) indicated in (A), and NBD pattern measured at point (b) indicated in (A).

Light transmittance though the bonded interface

The light transmittance through the bonded interface is very important for the practical application of lens bonding. Figure 10 shows the light transmittance for sapphire wafers bonded using ADB with various thickness Al films (0.15, 0.2, and 0.5 nm) on each wafer. The light transmittance was measured using spectrophotometry, as shown in the figure inset. The light transmittance of a double-side polished sapphire wafer was measured first, and the difference with this transmittance was defined as the transmittance through the bonded interface.

The transmittance though the bonded interface produced with the SAB method was approximately 100% over the entire range of wavelength from 250 to 350 nm, as we expected with the lack of a separate adhesive material. The transmittance for the bonded interface produced with the ADB method was dependent on the Al film thickness, and that for the interface bonded with 0.5 nm thick Al films (on each wafer) ranged from ca. 85% (350 nm) to ca. 90% (250 nm). However, the transmittance increased significantly as the Al thickness decreased, and reached 97–98% for 0.2 nm thick Al films and 98–99% for 0.15 nm thick Al films on each wafer.

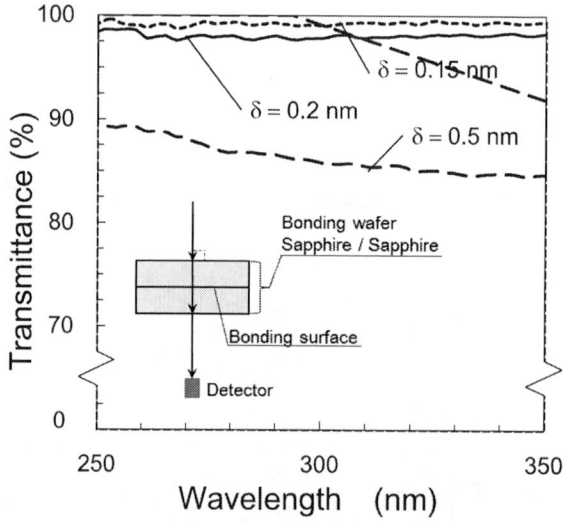

Figure 10. Light transmittance of sapphire wafers bonded using ADB with Al films of 0.15, 0.2, and 0.5 nm on each wafer.

Lens bonded DUV LEDs

DUV LED chips directly bonded to lenses were fabricated using both the ADB and SAB processes. The LED chip size was 1×1 mm^2, and the hemispherical lenses were made of sapphire. Two lens sizes with diameters of 3 and 5 mm were prepared, as these were the optimal diameters predicted by the Monte Carlo simulation.

Conventional DUV LED wafers emitting at 255 and 280 nm were fabricated. Special care was taken to polish the backside of the sapphire substrate using chemical mechanical polishing (CMP) to ensure atomically flat surfaces. The LED chips were then bonded with the lenses using the SAB or ADB process. The experimental conditions for these bonding processes were the same as those used for the fundamental studies. Thin Al films with a thickness of 0.2 nm on each side were used for the ADB process. The bonded chips were mounted on AlN sub-mount packages by soldering. The initial LED characteristics were measured with an integrating sphere under a pulsed current at room temperature. Lifetime tests were performed for the same packages under DC current conditions.

Figure 11 shows the output power L and η_{ex} for the 255 nm LED with a 3 nm diameter sapphire lens as a function of the forward current I from 20 to 1000 mA (11). The results for LEDs with lenses fabricated using both the SAB and ADB processes are shown, in addition to that for an LED without a lens. The value of L for the SAB device measured at 350 mA was 73.6 mW, which was 2.8 times larger than that for the LED without a lens (25.6 mW). η_{ex} for the SAB device increased with I and was a maximum of 4.56% at 100 mA. A further increase in the operating current led to a slight decrease in η_{ex}, presumably due to the thermal effects and carrier overflow.

L and η_{ex} for the ADB device were slightly smaller than those for the SAB device, although L was 2.5 times larger than that for the LED without a lens. These results are in good agreement with the slight reduction in transmittance through the ADB bonded interface with thin Al films (Figure 10).

Figure 12 shows L and η_{ex} for the 280 nm LEDs with 5 mm sapphire lenses as a function of I (11). L increased linearly with I up to 1 A. The value of L for the SAB device measured at 350 mA was 153.4 mW, which was 2.3 times larger than that for the conventional LED without a lens (66.0 mW). A maximum η_{ex} of 10.1% was obtained at I = 1 A. η_{ex} for the 280 nm LED did not show a maximum within the I range up to 1 A, which can be attributed to low forward voltages and weak thermal effects (11). Also L and η_{ex} for the ADB device were also smaller than those for the SAB device, although L was 1.9 times larger than that for the LED without a lens. It is noteworthy that the 280 nm LEDs with 3 mm diameter lens showed almost similar properties to those with the 5 mm diameter lenses, which is consistent with the simulation results. A lifetime test was also conducted for the 280 nm LED under application of 350 mA DC. The junction temperature was estimated to be 52 °C. After a 1000 h test, the output power remained above 90% of the original value, which is similar to that observed for conventional LEDs without lenses (11).

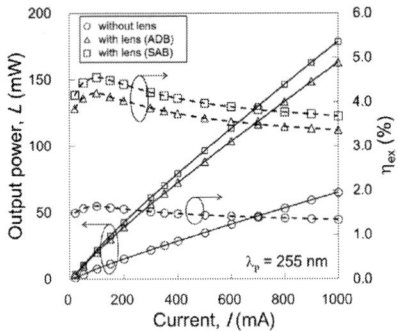

Figure 11. Output power L and η_{ex} for 255 nm LEDs with 3 nm diameter sapphire lenses as a function of the forward current I.

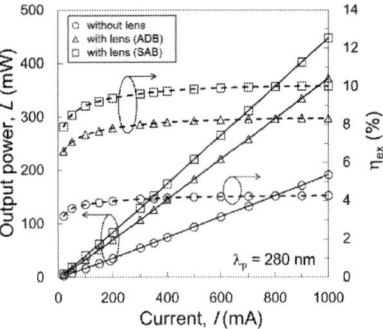

Figure 12. Output power L and η_{ex} for 280 nm LEDs with 3 nm diameter sapphire lenses as a function of the forward current I.

Conclusion

DUV LEDs with hemispherical inorganic material lenses were designed using Monte Carlo simulation and were then fabricated. Sapphire lenses were bonded to the sapphire substrates of LED chips using room temperature bonding methods. Total reflection did not occur at the bonded interface, because the refractive index of both substances was the same, and the DUV light exited from the LED chip to the lens without any losses. The LEDs fabricated with lenses exhibited high values of L and η_{ex}.

A fundamental study using the SAB method indicated that an amorphous layer with a total thickness of 7 nm was present at the bonded interface; however, the transmittance though the bonded interface was approximately 100% over the entire range of wavelength from 250 to 350 nm. The 250 nm LED with a sapphire lens fabricated using the SAB method had a high L value, which was 2.8 times larger than that for a conventional LED without a lens, and a maximum η_{ex} of 4.56%. The L value for the 280 nm LED with a lens was 2.3 times larger than that for a conventional LED without a lens.

L and η_{ex} for DUV LEDs bonded to sapphire lenses using the ADB method with 0.2 nm Al films on each side were slightly smaller than those fabricated using the SAB method. These results are in good agreement with the slight reduction of light transmittance through the bonded interface due to the thin Al films. This is a disadvantage of the ADB method for the fabrication of DUV LEDs with sapphire lenses. However, the simulation results indicated that any materials with n close to that of sapphire (1.75–2.0 at 280 nm) would also be good candidates for lenses to enhance the light extraction efficiency. The ADB process is potentially applicable to the bonding of lenses made of any material to the sapphire substrates of LED chips, and this is expected to extend the applications of LEDs with lenses, although more intensive efforts to enhance light transmittance at the bonded interface is required. The results presented here are expected to contribute not only to the research on DUV LEDs but also to research on visible light LEDs.

References

1. H. Hirayama, N. Maeda, S. Fujikawa, S. Toyoda, and N. Kamata, *Japan. J. Appl. Phys.*, **53**, 100209 (2014).
2. M. Kneissl, T. Kolbe, C. Chua, V. Kueller, N. Lobo, J. Stellmach, A. Knauer, H. Rodriguez, S. Einfeldt, Z. Yang, N. M. Johnson, and M. Weyers, *Semicond. Sci. Technol.*, **26**, 014036 (2011).
3. Y. Narukawa, M. Ichikawa, D. Sanga, M. Sano, and T. Mukai, *J. Phys. D.*, **43**, 354002 (2010).
4. A. Fujioka, K. Asada, H. Yamada, T. Ohtsuka, T. Ogawa, T. Kosugi, D. Kishikawa, and T. Mukai, *Semicond. Sci. Technol.*, **29**, 084005 (2014).
5. A. Fujioka, K. Asada, H. Yamada, T. Ohtsuka, T. Ogawa, T. Kosugi, D. Kishikawa, and T. Mukai, *Proc. SPIE Vol.* 9363, 93631L (2015).
6. M. Jo, N. Maeda, and H. Hirayama, *Appl. Phys. Express*, **9**, 012102 (2016).
7. T. Kinoshita, T. Obata, T. Nagashima, H. Yanagi, B. Moody, S. Mita, S. Inoue, Y. Kumagai, A. Koukitu, and Z. Sitar, *Appl. Phys. Express*, **6**, 092103 (2013).
8. J. R. Grandusky, J. Chen, S. R. Gibb, M. C. Mendrick, C. G. Moe, L. Rodak, G. A. Garrett, M. Wraback, and L. J. Schowalter, *Appl. Phys. Express*, **6**, 032101 (2013).

9. K. Yamada, Y. Furusawa, S. Nagai, A. Hirano, M. Ippommatsu, K. Aosaki, N. Morishima, H. Amano, and I. Akasaki, *Appl. Phys. Express*, **8**, 012101 (2015).
10. S. Inoue, T. Naoki, T. Kinoshita, T. Obata, and H. Yanagi, *Appl. Phys. Lett.*, **106**, 131104 (2015).
11. M. Ichikawa, A. Fujioka, T. Kosugi, S. Endo, H. Sagawa, H. Tamaki, T. Mukai, M. Uomoto, and T. Shimatsu, *Appl. Phys. Express*, **9**, 072101 (2016).
12. T. Suga, Y. Takahashi, H. Takagi, B. Gibbesch, and G. Elssner, *Acta Metall. Mater.*, **40**, s133 (1992).
13. T. Suga, K. Miyazawa, and Y. Yamagata, *MRS Internal Meeting on Advanced Materials*, Materials Research Society, **8**, 257 (1989).
14. E. Higurashi, T. Imamura, T. Suga, and R. Sawada, *IEEE Photonics Technol. Lett.*, **19**, 1994 (2007).
15. T. Shimatsu and M. Uomoto, *J. Vac. Sci. Technol.*, B **28**, 706 (2010).
16. T. Shimatsu and M. Uomoto, *ECS Trans.*, **33**(4), 61 (2010).
17. T. Shimatsu, M. Uomoto, and H. Kon, *ECS Trans.*, **64** (5) 317 (2014).
18. H. Hirayama, S. Fujikawa, N. Noguchi, J. Norimatsu, T. Takano, K. Tsubaki, and N. Kamata, *Phys. Status Solidi A.*, **206**, No. 6 (2009).
19. H. Hirayama, Y. Tsukada, T. Maeda, and N. Kamata, *Appl. Phys. Express*, **3**, 031002 (2010).
20. M. Shatalov, W. Sun, A. Lunev, X. Hu, A. Dobrinsky, Y. Bilenko, J. Yang, M. Shur, R. Gaska, C. Moe, G. Garrett, and M. Wraback, *Appl. Phys. Express*, **5**, 082101 (2012).
21. T. Inazu, S. Fukahori, C. Pernot, M. H. Kim, T. Fujita, Y. Nagasawa, A. Hirano, M. Ippommatsu, M. Iwaya, T. Takeuchi, S. Kamiyama, M. Yamaguchi, Y. Honda, H. Amano, and I. Akasaki, *Jpn. J. Appl. Phys.*, **50**, 122101 (2011).
22. M. Yamada, T. Mitani, Y. Narukawa, S. Shioji, I. Niki, S. Sonobe, K. Deguchi, M. Sano, and T. Mukai, *Jpn. J. Appl. Phys.*, **41**, L1431 (2002).
23. H. Takagi and R. Maeda, *J. Cryst. Growth*, **292**, 429 (2006).

Necessary Thickness of Au Capping Layers for Room Temperature Bonding of Wafers in Air using Thin Metal Films with Au Capping Layers

M. Uomoto[a] and T. Shimatsu[a,b]

[a] Frontier Research Institute for Interdisciplinary Sciences (FRIS),
Tohoku University, Sendai, 980-8578, Japan
[b] Research Institute of Electrical Communication (RIEC),
Tohoku University, Sendai, 980-8577, Japan

This study elucidated room temperature bonding of wafers in air using Ag, Al, and Cu films with thin Au capping layers. To avoid oxidation of 20-nm-thick of Ag and Cu films, a 2-nm-thick Au layer was found to be effective. Bonding was achieved in air over the entire area without a loading pressure. However, the necessary thickness of the Au capping layer for 20-nm-thick Al films was 6 nm or more, which is three times thicker than those for Ag and Cu films. The necessary Au layer thickness increased as the thicknesses of Al and Ag layers increased. Moreover, a loading pressure was necessary for bonding using thick films. These results were attributed to increased the surface roughness. However, the necessary Au capping layer thickness for bonding was 1/10 or less of the thickness of Al or Ag layers in the present thickness range of 20–500 nm.

Introduction

Bonding of two flat wafers using thin metal films is a promising process to achieve wafer bonding at room temperature (1–3) along with surface-activated bonding (4–7). Any mirror-polished wafer can be bonded using thin metal films. High surface energies of metal films and a large atomic diffusion coefficient at the grain boundaries and film surfaces enable bonding at room temperature without unusually high loading pressure.

We have two processes for bonding using thin metal films. One is bonding in vacuum, for which metal films are fabricated on two flat wafers' surfaces using sputter deposition with subsequent bonding of the two films on the wafers in vacuum. This process is similar to surface-activated bonding using nano-adhesive layers (8, 9). Bonding in vacuum can be done using almost any metal film (1, 2), including materials having high melting points such as tungsten. Wafers can be bonded even with film thickness of a few angstroms on each side (2, 3), which is useful to bond wafers having electrical devices on their surfaces while causing no marked damage to the electrical properties of the devices. Moreover, incident light can be transmitted through thin bonded films without marked reduction in their intensity.

Another bonding process is bonding in air. Room temperature bonding of two films on the wafers can be accomplished in air after sputter film deposition (2). This process is limited to use with Au and Au alloy films, but it is extremely convenient. For instance, bonding of patterned Au films can be done with sputter film deposition using metal masks. It is actually used for mass-production of new etalon filters for optical fiber communication (10). Moreover, bonding of wafers with mirror-polished metals was achieved using the bonding process (11), which is important for enhancing the heat dissipation efficiency for electrical power devices fabricated on wafers. During the bonding process using Au films in air, once the nucleation of recrystallization at Au–Au interface is formed by pushing some of the stacked wafers using tweezers, the recrystallization is propagated immediately. Then the bonding is completed over the entire bonded area (3, 12). Recrystallization at the Au–Au bonded interface occurs even with exposure time to air of 1 week, with bonding energy greater than the Au film surface energy (12).

It is noteworthy that the metal films used for bonding can also be used for electrical/thermal conduction and optical reflection or other purposes after bonding, which is useful to produce functional devices. Table I presents values of electrical resistivity ρ at 293 K (13), thermal conductivities (13), Young's modulus (13), calculated self-diffusion coefficients D at 300 K (14), free energies of formation of oxide compounds from the metals ΔG at 300 K (13) for Au, Ag, Cu, and Al. Also in this table, the required temperatures for bonding in air using these films (2) are shown.

The thermal and electrical conductivities for Ag and Cu are higher than those for Au. The Al Young's modulus is smaller than that for Au. Moreover, optical reflection coefficients for Ag and Al are well-known to be greater than that for Au. Furthermore, Ag, Cu, and Al are much less expensive than Au, underscoring the benefits of these films used for bonding for mass production.

From the perspective of bonding potential, the values of D for Au, Ag, and Cu are almost equal, in the range of 10^{-36}–10^{-41} m^2/s. Actually the bonded interface of these films disappears during the bonding process in vacuum (1, 2) because dynamic recrystallization occurs at the bonded interface. However, in air, the value of ΔG for Ag is slightly negative, implying that Ag forms oxide compounds to a slight degree at room temperature, although Au is not oxidized at room temperature (ΔG for Au is positive.). The value of ΔG for Cu is −144.9 kJ/mol, which is much smaller than that for Ag, implying that Cu forms oxide compounds much more readily than Ag at room temperature. Actually, room temperature bonding of wafers using Ag films in air requires a loading pressure in the bonding process. Moreover, the bonded Ag film structures is inhomogeneous with many vacancies (3). An increase in temperature up to 100 °C during the bonding process was necessary to bond wafers using Cu films in air. Moreover, Cu$_2$O forms at the bonded Cu–Cu interface (2). The value of D for Al is markedly larger than the other materials, indicating a high potential of recrystallization at the bonded film interface in vacuum. However, the value of ΔG for Al is −1584 kJ/mol, implying a high potential forming oxide compounds in air. It is reasonable to infer that bonding in air using Al films via recrystallization at the bonded interface is almost impossible.

Therefore, suppression of film surface oxidation is a key to bond wafers using Ag, Cu, and Al films in air. The use of a Au capping layer on these films was effective to avoid the oxidation of these films. Room temperature bonding of wafers was accomplished using these films with thin Au capping layers (15–17). All these films have an fcc crystal structure with (111) crystal plane parallel to the film plane. However, the surface energy of these films is material-dependent. Table II presents values of the surface free energy of (111) crystal lattice E_s (J/m^2) of Au, Cu, Ag, and Al (18–20). The values of differences from that of Au, E_s(Au), are also shown. Actually, E_s for Ag and Al, especially for Al, are lower than that of Au, suggesting that the two-dimensional continuity for the initial crystal growth of Au films on these films is poor (21). The necessary thickness of the Au

TABLE I. Values of electrical resistivity ρ at 293 K (13), thermal conductivities (13), Young's modulus (13), calculated self-diffusion coefficients D at 300 K (14), free energies of formation of oxide compounds from the metals, ΔG at 300 K (13) for Au, Ag, Cu, and Al. The temperatures required for bonding in air using these films are also shown.

Materials	Au	Cu	Ag	Al
ρ at 293 K (μΩcm)	2.20	1.694	1.63	2.67
Thermal conductivity (W/mK)	315.5	397	425	238
Young's modulus (GPa)	78.5	129.8	82.7	70.6
D at 300 K (m^2/s)	1.6×10^{-36}	1.4×10^{-41}	8.3×10^{-38}	3.3×10^{-29}
ΔG at 300 K (kJ/mol)	$0 < \Delta G$	-144.9 (Cu$_2$O)	-10.7 (Ag$_2$O)	-1584 (Al$_2$O$_3$)
Required temperature for bonding in air	room temperature	100 °C	room temperature	impossible

TABLE II. Values of the surface free energy of (111) crystal lattice, E_s (J/m^2) (18–20), and differences of E_s from that of Au, E_s(Au) (J/m^2)

Materials	Au	Cu	Ag	Al
Surface free energy of (111) crystal lattice, E_s (J/m^2)	1.50	1.81	1.25	1.15
Difference of E_s from E_s(Au) (J/m^2)	–	0.31	−0.25	−0.35

capping layer to avoid oxidation is expected to depend on film materials and film thickness. As described in this paper, this study examines room temperature bonding of wafers using Ag, Al, and Cu films with thin Au capping layers, with additional discussion of the necessary thickness of Au capping layer to bond wafers in air.

Experimental Procedure

Figure 1 presents a schematic illustration of the bonding process. For this study, 2-inch diameter micro-polished quartz glass wafers were used. An ultra-high vacuum (UHV) DC-magnetron sputtering system was used for film deposition. The vacuum chamber's base pressure was less than 2×10^{-6} Pa. Metal films were fabricated on two flat wafers' surfaces. The Ti films (1 nm or 5 nm thicknesses) were used as underlayers to enhance the adhesion strength of the deposited films on the wafer surface. 10–500-nm-thick Ag, Cu, Al films were deposited on Ti underlayers. Subsequently in the same vacuum, 2–20-nm-thick thin Au capping layers were deposited. After sputter film deposition on the surfaces of two wafers, bonding was accomplished in air of a clean room (20°C, 50% moisture level, class 100). No substrate heating was conducted during deposition or the bonding processes. The film surface was observed using atomic force microscopy (AFM). The bonded interface was observed using scanning acoustic tomography (SAT).

Results and Discussion

Bonding using 20-nm-thick Ag, Cu Films with 2-nm-thick Au Capping Layers

Bonding in air can be accomplished using 20-nm-thick Ag, Cu films with 2-nm-thick Au capping layers with no loading pressure: once the nucleation of recrystallization at the

Au–Au interface of capping layers was formed, recrystallization was propagated immediately. Then the bonding was completed over the entire bonded area.

Figure 2 portrays a high-resolution transmission electron microscopic (TEM) image of bonded Au(2 nm)/Ag(20 nm)/Ti(5 nm) films on each side. No interface corresponding to the original Au surfaces is visible. Crystal grains formed across the original Au surfaces of these films over the entire film thickness of 20 nm + 20 nm, indicating that recrystallization of Ag films occurred through the Au/Au connected capping layers. Figure 3 (A) presents a bright field scanning TEM image (BF-STEM) and the energy dispersive X-ray spectroscopy (EDX) mapping images of (B) Au and (C) Ag in the same area of (A). Image (B) revealed that Au diffused into Ag layers through recrystallization. Surprisingly, some Au reached the interfaces between Ag layers and Ti underlayers. These results indicate the substitution of some Ag for Au in crystal lattice propagated deeply into Ag films by the propagation of recrystallization. In the entier composition range, Au-Ag shows solid solution with fcc crystal structure. Moreover, the lattice constant of Au is almost equal to that of Ag. It is reasonable to infer that these properties of Au and Ag enhance propagation of the substitution of some Ag for Au in the crystal lattice.

Figure 4(A) presents a TEM cross-section image of bonded Au(2 nm)/Cu(20 nm)/Ti(5 nm) films on each side. A high-resolution image of the area shown as a dashed box in the image is also portrayed in Figure 4(B). No interface corresponding to the original film surface was observed. Lattice images shown in (B) were continuous from Au capping layers to Cu layers, indicating that recrystallization occurred at Au/Au

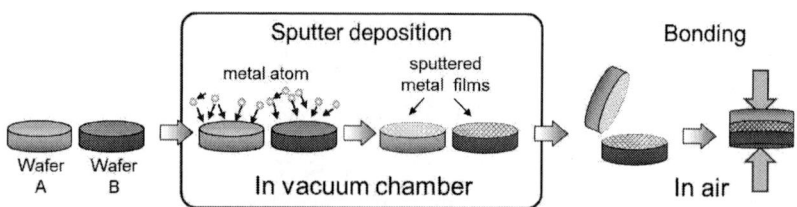

Figure 1. Schematic illustration of the bonding process.

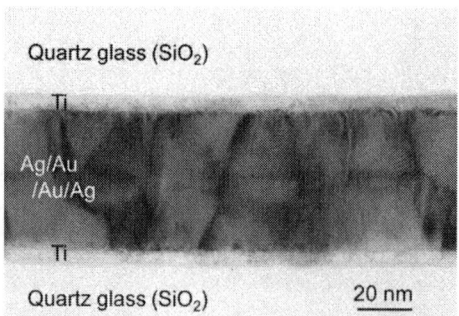

Figure 2. High-resolution TEM image for bonded Au(2 nm)/Ag(20 nm)/Ti(5 nm) films on each side.

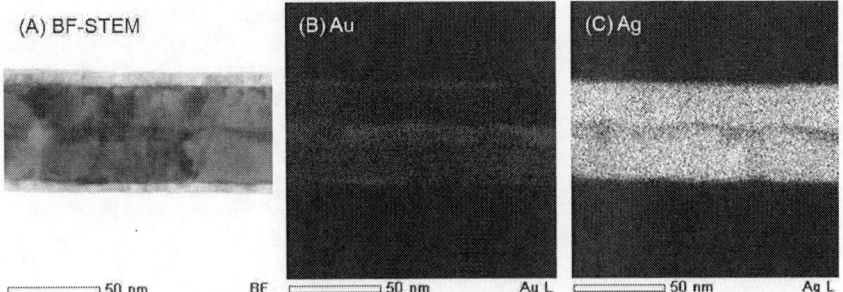

Figure 3. (A) Bright field scanning TEM image (BF-STEM) and energy dispersive X-ray spectroscopy (EDX) mapping images of (B) Au and (C) Ag in the same area of (A).

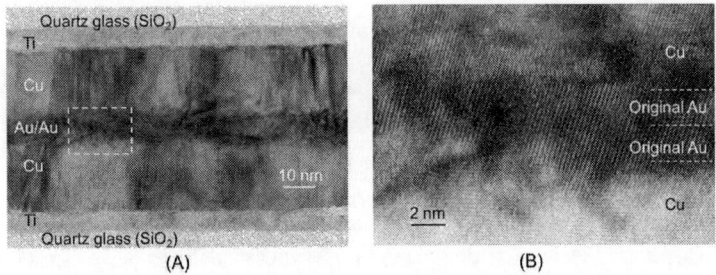

Figure 4. (A) TEM cross-section image of bonded Au(2 nm)/Cu(20 nm)/Ti(5 nm) films on each side. A high-resolution image of the area shown as a dashed box in the image is also shown in (B).

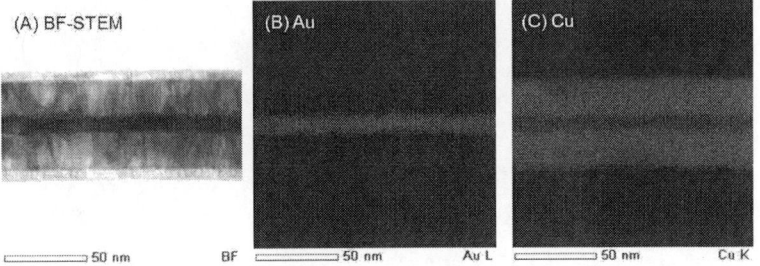

Figure 5. (A) A BF-STEM image, and EDX mapping images of (B) Au and (C) Cu in the same area of (A).

interface propagated into Cu layers. However, the propagation of recrystallization is not great compared to that for Au(2 nm)/Ag(20 nm) films. Figure 5(A) presents a BF-STEM image, with EDX mapping images of (B) Au and (C) Cu in the same area of (A). Image (B) showed that the distribution thickness of Au atoms was a few times thicker than the original Au capping layers. However, the diffusion depth into Cu layers was not remarkable compared to that into Ag layers shown in Figure 3(B). This might be attributable to a large misfit of the crystal lattice between Cu and Au (the lattice constant of Cu was 11.4% smaller than that of Au). Moreover, the value of D of Cu is slightly less than those of Ag and Au, which degrades the propagation further.

These results demonstrate that 2-nm-thick Au capping layer was effective to avoid oxidation of the surfaces of 20-nm-thick Ag and Cu films in air and to nucleate the recrystallization between the connected films in the bonding process. Actually the values of the surface free energy at the bonded interface γ, as estimated using blade-insertion method with Maszara's eq. (22) for these bonded wafers were greater than 2 J/m². Those values were larger than the surface energy of Au films (1.50 J/m²).

Necessary Thickness of Au Capping Layer for 20 nm-Thick Al Films

Bonding in air can also be accomplished using 20-nm-thick Al films with 2-nm-thick Au capping layers. However, the value of γ was about 0.3 J/m², which was one order lower than that of the surface energy of Au films. Figure 6(A) presents values of γ for bonded wafers using Au/Al(20 nm)/Ti(1 nm) films on each side as a function of the thickness of a Au capping layer. Bonding in air can be accomplished for all films shown in Figure 6(A) with no loading pressure. However, the values of γ were one order lower than the surface energy of Au in the range of Au film thickness lower than 5 nm. With increasing Au thickness from 5 nm, γ increased significantly beyond the surface energies of Au and Al. The value of γ at 10 nm thickness was not measured because the blade could not be inserted between the wafers. Necessary thickness of the Au capping layer for 20-nm-thick Al films was 6 nm or more, which was three times thicker than those for Ag and Cu films. Figure 6(B) presents an AFM image of the film surface of Au(10 nm)/Al(20 nm)/Ti(1 nm) film. Very fine island structure was observed. The surface roughness R_a was high at 1.41 nm. It is reasonable to infer that the small surface energy of Al films caused island-like initial film growth of Au films on Al films, resulting in an increase of the necessary Au layer thickness to avoid oxidation of the Al film surfaces. Au forms compounds with Al readily, which might enhance the necessary thickness.

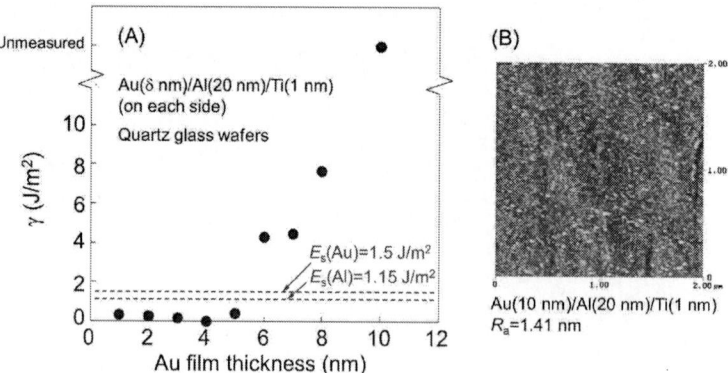

Figure 6. (A) Values of γ for bonded wafers using Au/Al(20 nm)/Ti(1 nm) films on each side as a function of Au capping layer thickness. (B) AFM image of Au(10 nm)/Al(20 nm)/Ti(1 nm) film.

Bonding Performance using Thick Al and Ag Layers with Thin Au Capping Layers

We examined bonding performance using thick Al and Ag films with Au capping layers. Figure 7(A) presents an AFM image of Au(10 nm)/Al(200 nm)/Ti(1 nm) film. The surface roughness of this film is 1.85 nm, which is greater than that for the film with Al(20 nm) shown in figure 6(B). The increase of Al layer thickness from 20 nm to 200 nm caused in a further increase of the roughness.

Bonding in air using this film thickness could not be accomplished without loading pressure. Therefore, after stacking two wafer's surfaces together, we tried to bond wafers by pushing the stacked wafers using tweezers. Figure 7(B) shows the SAT image of the bonded interface. Overlapping thick black lines, which are the bonded area, are observed in the image. Surprisingly, these lines correspond to the trace lines of tweezers on the stacked wafers with some forces. These results show that loading pressure is effective to bond wafers in air using thick Al films with Au capping layers to overcome the increased surface roughness of the films. It is noteworthy that bonding of wafers using the same film thickness in vacuum was accomplished with no loading pressure. Wafer and oxygen gases adsorbed onto the rough Au film surfaces are likely to have degraded the propagation of recrystallization at the Au/Au connected interface.

Figure 8(A) portrays an SAT image of the interface bonded using Au(20 nm)/Al(300 nm)/Ti(1 nm) film on each side with oading force F of 20 MPa. Figure 8(B) shows a TEM cross section image of the sample. Wafers were bonded over the entire area. A thick dark layer was observed at the original film surface in the TEM image, which corresponds to the Au diffusion layer. The thickness of this layer was four times greater than the original Au film thickness, indicating that Au diffused into the Al films significantly by the propagation of recrystallization at room temperature. These results suggest that a further increase in Al thickness will be acceptable for use in bonding.

Bonding using thick Ag films with Au capping layers was also examined. The necessary thickness of the Au capping layer increased as the Ag thickness increased, although the relation between these thicknesses has not been examined systematically. Figure 9 shows SAT images of the interfaces bonded using Au(20 nm)/Ag/Ti(1 nm) films on each side with F=20 MPa: (A) is with Ag(300 nm), and (B) with (500 nm). These images were obtained with blades inserted between the wafers used for the γ estimation. The dotted line corresponds to the blade edge. The blades were inserted at 1 hr after the

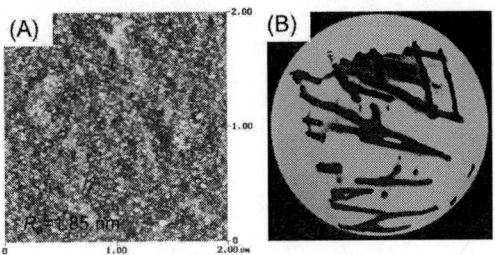

Au(10 nm)/Al(200 nm)/Ti(1 nm)

Figure 7. (A) AFM image of Au(10 nm)/Al(200 nm)/Ti(1 nm) film. (B) SAT image of the interface bonded using Au(10 nm)/Al(200 nm)/Ti(1 nm) film on each side.

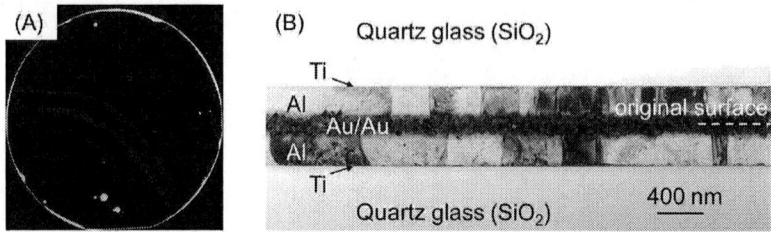

Au(20 nm)/Al(300 nm)/Ti(1 nm)

Figure 8. (A) SAT image of Au(20 nm)/Al(300 nm)/Ti(1 nm) film. (B) TEM image of the interface bonded using Au(20 nm)/Al(300 nm)/Ti(1 nm) film on each side.

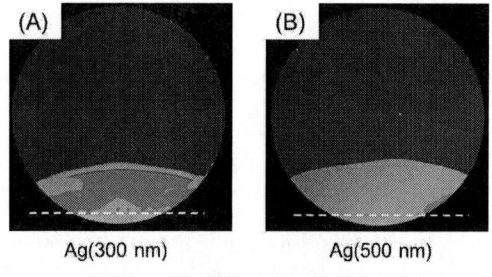

Au(20 nm)/Ag/Ti(1 nm), with F=20 MPa.

Figure 9. SAT images of the interfaces bonded using Au(20 nm)/Ag/Ti(1 nm) films on each side with F=20 MPa: (A) Ag(300 nm) and (B) Ag(500 nm).

bonding. Wafers were bonded over the entire area of wafers. However, the γ values estimated for these samples are low: only about 0.5 J/m². It is noteworthy that the γ values of wafers bonded in air using Ag films with Au capping layers showed a strong aftereffect: γ always increased as the time after the bonding increased. The γ values measured at 24 hr after the bonding were, respectively about 7.6 and 2.4 J/m² for samples with Ag(300 nm) and Ag(500 nm). Details of the aftereffects from using Ag films will be published in the near future.

Conclusion

Room temperature bonding of wafers in air using Ag, Al, and Cu films with thin Au capping layers was studied. To avoid oxidation of the surfaces of 20-nm-thick of Ag and Cu films in air, a 2-nm-thick Au capping layer was effective. Once nucleation of recrystallization at Au–Au interface was formed, the recrystallization propagated immediately throughout the entire bonding area. The substitution of some Ag or Cu for Au in crystal lattice propagated deeply into these films along with the propagation of recrystallization. However, the necessary thickness of the Au capping layer for 20-nm-thick Al films was 6 nm or more, which is three times thicker than those for Ag and Cu films.

Moreover, we examined the bonding performance using thick Al and Ag layers with Au capping layers. The necessary Au capping layer thickness increased as the film thickness increased. Moreover, loading pressure was necessary for bonding using thick Al and Ag layers films. These were caused mainly by an increase of the metal film surface roughness. However, even using 300-nm-thick Al films with a 20-nm-thick Au capping layer on each side, marked recrystallization occurred at the bonded interface. The necessary Au capping layer thickness for bonding was 1/10 or less of the thicknesses of Al or Ag layers, in the present thickness range of 20–500 nm.

Bonding in air using metal films with a thin Au capping layer is effective to use the bonding metal films used for electrical/thermal conduction and optical reflection or other purposes after the bonding. It is useful to produce functional devices.

Acknowledgments

The authors express their great appreciation to Mr. H. Yakumaru and Mr. K. Sahara (Hitachi Power Solutions Co. Ltd.) for their technical support using scanning acoustic tomography.

References

1. T. Shimatsu and M. Uomoto, *J. Vac. Sci. Technol.*, *B* **28**, 706 (2010).
2. T. Shimatsu and M. Uomoto, *ECS Transactions*, **33**(4), 61 (2010).
3. T. Shimatsu, M. Uomoto and H. Kon, *ECS Transactions*, **64** (5), 317 (2014).
4. T. Suga, Y. Takahashi, H. Takagi, B. Gibbesch, and G. Elssner, *Acta Metall. Mater.*, **40**, s133 (1992).
5. T. Suga, K. Miyazawa, and Y. Yamagata, *MRS Internal Meeting on Advanced Materials, Materials Research Society*, **8**, 257 (1989).
6. H. Takagi, K. Kikuchi, R. Maeda, T. R. Chung, and T. Suga, *Appl. Phys. Lett.* **68**, 2222 (1996).
7. E. Higurashi, T. Imamura, T. Suga, and R. Sawada, *IEEE Photonics Technology Letters*, **19**, 1994 (2007).
8. M. M. R. Howalder, H. Okada, T. H. Kim, T. Itoh, and T. Suga, *J. Electrochem. Soc.*, **151**, G461 (2004).
9. M. M. R. Howlader, T. Suga, and M. J. Kim, *IEEE Trans. Adv. Packag.*, **30**, 598 (2007).
10. T. Shimatsu, M. Uomoto, H. Kon, K. Wakabayashi, K. Oba and Y. Furukata, *Optronics*, **32**, 88 (2013).
11. H. Kon, M. Uomoto, and T. Shimatsu, *Conference on Wafer Bonding for Microsystems and Wafer Level Integration*, P18, Stockholm, Sweden, Dec. (2013).
12. H. Kon, M. Uomoto, and T. Shimatsu, *Journal of the Japan Institute of Electronics Packaging*, **17**, 431 (2014).
13. E. A. Brandes G. B. Brook, *Smithells Metals Reference Book, seventh Edition*, Table 8.1, Table 14.1, Table 15.1, Butterworth–Heinemann, Oxford (1992).
14. The Japan Institute of Metals, *Metal Data Book*, p20, Maruzen, Japan (1993).

15. M. Uomoto, H. Kon, and T. Shimatsu, *Proceedings of 29th Spring Meeting of Japan Institute of Electronics Packaging*, 202 (2015).
16. H. Kon, M. Uomoto, and T. Shimatsu, *Proceedings of 29th Spring Meeting of Japan Institute of Electronics Packaging*, 212 (2015).
17. T, Shimatsu, M. Uomoto, and H. Kon, *Conference on Wafer Bonding for Microsystems and Wafer Level Integration*, S7-3, Braunschweig, Germany, Dec. (2015).
18. L. Vitos, A.V. Ruban, H.L. Skriver, and J. Kolla´r, *Surface Science*, **411**, 186 (1998).
19. W.R. Tyson, W.A. Miller, *Surf. Sci.*, **62**, 267 (1977).
20. F.R. de Boer, R. Boom, W.C.M. Mattens, A.R. Miedema, A.K. Niessen, *Cohesion in Metals*, North-Holland, Amsterdam (1988).
21. O. Kitakami, S. Okamoto, and Y. Shimada, *J. Appl. Phys.*, **79**, 6880 (1996).
22. M. P. Maszara, G. Goetz, A. Cavigila, and J. B. McKitterick, *J. Appl. Phys.*, **64**, 4943 (1988).

Direct Wafer Bonding of SiC-SiC at Room Temperature by SAB Method

F. Mu[a], K. Iguchi[b], H. Nakazawa[b], Y. Takahashi[b], M. Fujino[a], and T. Suga[a]

[a] Department of Precision Engineering, School of Engineering, The University of Tokyo, Hongo, Tokyo 113-8656, Japan
[b] Fuji Electric Co., Ltd., Matsumoto, Nagano 390-0821, Japan

> In this work, direct wafer bonding of SiC-SiC was accomplished by surface activated bonding (SAB) method at room temperatrue. The bonding energy of ~1.4 J/m^2 was obtained without orientation dependence, which is much weaker than bulk SiC strength. Correspondingly, the tensile strength of bonding interface is ~12.2 MPa. The bonding mechanisms were investigated through Monte Carlo simulation and interface analysis by transmission electron microscopy (TEM) and energy dispersive X-ray spectroscopy (EDX). The formation of amorphous SiC during surface activation may eliminate the affects of orientation, resulting in the disappearance of orientation dependence. In addition, the possible Si preferentially sputtering during surface activation is assumed to be the reason why direct wafer bonding of SiC-SiC cannot reach bulk SiC strength. After a rapid thermal annealing at 1273 K for 180 s in Ar, the tensile strength of bonding interface could be improved to the values higher than 21.6 MPa.

Introduction

SiC has been attracting much attention in power device field owing to its superior properties like wider band-gap, higher electric breakdown field, higher carrier saturation velocity and higher thermal conductivity compared with silicon. These properties can make the devices with smaller size, lower losses, higher operation temperature and frequency, and simpler heat sink. Now 6-inch single crystalline 4H-SiC wafer and many kinds of SiC devices such as schottky diodes, bipolar junction transistors, MOSFETs and JFETs have been available (1, 2). With the further development of SiC devices, SiC-SiC wafer bonding will be indispensable in SiC device fabrication and integration.

As far as we know, only a few reports about SiC-SiC bonding have been published (3, 4). G. Yushin et al. accomplished SiC-SiC wafer bonding in ultrahigh vacuum (UHV) at a temperature range of 1073-1273 K for 15 hours under 20 MPa (3). Grekhov et al. also bonded 6H-SiC in chip size in water, followed by heat treatment at 368-373 K for 4 hours and at 1523 K for 2 hours under ~50 kPa (4). Both of their works didn't report the bonding strength and needed an annealing at a very high temperature for a long time, which would cause a huge energy waste and a low yield.

SAB is a promising method to achieve bonding at room temperature, which has apparent advantages compared with conventional high temperature bonding methods (5). In previous studies, SiC-SiC strong bonding at room temperature by modified SAB with Fe-Si deposited layer or Si deposited layer have been demonstrated or proved to be feasible, respectively (6, 7). Although a deposited intermediate layer is very helpful for SiC-SiC bonding, a direct wafer bonding of SiC-SiC without any non-SiC interfacial layer will also be desired in some applications, which may need to consider the interface reliability in extreme conditions. Therefore, direct wafer bonding of SiC-SiC by SAB method was investigated in this paper.

Experimental Procedure

Wafer bonding was performed in a UHV-bonding machine, which consists of a load-lock chamber and a processing-bonding chamber. At the beginning, wafers were set into the load-lock chamber and then transferred to the processing-bonding chamber, in which an Ar ion beam was set to remove the oxide layer and contaminations on SiC wafer surfaces. After surface activation, two SiC wafers were bonded directly at room temperautre under ~2.5 MPa for 180 s. The base pressure in the processing-bonding chamber is 5.0×10^{-6} Pa. The accelerating voltage and current of the Ar ion beam source were 1 kV and 100 mA, respectively.

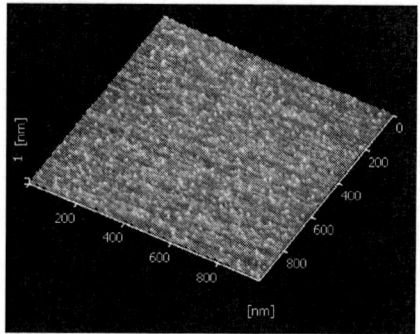

Figure 1. Typical DFM image of SiC surface after Ar ion beam irradiation.

The used specimens are n-type, 4-inch, 4° off-axis 4H-SiC wafers with a thickness of ~350 μm. Both the Si-face and C-face of 4H-SiC wafers, polished by chemical mechanical method, were selected as bonding surfaces to study the orientation dependence for bonding. Their root-mean-square (rms) roughnesses before and after acativation were measured by a dynamic force microscopy (DFM, NanoNavi/L-trace II; Hitachi High Tech Science). The roughnesses of Si-face and C-face were almost same, regardless of before or after surface activation. The measurements revealed that the bonding surfaces before and after surface activation have an rms roughness of ~0.18 nm and ~0.25 nm, respectively. The typical DFM image of SiC wafer surface after surface activation is shown in Fig. 1. Since there are two kinds of bonding surfaces, two pairs wafer, SiC (Si-face)-SiC (Si-face) and SiC (C-face)-SiC (C-face) were bonded for comparison to study the orientation dependence. After bonding, the bonded wafers were observed by scanning acoustic microscopy (SAM, Hitachi FineSAT FS300) to examine the un-bonded areas. The bonding energy was measured by razor blade method in air at room temperature (8). The Young's modulus of single crystalline 4H-SiC was chosen as 530 GPa (9). Since the rapid thermal annealing at 1273 K in inert gas is widely used for the ohmic contact formation of SiC device, the affect of such an annealing on bonding strength was also investigated by pulling test for the samples before and after annealing. In the pulling test, the 9×10 mm^2 chips without large voids (diced from bonded wafers) were fixed between two copper jigs, which would be pulled by a tensile pulling tester (Shimadzu AG-X), by epoxy.

Since preferentially sputtering might happen for SiC during Ar ion beam bombardment, simulation was performed to investigate the surface activation process by Monte Carlo code SRIM2013 using the model of SiC compound number 590 from the ICRU library (10). Then, TEM (JEM 2010F, JEOL) and EDX at 200 kV were used to investigate the structure and composition of the bonding interfaces.

Results and Discussion

<u>Bonding Characterization</u>

SAM images of the bonded wafers are shown in Fig. 2 (a) and (b). Both of them were bonded except some voids and un-bonded edge area. It is assumed that the voids were caused by particle contaminations since the bonding experiments were not performed in a clean room. By magnifying some of these voids such as the rectangular area in Fig. 2 (a) and the circle area in Fig. 2 (b), the particles were confirmed, as shown in Fig. 2 (c) and (d). It can be expected that this problem of bonding void would be solved when the bonding is conducted in a clean room. The un-bonded edge area may be caused by the misalignment during bonding.

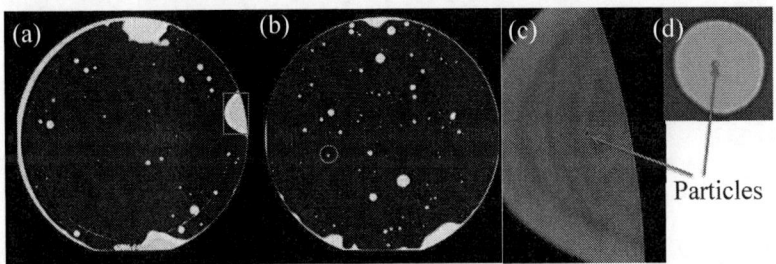

Figure 2. SAM images of SiC-SiC bonded wafers: (a) SiC (C-face)-SiC (C-face) and (b) SiC (Si-face)-SiC (Si-face), (c) the magnified rectangular area in (a), and (d) the magnified circle area in (b).

The bonding energies of these two bonded wafers were compared in Fig. 3. Both of the average bonding energies of SiC (C-face)-SiC (C-face) and SiC (Si-face)-SiC (Si-face) are ~1.4 J/m^2, much less than bulk SiC strength (11). The average tensile strengths of the two bonding interfaces are also same, ~12.2 MPa. The fractures always happened at the interface, as shown in Fig. 4. According to the similar bonding strength of the bonding pairs with different orientation, it can be inferred that there is no orientation dependence between Si-face and C-face for SAB method. After rapid thermal annealing at 1273 K in Ar gas for 180 s, all the average tensile strengths of the bonding interfaces were improved to ~21.6 MPa. However, the fracture during pulling test always happened in the epoxy. This indicates the tensile strengths of the bonding interfaces after rapid thermal annealing should be higher than 21.6 MPa, which is much higher than the bulk Si strength tested in our previous research (7).

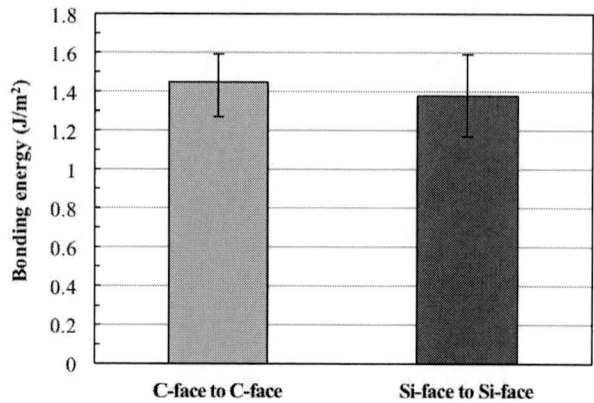

Figure 3. Comparison of the bonding energies of SiC (C-face)-SiC (C-face) and SiC (Si-face)-SiC (Si-face) bonded wafer.

Figure 4. Typical fractured SiC-SiC bonded chip after pulling test.

Bonding Mechanism

Through the Monte Carlo simulation, it was found that the sputtering yield of C and Si are 0.65 atoms/ion and 1.14 atoms/ion, respectively, which means Si was preferentially sputtered. In addition, the projected range of Ar ions was 1.8 nm (straggling ΔRp=0.9 nm). According to the simulation results, the activated SiC surface layer should consist of Si, C and Ar, but be Si-less or C-rich.

Furthermore, the bonding interfaces of SiC (C-face)-SiC (C-face) and SiC (Si-face)-SiC (Si-face) were analyzed by TEM and EDX. Fig. 5 shows their TEM images. It can be seen that there is a uniform ~4 nm amorphous layer at both of the two interfaces. Since the bonding interfaces were symetrical, it can be inferred that a ~2 nm amorphous SiC layer formed on both Si-face and C-face after Ar ion beam irradiation. The formation of amorphous surface layer by surface activation may eliminate the affects of orientation, which can explain the similar bonding strength of the two bonding pairs with different orientation.

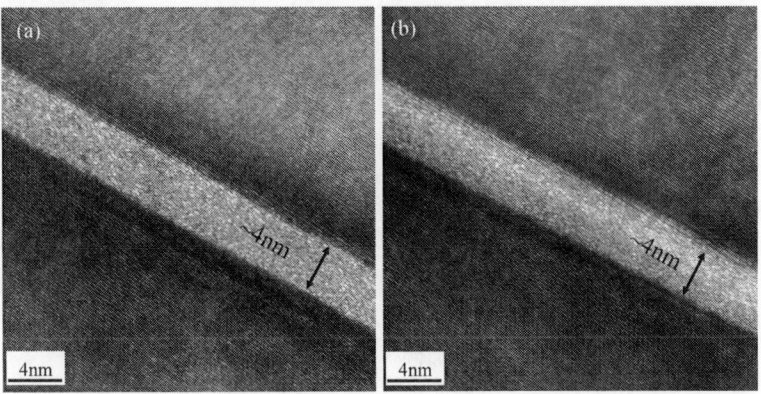

Figure 5. TEM images of the bonding interfaces of (a) SiC (C-face)-SiC (C-face) and (b) SiC (Si-face)-SiC (Si-face).

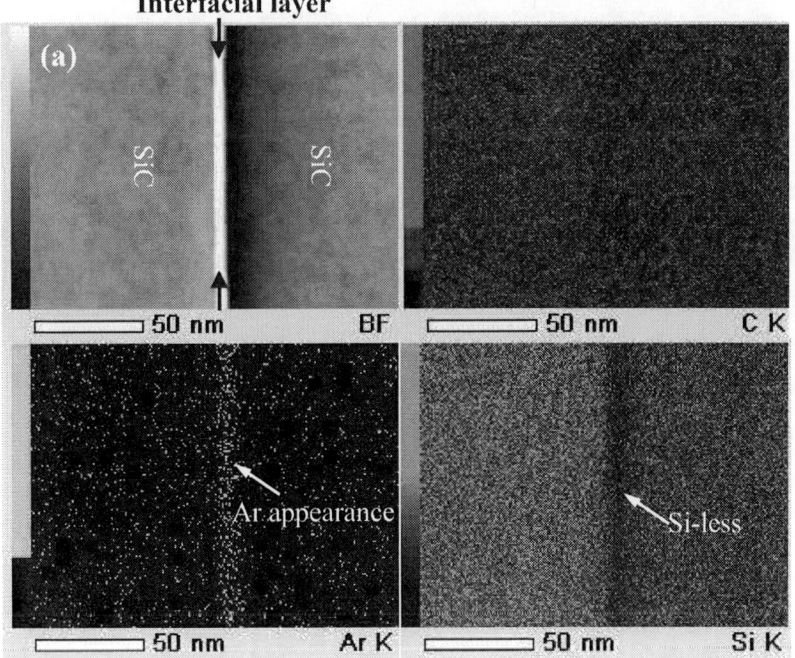

Figure 6. EDX element (C, Ar and Si) mapping of the bonding interfaces of SiC (C-face)-SiC (C-face).

Since the two bonding interfaces have similar results of EDX analysis, only EDX element (C, Ar and Si) mapping of the SiC (C-face)-SiC (C-face) bonding interface was presented in Fig. 6. No difference of the C amount in the SiC substrate and the interfacial layer could be distinguished; while the reduction of Si amount and appearance of Ar in the interfacial layer could be found by comparing with the adjacent SiC substrate. This agrees well with the results of Monte Carlo simulation. According to the molecular dynamic study of diamond bonding in UHV at room temperature, it could be assumed that the formation of graphite-like structure on the surfaces will be harmful for bonding (12). In case of SiC-SiC direct bonding by SAB, some C-C bonds or graphite-like structure might also be formed in the Si-less or C-rich activated surface layer caused by the Si preferentially sputtering during surface activation as inferred from Monte Carlo simulation and EDX analysis. This might be the reason why direct wafer bonding of SiC-SiC cannot reach bulk SiC strength.

Conclusion

By SAB method, direct wafer bonding of SiC-SiC has been achieved with a bonding energy of ~1.4 J/m^2 without orientation dependence. The formation of ~2 nm amorphous SiC layer on both Si-face and C-face during surface activation may eliminate the affects of orientation on bonding. In addition, the amorphous SiC layer being Si-less or C-rich caused by possible Si preferentially sputtering might result in the formation of C-C bonds or graphite-like structure in the activated SiC surface layer, which is assumed to be the reason why the direct wafer bonding of SiC-SiC cannot be as strong as bulk SiC. A rapid thermal annealing treatment at 1273 K for 180 s in Ar gas, which is widely used for ohmic contact formation on SiC device, could improve the tensile strength of bonding interface from ~12.2 MPa to the value higher than 21.6 MPa. Such a direct wafer bonding of SiC-SiC by SAB method at room temperature is expected to be useful for the fabrication of SiC power device in the future.

Acknowledgments

This research was partially supported by a Grant-in-Aid for Scientific Research (A), 2011, 23246125 from the Ministry of Education, Culture, Sports, Science and Technology.

References

1. T. Kimoto and J. A. Cooper, Fundamentals of Silicon Carbide Technology Growth, Characterization, Devices and Applications, John Wiley & Sons, New York (2014).
2. H. Okumura, *Mater. Res. Soc. Bull.*, **40**, 439 (2015).
3. G. N. Yushin, A. V. Kvit and Z. Sitar, *J. Mater. Sci.*, **40**, 4369 (2005).
4. I. V. Grekhov, L. S. Kostina, T. S. Argunova, E. I. Belyakova, J. H. Je, P. A. Ivanov, and T. P. Samsonova, *Tech. Phys. Lett.*, **32**, 453 (2006).
5. H. Takagi, K. Kikuchi, R. Maeda, T. R. Chung, and T. Suga, *Appl. Phys. Lett.*, **68**, 2222 (1996).

6. T. Suga, F. Mu, M. Fujino, Y. Takahashi, H. Nakazawa, and K. Iguchi, *Jpn. J. Appl. Phys.*, **54**, 030214-1 (2015).
7. F. Mu, K. Iguchi, H. Nakazawa, Y. Takahashi, M. Fujino, and T. Suga, *Jpn. J. Appl. Phys.*, **55**, 04EC09-1 (2016).
8. W. P. Maszara, G. Goetz, A. Caviglia, and J. B. McKitterick, *J. Appl. Phys.*, **64**, 1988, 4943 (1988).
9. F. Zhao, W. Du, and C.-F. Huang, *Microelectron. Eng.*, **129**, 53 (2014).
10. [Online]. Available: http://www.srim.org.
11. H. Kikuchi, R. K. Kalia, A. Nakano, P. Vashishta, P. S. Branicio, and F. Shimojo, *J. Appl. Phys.*, **98**, 103524-1 (2005).
12. D. Conrad, K. Scheerschmidt, and U. Gösele, *Appl. Phys. Lett.*, **77**, 49 (2000).

84

Chapter 2

Heterogeneous & Photonic Integration

Diverse Accessible Heterogeneous Integration (DAHI) Foundry at Northrop Grumman Aerospace Systems (NGAS)

A. Gutierrez-Aitken, D. Scott, K. Sato, B. Poust, E. Nakamura, K. Thai, W. Chan,
E. Kaneshiro, C. Monier, I. Smorchkova, N. Lin, D. Ferizovic, X. Zeng,
A. Oki, R. Kagiwada

Northrop Grumman Aerospace Systems
One Space Park, D1-1302J, Redondo Beach, CA 90278

The performance, size, weight and power requirements for future systems are increasingly demanding. These can be met by intimate integration of lower power and higher performance scaled CMOS and compound semiconductor technologies into smaller areas and volumes. The more adaptable and more intimate is this integration between two or more technologies, the more flexibility is given to the designer for the selection of the technology for a specific function, or even better, for an optimum combination of different transistor technologies in the same function or cell in the design. Northrop Grumman Aerospace Systems (NGAS) under the Diverse Accessible Heterojunction Integration (DAHI) DARPA program is developing integration processes, design kits and thermal simulation tools to integrate submicron CMOS, InP HBT, GaN HEMT and high-Q passive technologies for advanced DoD systems. We have demonstrated integration of NGAS' InP HBT and GaN HEMT technologies on 65nm and 45nm CMOS wafers.

DAHI Heterogeneous Integration Technology

Advantages of DAHI Integration Technology

The NGAS DAHI integration process has several advantages. It is scalable to 200 mm and 300 mm CMOS wafers and fully compatible with all CMOS technologies. All the technologies to be integrated are independently fabricated in parallel for optimum cycle time and decoupled line yield. The NGAS DAHI integration approach also offers advantages from the thermal management point of view. The low thermal resistance heterogeneous interconnects (HICs) between the technologies enable flexible electrical and thermal routing to the CMOS or to other semiconductor technologies. The GaN HEMT is integrated with its backside facing the CMOS front side to allow better intra-chiplet heat spreading and large inter-chip HIC heat sink arrays. The thickness of the Silicon Carbide (SiC) substrate in the GaN HEMT chiplet is designed to be thick enough to allow optimum lateral heat spreading before reaching the thermal HIC interface. The TF InP HBT technology is integrated with its front side facing the CMOS front side for maximum density of interconnects. An additional very important task in the development of the DAHI technology is the creation of an integrated DAHI process design kit (PDK). The objective of the DAHI PDK development is to have a modular and flexible design environment that can accommodate schematics, layouts and simulations of integrated DAHI circuits with device-level integration designs of a wide variety of technologies.

Heterogeneous Integration Approaches

There are several approaches to perform heterogeneous integration. At NGAS we have developed in the past several years selective epitaxial growth, metamorphic growth, wafer level packaging (WLP), and DAHI heterogeneous integration processes (1) and demonstrated state of the art circuit performance (2)-(3). Figure 1 describes schematically these integration approaches.

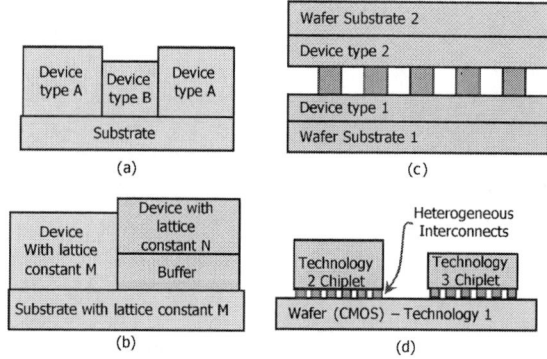

Figure 1. Heterogeneous integration methods used or being developed at NGAS. (a) Selective epitaxy, (b) metamorphic growth, (c) wafer level packaging and (d) DAHI integration.

In the selective epitaxy heterogeneous integration technique (Figure 1a), two (or more) different types of devices (typically HBT and HEMT) are grown on the same substrate (For example InP or GaAs) either sequentially, one on top of the other, or grown, etched back selectively and grown again to create the cross-section shown in Figure 1a. Metamorphic growth combined with selective epitaxy enables the integration of semiconductor devices of one material system on a substrate of a different material system where the device material system lattice constant is significantly different than the substrate lattice constant (Figure 1b). There are several approaches to achieve this that in general use one or more buffer layers between the substrate and the device. The key factor in this approach is to design and implement a buffer layer that provides a high quality "virtual substrate" on which to grow the device. The buffer layer should be carefully designed to minimize the increase on thermal resistance due to this added layer.

In the wafer level packaging approach (Figure 1c) the wafers of the individual technologies to be integrated are fully fabricated and bonded together at the wafer level using a metal to metal thermo-compression process. This technology, also referred as wafer scale assembly (WSA), has several advantages that include no device performance degradation, no change in device fabrication process and preservation of existing high-reliability production processes. The DAHI heterogeneous integration process (Figure 1d) developed by NGAS under the DAHI DARPA program is a chiplet to wafer metal to metal thermo-compression bonding with highly scaled heterogeneous interconnects (HICs).

DAHI Integration Process

The DAHI process combines the desired characteristics of low temperature bonding with high yield, low electrical and thermal resistance heterogeneous interfaces. Figure 2(a) shows a section of a DAHI integrated wafer of face-up GaN HEMT and face-down InP HBT chiplets on CMOS. Figure 2(b) shows a schematic cross-section of the integration illustrating some details of the electrical HICs and thermal HICs. Figure 2(c) shows a scanning electron microscope (SEM) picture of a focused ion-beam (FIB) cross-section of the DAHI HIC.

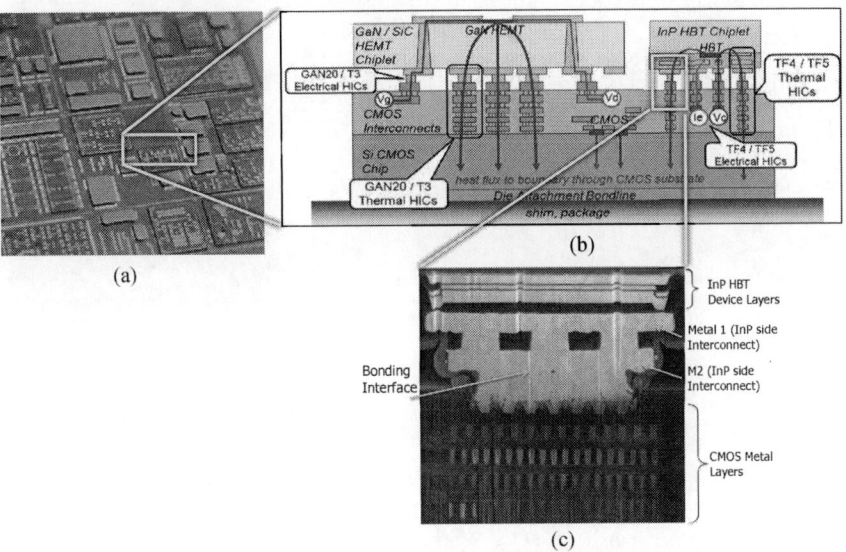

Figure 2. SEM picture of a FIB cross-section of the DAHI heterogeneous interconnect.

For the DAHI Phase 1 of the program, the selected technologies for integration were IBM's 65nm CMOS, NGAS' TF4 InP HBT and GaN20 GaN on SiC HEMT technologies. A top level integration process flow is illustrated in Figure 3. Commercial CMOS wafers are obtained through foundry suppliers without alterations to the CMOS process. Once delivered, the CMOS wafers are cored to 100mm diameter for compatibility with NGAS' existing compound semiconductor toolset, and the HIC metal is applied. While the CMOS is being fabricated at the silicon foundry, the InP and GaN wafers are fabricated at NGAS. The compound semiconductor wafers follow existing frontside fabrication procedures and enter backside fabrication where minor changes to final wafer thickness and metallization processes are performed. Once fabrication of the GaN and InP wafers is complete, the thinned wafers are mounted to carriers and singulated into chiplets before alignment and integration to the prepared CMOS wafer is performed.

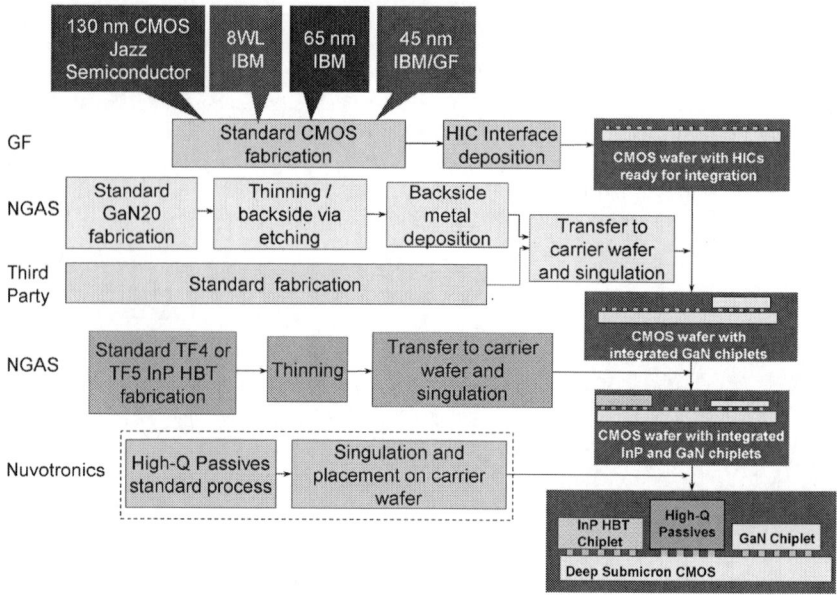

Figure 3. NGAS DAHI integration sequence.

DAHI Technologies.

A cross-sectional view of the InP Heterojunction Bipolar Transistors (HBT) technology available for DAHI integration is shown in Figure 4(a). It offers NGAS advanced InP bipolar transistors (TF), Schottky diodes, high-precision thin film resistors and metal-insulator-metal capacitors. The backend interconnect fabrication uses three levels of interconnect metal with low dielectric constant interlayer. Heterogeneous interconnects (HIC) are deposited onto the TF most top metal to serve as the interface to the CMOS wafer. After front-side fabrication, InP wafers are thinned down to 75 μm thickness during backside process.

Two distinct TF transistor technologies (TF4, TF5) of different DC and RF characteristics Figure 4(b) are available for DAHI integration. Both TF4 and TF5 devices use a MBE-grown InP/InGaAs double HBT structure. TF4 is based on 0.65 μm emitter width, with peak cutoff frequency (ft) of 150 GHz at 1.5 mA/μm^2 maximum current density. TF5 transistors use a more advanced epilayer profile and device scaling with a minimum emitter width of 0.25 μm for superior RF performance (ft > 350 GHz at 5 mA/μm^2 current density). The measured RF performance if these technologies are presented in Figure 4(c).

The GaN HEMT technology is fabricated using NGAS' T-gate GaN20 on SiC developed for use up to 60GHz. The key features of the technology are shown in the cross-sectional drawing in Figure 4(d). GaN HEMT epitaxial layers for NGAS process are grown by molecular beam epitaxy on 100mm high purity, semi-insulating 4H-SiC substrates. The device epitaxial profile used in this technology provides excellent carrier confinement and minimizes short channel effects. The HEMT device utilizes 200 nm T-gate formed using electron beam lithography. The process has two levels of interconnect metals to form air bridges, and includes NiCr thin film resistors (TFRs), metal-insulator-metal (MIM) capacitors and inductors. 15x25 μm^2 through-wafer vias are etched in the 50 μm-thick SiC to form electrical connections to the backside of the wafer.

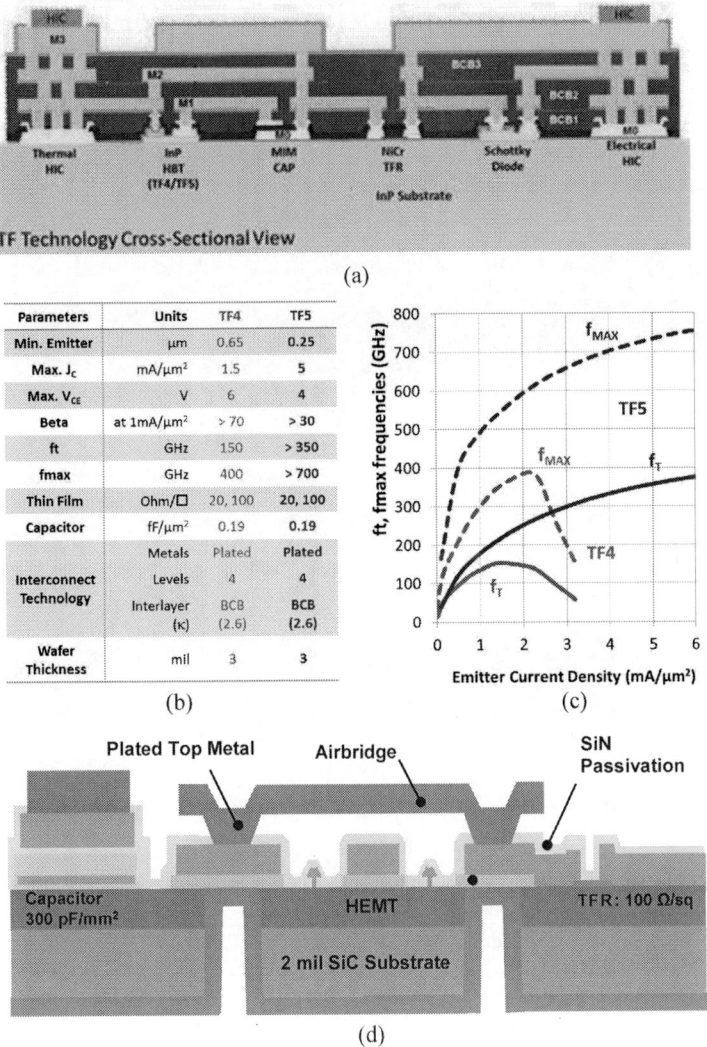

Figure 4. DAHI InP HBT and GaN HEMT Technologies.

DAHI Process Design Kits and Thermal Analysis Tool

DAHI Process Design Kit (PDK)

The DAHI PDK software was developed by NGAS to handle the creation of heterogeneous circuits. Creation of these device-level heterogeneous circuits follows the basic design stages of design capture, layout, simulation, thermal analysis, and system level design. The DAHI PDK is designed to be flexible and modular to accommodate a wide variety of technologies and their corresponding native stand-alone PDKs. Circuit designers then selectively choose the best technology offering within the DAHI PDK and merge them at the device level to implement a higher performing electronic function. The DAHI PDK was created to integrate multiple technology PDKs into a single and complete platform capable of performing various types of simulations that include time-domain, frequency-domain, and thermal simulations as shown in Figure 5

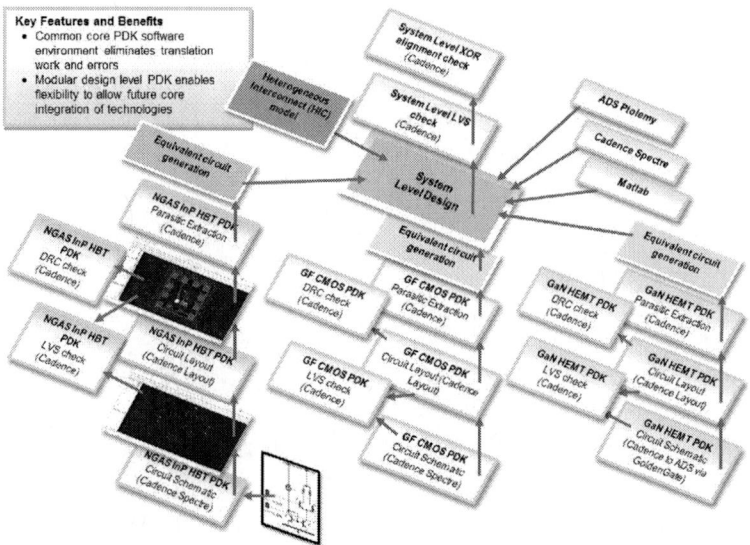

Figure 5. Modular NGAS DAHI PDK hierarchical design environment.

Other important functions of the DAHI PDK consist of layout, design rule check (DRC) and layout versus schematic (LVS) validation at different hierarchical levels, and parasitic extraction. The DAHI PDK allows for integration of additional third-party (non-CMOS) PDKs presently. In the near future, integration of multiple CMOS node PDKs will be addressed such that the DAHI PDK can continue to expand and offer more design capability and flexibility.

PDK compatibility is very important in maintaining a modular technology infrastructure and future expansion. Using unaltered PDKs from other vendors aligns itself to maintaining modularity within the DAHI PDK as well as remaining compatible with industry standard electronic design automation (EDA) tools. This approach helps to eliminate translation work and errors while keeping all the validation of the individual

PDKs and their models intact. The Cadence design environment was chosen for the PDK development since it is available for most technologies including many commercial CMOS foundry offerings. Adopting Cadence also allows circuit designers access to other software such as ADS through Cadence Dynamic Link, Ptolemy, MATLAB, and GoldenGate to help design and analyze DAHI circuits. NGAS developed the DAHI translator software to provide all the necessary links between the individual PDKs within the DAHI PDK software. The DAHI translator is also necessary to combine individual technologies into a DAHI layout which can then be LVS checked against its corresponding DAHI schematic as shown in Fig.13.

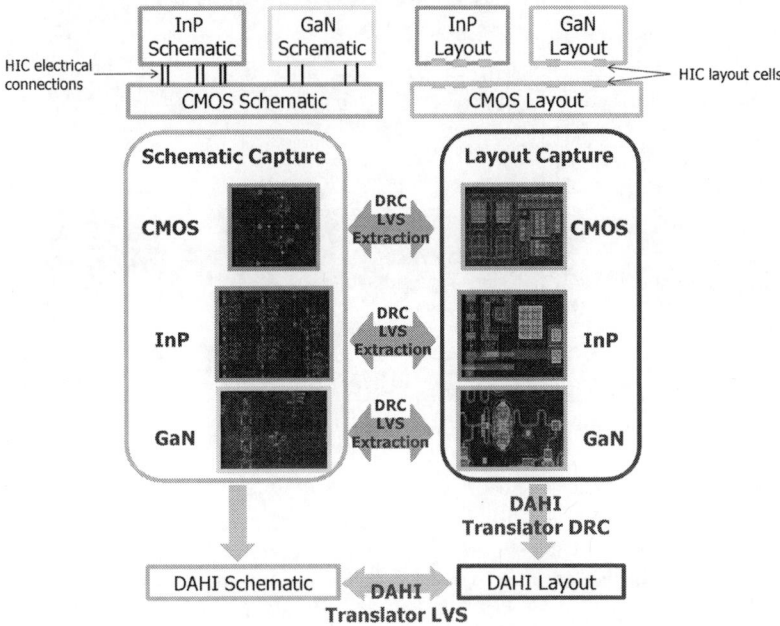

Figure 6. Typical DAHI Design Flow.

DAHI Compact Analytical Thermal Solver (DAHICATS) and Finite Element Model Automation (DAHIFEMA)

Semiconductor device failure mechanisms in general are accelerated by operation at high temperature. Foundries typically perform multiple-temperature accelerated life tests to establish the maximum operating temperature at which the projected mean time to failure meets or exceeds mission requirements. Circuit designers must ensure devices will not exceed reliability limited maximum operating temperatures under worst case operating and ambient conditions.

DAHI circuit design and heterogeneous integration in general complicate the thermal management challenge. Circuit designers must provide low thermal resistance paths for waste heat generated in DAHI chiplets to flow into and through the underlying CMOS substrate. This is accomplished by judicious use and placement of DAHI thermal HICs in the circuit layout. The simple self-heating models typically included in process design kits do not address this need, and iteration between circuit designers and thermal analysts is impractical due to problem complexity.

NGAS' DAHI process design kit offers a suite of built-in thermal analysis tools developed specifically to address the thermal management challenge. Accessibility to the circuit designer is ensured by a simple set of user interfaces and fully automated model extraction, analysis, and post-processing flow as shown in Figure 6.

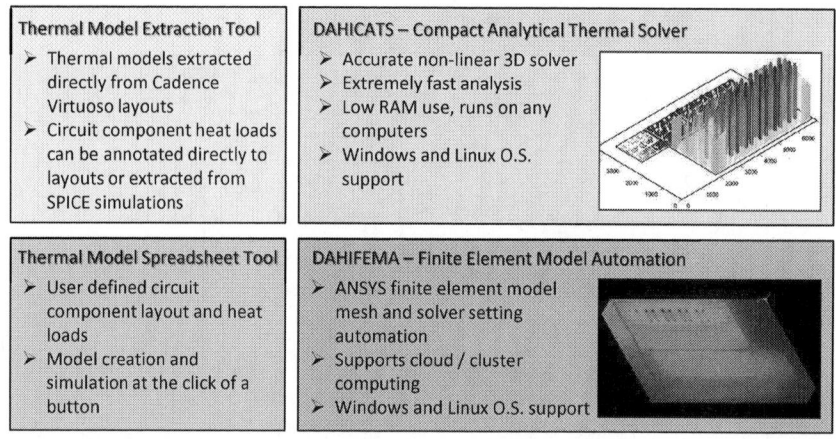

Figure 7. DAHI thermal tool suite makes complex DAHI thermal analysis accessible to circuit designers.

The DAHICATS tool is a proprietary non-linear 3D thermal solver developed specifically to handle DAHI thermal analysis problems efficiently. Significant solution time and memory reduction are offered over more traditional computation methods such as thermal finite element analysis (FEA), as shown in the comparison study presented in Figure 8.

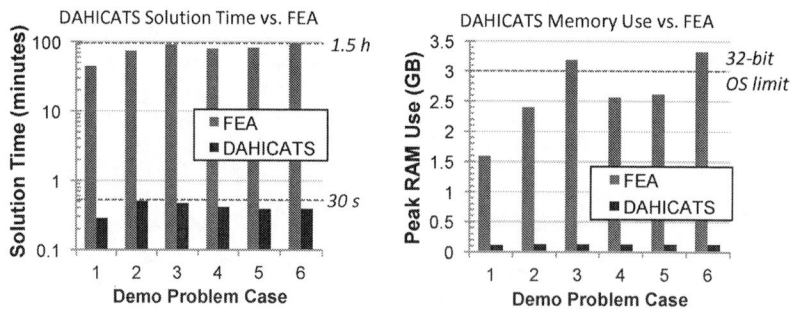

Figure 8. DAHI thermal tool suite makes complex DAHI thermal analysis accessible to circuit designers.

Thermal model automation and thermal solver efficiency improvements reduce the level of effort for the circuit designer / thermal analyst to a few clicks and minutes of turn-around time, as illustrated in Figure 9.

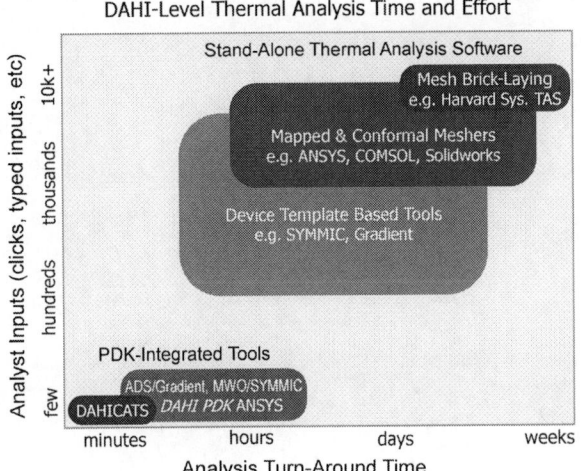

Figure 9. DAHI thermal tool suite makes complex DAHI thermal analysis accessible to circuit designers.

Acknowledgments

This work is supported by DARPA DAHI Program under AFRL contract No. FA8650-13-C-7312, Daniel Green DARPA program manager and Robert Fitch AFRL COTR. The views expressed are those of the authors and do not reflect the official policy or position of the Department of Defense or the U.S. Government..

References

1. A. Gutierrez-Aitken, P. Chang-Chien, D. Scott, K. Hennig, E. Kaneshiro, P. Nam, N. Cohen, D. Ching, K. Thai, B. Oyama, J. Zhou, C. Geiger, B. Poust, M. Parlee, R. Sandhu, W. Phan, A. Oki, R. Kagiwada, IEEE Compound Semiconductor Integrated Circuit Symposium (CSICS), Monterey, CA, Oct 5, 2010.
2. B. Oyama, D. Ching, K. Thai, A. Gutierrez-Aitken, V.J. Patel, IEEE Journal of Solid-State Circuits, Vol. 48 , No.10, Oct. 2013, pp. 2265 - 2272 .
3. P. Chang-Chien, X. Zeng, K. Tornquist, M. Nishimoto, M. Battung, Y. Chung, J. Yang, D. Farkas, M. Yajima, C. Cheung, K. Luo,D. Eaves, J. Lee, J. Uyeda, D. Duan, O. Fordham, T. Chung, R. Sandhu, R. Tsai, IEEE 19th International Conference on Indium Phosphide and Related Materials (IPRM), May 2007, pp. 14 – 17.

Suppressed Self-Heating in Multi-Finger InP-Based DHBTs with Au Subcollector Fabricated on SiC Substrate by Surface-Activated Bonding

Y. Shiratori[a], T. Hoshi[a], N. Kashio[b], K. Kurishima[a], E. Higurashi[c], and H. Matsuzaki[a]

[a] NTT Device Technology Laboratories, NTT Corporation, Atsugi, 243-0198, Japan
[b] NTT Device Innovation Center, NTT Corporation, Atsugi, 243-0198, Japan
[c] The University of Tokyo, Tokyo, 113-8656, Japan

> InP-based multi-finger double-heterojunction bipolar transistors (DHBTs) with an Au subcollector were fabricated on a highly-thermal-conductive SiC substrate by surface-activated bonding (SAB). The fabricated DHBTs have good electrical properties without any degradation due to the wafer-bonding process. They also show 68% reduction of thermal resistance (R_{th}) compared with a conventional DHBT on an InP substrate. Reduced R_{th} enables us to increase current density while maintaining the junction temperature of DHBTs, which boosts their operating speed and enhances their integration density. Therefore, multi-finger DHBTs on a SiC substrate fabricated by SAB is very useful for reducing IC chip size and increasing the operation speed.

Introduction

InP-based double-heterojunction bipolar transistors (DHBTs) have demonstrated excellent high-frequency performance suitable for submillimeter-wave and telecommunications integrated circuits (ICs). For these ICs, increasing the current density of the DBHTs is very effective for improving their high-frequency performance because of reducing the charging time. Multi-finger transistor design is also useful in reducing IC chip size because it reduces the total area and the number of DHBTs in ICs needed in order to obtain the required output current. On the other hand, the increased current density and multi-finger design both result in higher power consumption per unit area, leading to an increase in junction temperature and degradation of the current gain of DHBTs. To suppress these undesirable effects, effective heat transfer from inside the DHBTs to the backside of the substrate is required. However, this is challenging because InP and InGaAs, which are used for the substrate and subcollector materials in InP-based DHBTs, have low thermal conductivities (κ) of 68 and 5 $Wm^{-1}K^{-1}$, respectively.

In order to improve heat transfer, we propose the multi-finger DHBT with an Au subcollector fabricated on a SiC substrate by surface-activated bonding (SAB). The measured thermal resistance (R_{th}) shows the significant impact of the proposed DHBT structure on suppressing self-heating, thanks to the high thermal conductivity of the SiC substrate and the simultaneously introduced highly thermal-conductive Au subcollector.

Device Structure and Fabrication Process

Multi-Finger DHBT Structure with an Au Subcollector on a SiC Substrate

Figure 1 shows a cross-sectional view of a multi-finger DHBT. The DHBT structure is formed on a SiC substrate ($\kappa = 490$ Wm^{-1}K^{-1}) through an Au subcollector ($\kappa = 320$ Wm^{-1}K^{-1}). The epitaxial-layer structure is designed for high-current-density operation (> 10 mA/μm^2). The emitter is 20-nm-thick InP. The InGaAsSb base is 20-nm thick and highly C-doped at 7×10^{19} cm^{-3}. The collector is 60-nm-thick InP. For the contact to the subcollector, 10-nm-thick, highly Si-doped InGaAs is used. Emitter fingers and base electrodes are formed on one Au subcollector. We prepared nine types of multi-finger DHBTs with combinations of four to six emitter fingers and emitter-finger spacing of 0.9, 1.4, and 3.4 μm. For single-finger devices, we have already demonstrated a significant improvement in heat dissipation [1]. To compare the R_{th} between single-finger and multi-finger DHBTs, single-finger DHBTs were also fabricated.

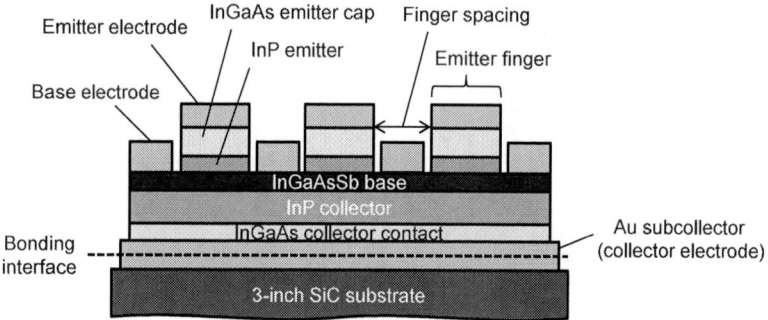

Figure 1. Cross-sectional view of a multi-finger DHBT with an Au subcollector on a SiC substrate.

Fabrication Process

The DHBT fabrication process started with the epitaxial growth of DHBT layers on a 3-inch InP substrate by metal-organic chemical vapor deposition, as shown in Fig. 2. After Au had been deposited as an adhesive and subcollector material on both the DHBT layers and 3-inch SiC substrate, they were bonded by SAB. This bonding technique prevents thermal degradation of the epitaxial layers because of its sufficiently low bonding temperature (~150°C) [2]. Note that the growth sequence of the epitaxial-layer structure is inverted for the wafer bonding, which started from the emitter cap layer.

Figure 3(a) shows an acoustic microscope image of the bonded wafer. The InP and the SiC substrates were bonded to each other with almost no voids except for at the wafer edge. After the removal of the InP substrate, DHBTs with emitter widths of 0.25, 0.35 and 0.5 μm were fabricated using the self-aligned process [3]. Figure 3(b) shows a scanning electron microscope image of a fabricated four-finger DHBT with an emitter-finger size of 0.5 μm x 4.0 μm. Multi-finger DHBTs were successfully fabricated on a SiC substrate without any serious problems related to the bonding process.

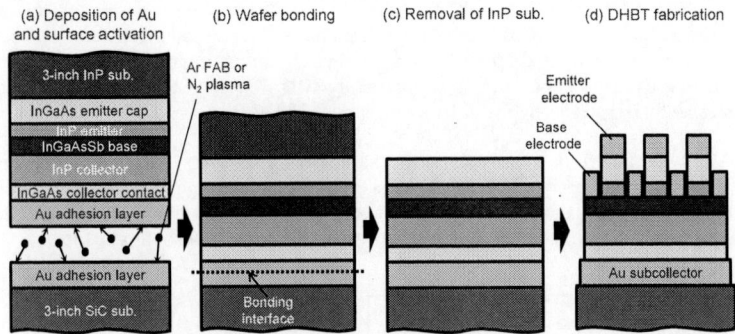

Figure 2. Fabrication steps for the multi-finger DHBT.

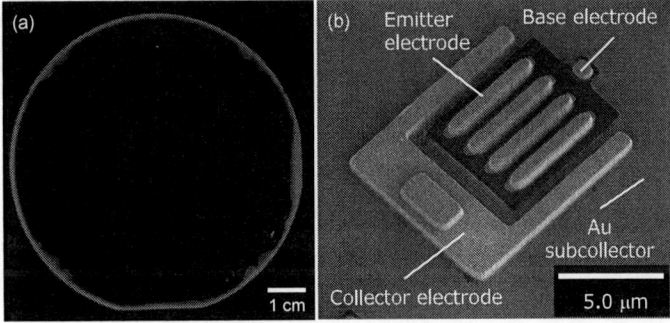

Figure 3. (a) Scanning acoustic microscope image of a 3-inch bonded wafer with an InP substrate and SiC substrate. (b) Scanning electron microscope image of a fabricated four-finger DHBT with emitter-finger size of 0.5 μm x 4.0 μm on an Au subcollector.

Device Characteristics

Electrical Characteristics

We measured the electrical characteristics of fabricated DHBTs to investigate the influence of the bonding process. Figure 4(a) shows Gummel plots of a four-finger DHBT on a SiC substrate (hereinafter referred to as "DHBT on SiC") with the emitter-finger size of 0.25 μm x 4.0 μm. For comparison, Gummel plots of a conventional DHBT with the same epitaxial-layer structure on an InP substrate (hereinafter referred to as "DHBT on InP") are shown by dotted lines. As seen from the figure, there is little difference in the current-transfer characteristics between the two kinds of DHBTs in the region of high base-emitter voltage ($V_{DE} > 0.7$ V). In addition, an increase in base and collector leakage current is not observed for the DHBT on SiC in the low-V_{BE} region (< 0.7 V). These results indicate that the emitter-base and base-collector junctions did not degrade after the bonding process.

We also performed on-wafer measurement of S-parameters with an HP8510C network analyzer. Figure 4(b) shows current gain ($|h_{21}|^2$) and Mason's unilateral gain (U) as a function of frequency for both DHBTs at collector current density (J_C) of 15 mA/μm^2. Cutoff frequency (f_t) and maximum oscillation frequency (f_{max}) for the DHBT on SiC are 502 and 345 GHz, respectively, at collector-emitter voltage (V_{CE}) of 1.6 V. These values are slightly different from the DHBT on InP (f_t of 571 GHz and f_{max} of 312 GHz at V_{CE} of 1.0 V). According to delay time analysis, the difference in f_t between the two kinds of DHBTs can be explained by the difference in their extrinsic collector capacitance, and the intrinsic transit time across the base and collector is almost the same. Consequently, the bonding process essentially has very little impact on the carrier transport properties.

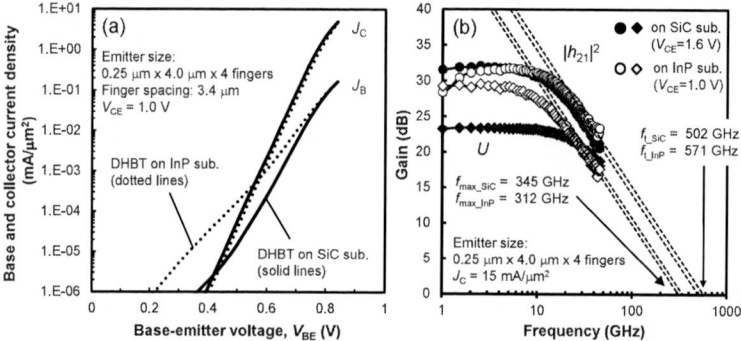

Figure 4. DC and RF characteristics of fabricated four-finger DHBTs with the emitter size of 0.25 μm x 4.0 μm. (a) Gummel plots and (b) current gain ($|h_{21}|^2$) and Mason's unilateral gain (U) as a function of frequency at J_C of 15 mA/μm^2.

Thermal Characteristics

We estimated the R_{th} for the DHBTs from electrical characteristics in a common-base configuration, which is expressed by the following equation [4].

$$R_{th} = \frac{1}{\phi} \frac{\Delta V_{BE}}{\Delta P}$$ (1)

Here, ΔV_{BE} is a shift in base-emitter voltage for a given collector current when collector-base voltage increases. ΔP is an associated increase in power dissipation, and ϕ is a thermo-electro feedback coefficient, experimentally obtained from the Gummel plots at various ambient temperatures. For the DHBTs, the measured ϕ is about 0.9 mV/K. In this paper, the R_{th} is normalized by the total emitter junction area.

Figure 5 shows R_{th} as a function of the emitter width of DHBTs. For both DHBTs on SiC and InP, the R_{th} decreases with the emitter width due to the increased importance of heat diffusion from the periphery of DHBTs. In addition, the DHBTs on SiC show over 62% reduction of the R_{th} compared to the DHBTs on InP, independently of the emitter width. Thus, the lower R_{th} of the DHBT on SiC can be maintained even when the emitter width is further scaled down ($< 0.25\ \mu m$).

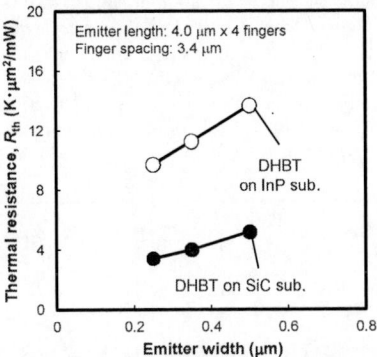

Figure 5. Emitter-width dependence of thermal resistance for four-finger DHBTs fabricated on a SiC and InP substrate with the emitter length of $4.0\ \mu m$.

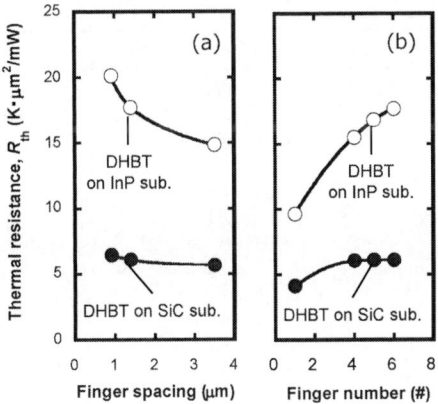

Figure 6. (a) Finger-spacing dependence and (b) finger-number dependence of thermal resistance for DHBTs fabricated on a SiC and InP substrate with the emitter-finger size of $0.5\ \mu m \times 4.0\ \mu m$

Figure 6 shows the finger-spacing and the finger-number dependence of R_{th} for the DHBTs with the emitter-finger size of $0.5\ \mu m \times 4.0\ \mu m$. For the DHBTs on InP, decreasing the emitter-finger spacing or increasing the emitter-finger number results in higher heat density and, consequently, a higher R_{th}. On the other hand, such abrupt increases are not observed for the DHBTs on SiC. These results clearly show the combination of the Au subcollector and SiC substrate effectively transfers the generated heat from inside the DHBT to the substrate. The proposed DHBTs exhibit a 68% reduction in the R_{th}, compared with the DHBTs on InP, when the emitter-finger spacing is 0.9 μm.

Figure 7 shows collector I-V curves for the four-finger DHBTs with a 0.25-μm-wide emitter on InP and SiC. Junction temperatures (T_j) in each device were estimated from measured R_{th} and are plotted by dotted curves. By reducing R_{th}, T_j of the DHBT on SiC become less than half of that of the DHBT on InP. Owing to the maintained low T_j, the DHBT on SiC successfully suppresses the current-gain lowering in the high-V_{CE} and high-I_C bias regions (~20 mW/μm²). This result confirms that the proposed DHBT structure enables to high-current-density operation.

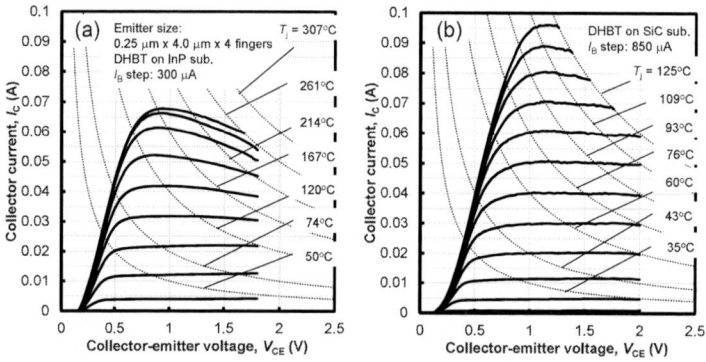

Figure 7. I_C-V_{CE} characteristics of the four-finger DHBTs with an emitter-finger size of 0.25 μm x 4.0 μm fabricated on (a) an InP substrate and (b) SiC substrate.

Conclusion

We fabricated multi-finger InP DHBTs with an Au subcollector on a SiC substrate using the SAB technique for high-power and high-speed operation. The measured thermal resistance is almost completely independent of the emitter-finger spacing and emitter-finger number. In addition, the DHBTs on a SiC substrate exhibit a 68% reduction of R_{th} compared with those on an InP substrate. Reducing R_{th} can contribute to enhancing the operating speed and integration density of DHBTs. The proposed multi-finger DHBT is very promising for boosting the operation speed and reducing the chip size of ICs.

References

1. Y. Shiratori, T. Hoshi, N. Kashio, K. Kurishima, E. Higurashi, and H. Matsuzaki, *IPRM 2015*, Santa Barbara, E2.5 (2015).
2. Y.-H. Wang, J. Liu, and T. Suga, *Proc. ICEPT-HDP '09*, pp. 516-519 (2009).
3. N. Kashio, T. Hoshi, K. Kurishima, M. Ida, and H. Matsuzaki, *IEEE Electron Device Lett.*, **35**, pp. 1209-1211 (2014).
4. W. Liu, *Handbook of III-V Heterojunction Bipolar Transistors*, p. 363, Wiley-Interscience, New York (1998).

Au/SiO$_2$ Hybrid Bonding with 6-μm-Pitch Au Electrodes for 3D Structured Image Sensors

Yuki Honda[a], Kei Hagiwara[a], Masahide Goto[a], Toshihisa Watabe[a], Masakazu Nanba[a], Yoshinori Iguchi[a], Takuya Saraya[b], Masaharu Kobayashi[b], Hiroshi Toshiyoshi[b], Eiji Higurashi[b], and Toshiro Hiramoto[b]

[a]NHK Science and Technology Research Laboratories, Tokyo, Japan
[b]The University of Tokyo, Tokyo, Japan

A 3D structured CMOS image sensor that can process signals pixel-wise in parallel is investigated as a means to simultaneously increase the pixel count and the frame rate. To make the pixel pitch finer, the Au electrode dimension is reduced to a 6-μm pitch that is electrically interconnected using a direct-bonding technique. We also develop a daisy-chain test device to examine the numerous interconnections. The experimental results show that the series of electrical interconnections exceeds 230,000 contacts, with the Au contact resistance for a single connection being ~23.6 mΩ.

Introduction

Pixel counts and frame rates of broadcasting TV cameras are increasing to meet the escalating demands for high quality pictures. The pixel count could be as high as 133 million [1] in the state-of-the-art CMOS video image sensor, and the frame rate could be 120 frames per second (fps) [2], which will further keep increasing in coming years. Owing to large pixel count, the sensors are designed to share the signal processing circuits in a form such as column-parallel processing [2], and hence the conventional image sensors face difficulties in processing the signals at higher frame rate. In other words, the signal processing speed at each pixel is the bottleneck that limits the pixel count and the frame rate. To overcome this problem, we developed a 3D structured CMOS image sensor, as shown in Fig. 1 [3]. This sensor can process signals pixel-wise in parallel as a means to simultaneously increase the pixel count and the frame rate. The image sensor uses stacked signal processing circuits for each pixel by placing the circuits immediately below the photodiodes. Compared with conventional image sensors, the new structure enables us to maintain a higher frame rate even when the pixel count increases.

Pixel–pitch interconnection is a key technology in 3D structured CMOS image sensors. We have previously developed a prototype 3D structured CMOS image sensor (8 × 8 pixels) with 40-μm-pitch Au interconnections [3]. To further enhance the resolution, we have recently reduced the Au electrode dimension to a 6-μm pitch that could be electrically interconnected using a direct-bonding technique. Moreover, we have developed a daisy-chain test device to examine the numerous interconnections.

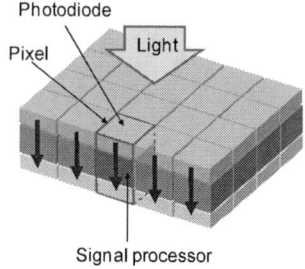

Figure 1. Schematic of a 3D structured CMOS image sensor.

Fabrication

The device is fabricated using the process shown in Fig. 2, which comprises the following steps: (a) an Al wiring layer is formed on a fully depleted silicon-on-insulator wafer; (b) via holes are formed in the top intermediate SiO_2 layer by dry etching; (c) Ti and Au seed layers are deposited by sputtering, and then the via holes are embedded with an electroplated Au layer; (d) an Au/SiO_2 bonding surface is formed by chemical mechanical polishing, and the wafer is diced into chips of 20×20 mm^2 and 18×18 mm^2 each; (e) the bonding surface is sequentially activated by Ar and O_2 plasmas; (f) two chips are directly bonded by 2,000 N for 60 min at 200 °C.

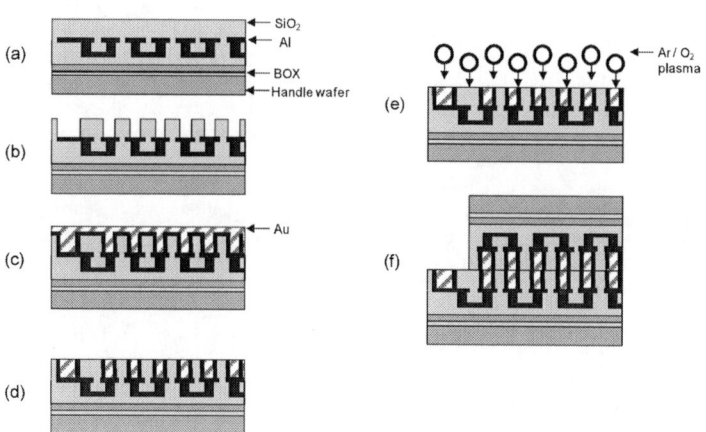

Figure 2. Process flow for fabrication of the daisy-chain test device.

Fig. 3 shows an atomic force microscope (AFM) image of the chip surface. The height of the Au electrode above the SiO_2 surface is ~18 nm. The average roughnesses (Ra) of the polished SiO_2 and the Au surfaces are <0.4 nm and ~1 nm, respectively. The diameter and the pitch of the Au electrode are 3 μm and 6 μm, respectively.

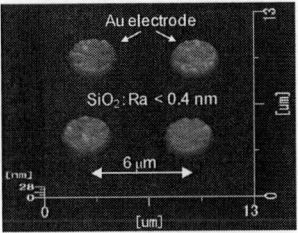

Figure 3. AFM image of a polished chip surface.

Measurement Results

Fig. 4 shows a cross-sectional scanning electron microscope (SEM) image of the bonded electrodes. No voids were observed at the bonded interface. Fig. 5 shows the measured daisy-chain resistance. We confirmed that the series of electrical interconnections exceeded 230,000 contacts and that the chain resistance was proportional to the number of electrodes. The Au contact resistance for a single connection was ~23.6 mΩ, which is considerably smaller than that of the plug electrodes of LSI circuits. Fig. 6 shows the alignment errors at 5 positions of 2 bonded daisy-chain test devices. The alignment errors were measured by observing alignment marks of the bonded chips with a transmission infrared microscope. Measured alignment errors between the upper and lower chips were less than 1.0 μm, which is sufficiently smaller compared to an Au electrode diameter.

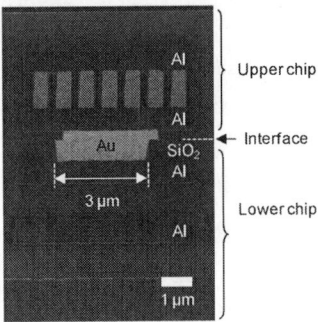

Figure 4. Cross-sectional SEM image of the bonded daisy-chain test device.

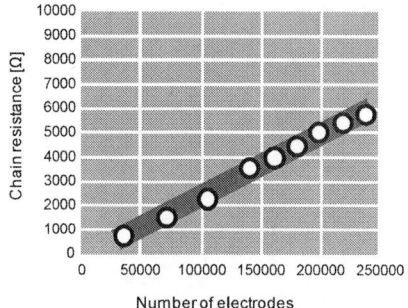

Figure 5. Measured chain resistance of the bonded daisy-chain test device.

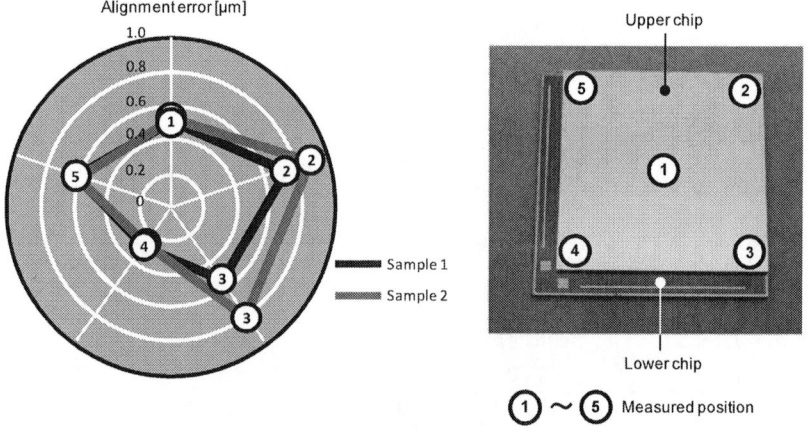

Figure 6. Alignment errors of 2 bonded daisy-chain test devices.

Conclusions

To summarize, a daisy-chain test chip with 6-μm-pitch Au electrode interconnections was successfully fabricated using Au/SiO$_2$ hybrid bonding. The results were promising for realizing a 3D structured CMOS image sensor for pixel-parallel signal processing.

References

1. R. Funatsu et al., *ISSCC 2015*, pp. 112-113, (2015).
2. K. Kitamura et al., IEEE Trans. Electron Devices, vol. 59, No. 12, pp. 3426-3433, (2012).
3. M. Goto et al., IEEE Trans. Electron Devices, vol. 62, No. 11, pp. 3530-3535, (2015).

Chapter 3

Hybrid Bonding by Surface Activation

Cu-Cu Die to Die Surface Activated Bonding in Atmospheric Environment using Ar and Ar/N₂ Plasma

S. L. Chua and C. S. Tan

School of Electrical and Electronics Engineering, Nanyang Technological University, 50 Nanyang Avenue, Singapore 639798

> Surface activated bonding (SAB) method is an attractive option for Cu-Cu bonding as the force application duration is considerably shorter as compared to thermocompression bonding. Die to die SAB allows 3D heterogeneous integration to be done with lesser energy consumption compared to thermocompression. In this report, blanket and patterned Cu-Cu die to die SAB are demonstrated. Two different types of plasma were used for the surface activation of Cu. Both Ar and Ar/N₂ plasma activation were performed on the Cu surfaces enabling direct bonding at room temperature, ambient air, and atmospheric pressure without any wet chemical process.

Introduction

System performance can be improved by stacking and vertically interconnecting distinct device layers by replacing long two dimensional (2D) interconnects with shorter vertical interconnects. This reduces power loss across interconnects while allowing the design of high bandwidth parallel communication between the two dies which is challenging in a side by side placement. In order to allow communications between stacked dies, metallic connections are required. Cu-Cu solid state direct bonding is ideal to form the required metallic connections if the bonding process is compatible to semiconductor manufacturing. Solid state direct bonding allows reduction in pitch size required for the bonding of interconnects from non-solid state bonding methods (1).

Surface activated bonding (SAB) method is an attractive option for Cu-Cu bonding as the force application duration is considerably shorter as compared to thermocompression bonding, and multiple wafers or dies can be batch processed in the annealing process at less than 400°C (2). Cu-Cu wafer to wafer SAB using Ar/N₂ plasma for activation has been shown, using transmission electron microscopy (TEM), that it was able to form zig-zag bonding interface (3). However, SAB is more challenging for Cu-Cu die to die bonding as particles or edge defects from the dicing process could negatively affect the bonding quality. Die to die bonding will allow flexibility and optimization in the design of the individual die in the case of heterogeneous integration.

In this report, blanket and patterned Cu-Cu die to die SAB are demonstrated. Two different types of plasma were used for the surface activation of Cu surfaces. Both Ar and Ar/N₂ plasma activation were performed on the Cu surfaces enabling direct bonding at room temperature, ambient air, and atmospheric pressure without any wet chemical process.

Experiment and Discussion

In this report, Ar and Ar/N_2 plasma were investigated for Cu-Cu SAB. It have been shown that it is possible to bond Si and SiO_2 using only inert plasma gases for surface activation due to surface modification by physical bombardment (4). Hence Ar and Ar/N_2 gases were used in this work for activation. Two 6 inch Si wafers were deposited with 10nm of Ti as adhesion/barrier layer and 100nm of Cu using electron beam physical vapor deposition (e-beam PVD). After the metal layers were deposited, the wafers were sent to a surface activation process which uses either Ar or Ar/N_2 gas plasma for activation. The activation process includes two steps. The first step cleans the surface with fast atom bombardment using Ar which removes contaminants and native oxides from the Cu surface on the wafer. The second step uses either Ar or Ar/N_2 plasma for the activation of the Cu surface. The usage of Ar/N_2 plasma for activation was observed to cause nitridation on the Cu surface. However the effects of the nitridation are still under investigation and are inconclusive at the time of this report. The inclusion of the second step was intended for the investigation of the effects of the nitridation by Ar/N_2 plasma. Immediately after activation, the wafers were placed face to face and bonded together in a double sided wafer aligner with the vacuum contact mode for 4 minutes. After the wafers were bonded together, they were transferred to a furnace for annealing at 300°C for 1 hour.

The annealed wafers were diced and prepared with focus ion beam (FIB) for transmission electron (TEM) and electron energy loss spectroscopy (EELS) to observe the bonding interface. Both Ar and Ar/N_2 plasma activated samples showed bonding interface with no voids in the TEM images which were shown in Figure 1 and 2. O and N were not observed in the EELS mapping shown in Figure 3 and 4 at the bonding interface. This showed that both Ar and Ar/N_2 plasma activation allowed good SAB of Cu without noticeable oxidation or nitridation at the bonding interface.

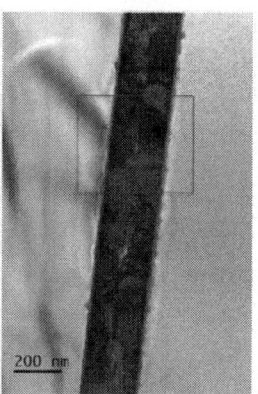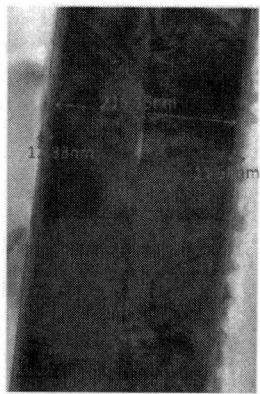

Figure 1. TEM image of Ar plasma activated sample (left) and higher magnification (right)

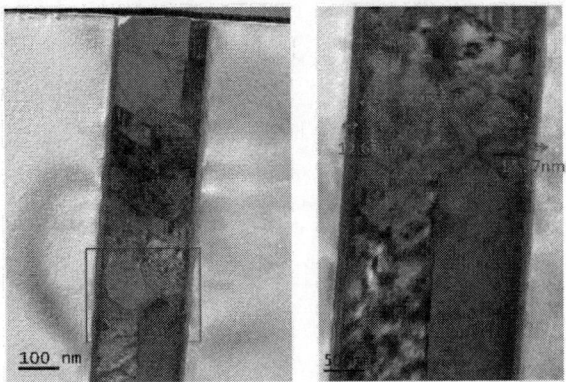

Figure 2. TEM image of Ar/N$_2$ plasma activated sample (left) and higher magnification (right)

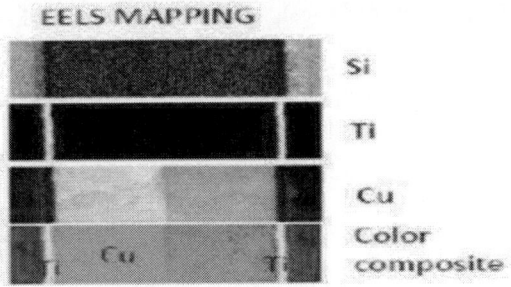

Figure 3. EELS mapping of Ar plasma activated sample

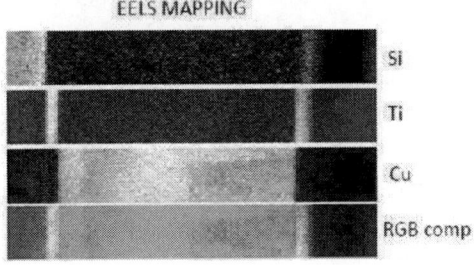

Figure 4. EELS mapping of N$_2$ plasma activated sample

As TEM/EELS observations confirm successful Cu-Cu bonding with SAB at wafer level bonding, Cu-Cu die to die bonding using the surface activation process was then attempted. Preparation of the die to die SAB samples started with deep reactive-ion etching (DRIE) 8μm deep trenches on a 6 inch Si wafer to form 2mm by 2mm square island features. 200nm of SiO_2 were then deposited on the etched wafer by tetraethoxysilane (TEOS) plasma enhanced chemical vapor deposition (PECVD) followed by 10 nm of Ti as adhesion/barrier layer and 100nm of Cu using e-beam PVD. Dicing was carried out on the trenches to form 3mm by 3mm dies with the 2mm by 2mm square island features in the center. The trenches will prevent the unevenness of the dicing edge from affecting the bonding. The corresponding dies for the bonding samples were fabricated with a similar Si wafer and similar deposition methods to form the same thickness of SiO_2, Ti and Cu. The wafer was diced into dies bigger than the 2mm by 2mm square island features. Schematics of the SAB samples were shown in Figure 5.

After both the top and bottom dies have been diced, the dies were processed with the surface activation process by either Ar or Ar/N_2 plasma. Atomic-force microscopy (AFM) has been performed on the surface activated dies and surface roughness have been characterized. Surface roughness measurements of the samples were of acceptable level required for SAB. Figure 6 shows that the AFM result of an Ar/N_2 plasma activated sample with root mean squared surface roughness of 0.77nm measured on a 5μm by 5μm area. The activated dies were placed face to face immediately after activation in a bonder in ambient air, at atmospheric pressure and room temperature. The bonder applied a bonding force of 500N for each sample for 5 minutes. After the force was removed, annealing was carried out after multiple purges to form a N_2 environment in the bonder chamber. Annealing was carried out at 300°C for 1 hour with a slow temperature ramp rate. After annealing, the bonded dies were shear strength tested by a shear test machine. The shear test was conduct in accordance to MIL-STD-883E method 2019.5 (5).

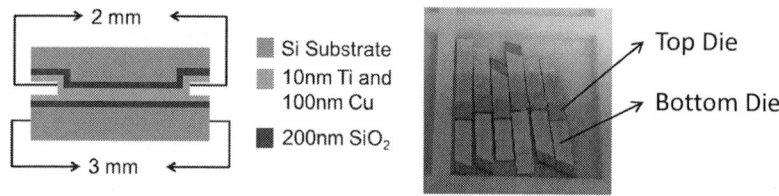

Figure 5. Schematics (left) and photo (right) of the bonded samples

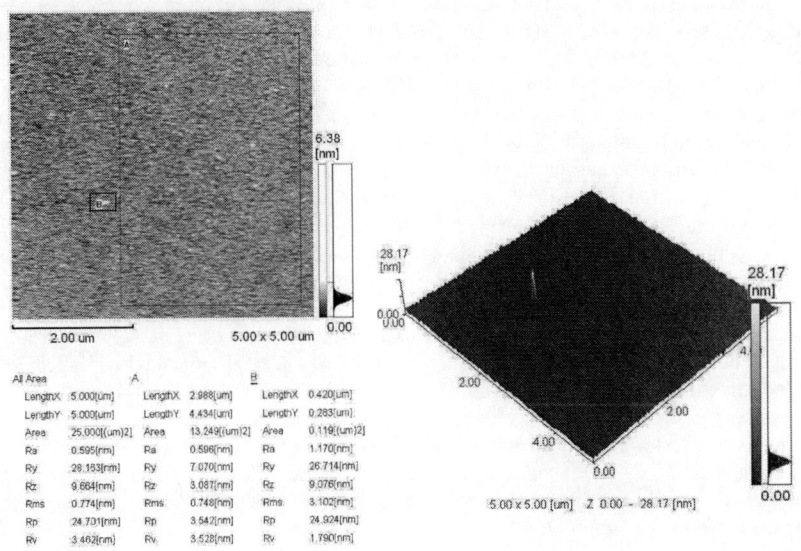

Figure 6. AFM roughness (left) and 3D profile (right) of Ar/N$_2$ plasma activated sample

The shear strength results for the SAB with Ar plasma and Ar/N$_2$ plasma were shown in Figure 7. The mean shear stress measured for the Ar plasma SAB samples is 18.85MPa and 18.37MPa for the Ar/N$_2$ samples. Most samples failed at the Si substrate.

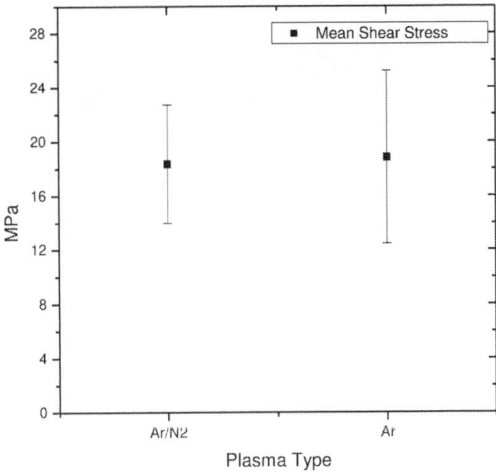

Figure 7. Chart showing the mean shear strength of Ar and Ar/N$_2$ plasma SAB samples

Patterned dies with cross bridge kelvin resistor (CBKR) structure were also bonded together successfully. The CBKR dies were prepared by DRIE for 8μm to form the CBKR structures. 200nm SiO₂ were then deposited by PECVD and 10nm of Ti as adhesion/barrier layer and 100nm of Cu were deposited using e-beam PVD. Metal lift off were done to remove the unwanted Ti and Cu layers outside the CBKR structures. The DRIE of 8μm depth around the CBKR structures will prevent the residues formed from the metal lift off process and the dicing edges from affecting the bonding. The cross section and a section of the mask layers of the CBKR samples are shown in Figure 8. There are four CBKR structures on each die, one at each corner for testing the uniformity of the bonding across the entire die. The contact area for the kelvin resistor is 100 μm by 100 μm.

Each pair of top and bottom dies was sent for surface activation process simultaneously. After the plasma activation, the dies were immediately aligned and bonded together using a commercial die bonder. The bonding force used in the bonding process was 210N for 4 minutes. After bonding a few set of dies, the samples were transferred into a furnace for a batch annealing process. Annealing was carried out in N₂ environment at 300°C for 1 hour with a slow ramp rate. After annealing, the bonded samples were preliminarily verified by shaking the dies in a waffle tray. The bonded samples are shown in Figure 9.

The dies were then tested electrically using a four point probe station to ascertain that all CBKR structures on each sample were bonded. A sweeping current from -0.05A to 0.05A were supplied into the CBKR and voltmeter units in the probe station were used to measure the voltage difference from the top and the bottom bonded metal layer. The measured voltage differences were shown in Figure 10. The measured voltage difference increased linearly with the current input which shows that the bonded interface has an ohmic contact. All four CBKR from each dies were of comparable resistances which suggest that the die have been bonded uniformly. An average resistance of 0.035 Ω is obtained from the samples. Cu-to-Cu bonding with Ti passivation at 180 °C samples in a study shows measured resistances of between 1 to 1.5 Ω before and after current stressing (6). The specific contact resistance is calculated to be 3.5×10^{-6} Ωcm² which is below the specific contact resistance of 7×10^{-4} Ωcm² for Cu-to-Cu bonding with Ti passivation at 180 °C and higher than 2×10^{-7} Ωcm² for Cu–Cu Bonding at 150 °C with Pd Passivation (7).

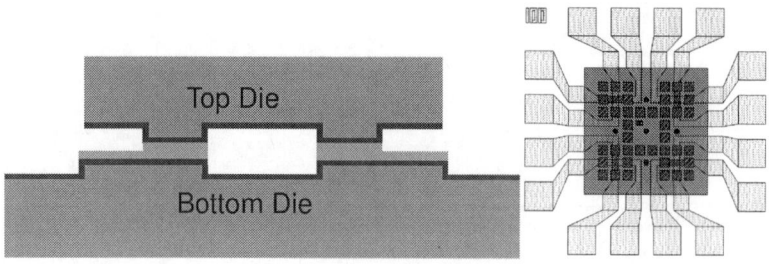

Figure 8. Cross section representation (left) and a section of the mask layers (right) of CBKR die samples

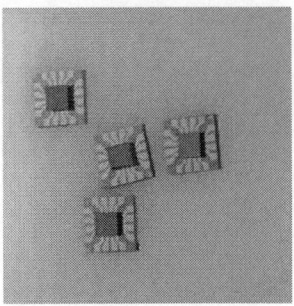

Figure 9. Photo of bonded dies

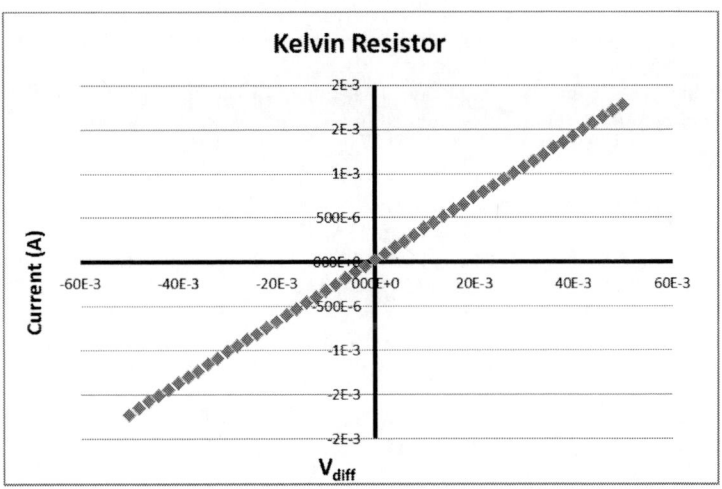

Figure 10. Chart showing voltage difference against a sweeping current input for a CBKR in a die to die SAB sample

Conclusion

Surface activation process using Ar and Ar/N_2 plasma for Cu have shown promising bonding capabilities based on TEM/EELS analysis. Blanket Cu-Cu die to die SAB have been demonstrated with the surface activation process. Good shear stress results have been achieved from the bonded samples. CBKR structures patterned dies were also successfully bonded together with ohimc contact at the bonding interface. This shows that both Ar and Ar/N_2 plasma activation enables patterned die to die direct bonding at room temperature, ambient air, and atmospheric pressure without any wet chemical process.

Acknowledgments

The authors will like to thank Nanyang NanoFabrication Centre for the support provide in the experiments and SIMTech for providing the equipment for shear force test.

References

1. T.Suga, Feasibility of Surface Activated Bonding for Ultra-fine Pitch Interconnection-A New Concept of Bump-less Direct Bonding for System Level Packaging, pp 702 – 705, *Electronic Components and Technology Conference*, Las Vegas NV (2000)
2. T.Suga, Cu-Cu Room Temperature Bonding - Current Status of Surface Activated Bonding(SAB), *ECS Transactions*, 3(6), 155-163 (2006).
3. S.L.Chua, G.Y.Chong, Y.H.Lee and C.S.Tan, Direct Copper-Copper Wafer Bonding with Ar/N2 Plasma Activation, p 134-137, *Electron Devices and Solid-State Circuits (EDSSC)*, Singapore (2015)
4. T.Suni, K.Henttinen, I.Suni and J.Mäkinen, Effects of Plasma Activation on Hydrophilic Bonding of Si and SiO_2, *J. Electrochem. Soc*, 149(6), G348-G351 (2002).
5. Die shear strength, 2019.5, MIL-STD-883E (1996).
6. Y.P.Huang, Y.S.Chien, R.N.Tzeng, M.S.Shy, T.H.Lin, K.H.Chen, C.T.Chiu, J.C.Chiou, C.T.Chuang, W.Hwang, H.M.Tong and K.N.Chen, Novel Cu-to-Cu Bonding With Ti Passivation at 180 °C in 3-D Integration, *IEEE Electron Device Letters*, 34(12), pp. 1551–1553 (2013).
7. Y.P.Huang, Y.S.Chien, R.N.Tzeng and K.N.Chen, Demonstration and Electrical Performance of Cu-Cu Bonding at 150 °C With Pd Passivation, *IEEE Transactions on Electron Devices*, 62(8), pp. 2587–2592 (2015).

Combined Surface Activated Bonding Technique for Hydrophilic SiO₂-SiO₂ and Cu-Cu Bonding

Ran He[a], Masahisa Fujino[a], Akira Yamauchi[b], and Tadatomo Suga[a]

[a] Department of Precision Engineering, The University of Tokyo, Tokyo 113-8656, Japan
[b] Bondtech Co., Ltd., Kyoto 611-003, Japan

Wafer bonding with high quality (such as high strength, void-less interface, and low-resistance vertical electrical interconnects) obtained at low temperatures is highly desired for 3D integration of microsystems. This paper reports a combined surface activated bonding (SAB) technique for improvement of the bonding quality of SiO_2-SiO_2 pairs bonded at 200°C. The present bonding technique employs a combination of surface irradiation using a Si-containing Ar beam and prebonding attach-detach procedure prior to bonding in vacuum, followed by postbonding annealing in ambient pressure. The bonding method has also been found effective for SiO_2-SiN_x and Cu-Cu bonding at 200°C. Results of bonding strength measurements, transmission electron microscopy (TEM) and energy-dispersive X-ray spectroscopy (EDS) observations are reported and discussed to understand the present technique. The feasibility of this bonding technique for Cu/SiO_2 and $Cu/SiO_2/SiN_x$ hybrid bonding is also discussed.

Introduction

Low-temperature wafer bonding is an important enabling technology for the emerging three-dimensional (3D) integration of microsystems. Among many bonding schemes, low-temperature SiO_2-SiO_2 bonding has been extensively studied because of its potentials for high bonding quality and good compatibility with the complementary metal-oxide-semiconductor (CMOS) technology. The SiO_2-SiO_2 bonding is also the basis for the Cu/SiO_2 hybrid bonding, in which the SiO_2-SiO_2 and Cu-Cu bonding is obtained at the same time between two wafers. The Cu/SiO_2 hybrid bonding promises both direct metal interconnections and enhanced thermal/mechanical stability with a seamless bonding structure during the single bonding process (1,2).

However, the SiO_2-SiO_2 bonding strength is still low by using the popular plasma activated wafer bonding (3,4) and the CMP (chemical mechanical polishing) activation method (4). The possible reason may be related to the presence of excess interfacial H_2O. It has been reported that nano-sized voids are generated owing to the excess H_2O at the SiO_2-SiO_2 bonding interface (5,6). Furthermore, F. Fournel et al. reported that the SiO_2-SiO_2 bonding strength can be decreased by the water stress corrosion effect induced by the interfacial H_2O, which is difficult to remove by annealing at below 500-600 °C (7).

To improve the SiO_2-SiO_2 bonding, it is of great important to reduce the number of excess interfacial H_2O. A main origin of the excess H_2O is trapping of H_2O that is covered on the hydrophilic wafers and from the humid air. Wafer bonding in vacuum is a promising way to minimize the interfacial H_2O by enabling desorption of excess H_2O from the hydrophilic surfaces in vacuum. However, the SiO_2-SiO_2 bonding conducted in vacuum usually yields even lower bonding strength compared to bonding in air (3,7–10).

In this paper, we developed a combined SAB technique to improve SiO_2-SiO_2 bonding at 200 °C (11,12). The combined SAB method involves a combination of surface irradiation using a Si-containing Ar beam and prebonding attach-detach process prior to bonding in vacuum. The Si atoms are added in the Ar beam to increase the number of reactive Si sites on SiO_2 surface, while the prebonding attach-detach process is used to enhance the OH adsorption and to remove excess H_2O prior to bonding. The developed combined SAB was also found to be effective for SiO_2-SiN_x and Cu-Cu bonding. Mechanisms of the present combined SAB are discussed based on the experimental results of bonding strength measurements by crack opening tests and interface observation by transmission electron microscopy (TEM) and energy-dispersive X-ray spectroscopy (EDS). The feasibility of the present bonding technique for low-temperature Cu/SiO_2 and $Cu/SiO_2/SiN_x$ hybrid bonding is also discussed.

Experimental Methods

The process flow of the bonding experiments is shown in Figure 1. Wafers were loaded into a load-lock chamber and then transported into an ultra-high vacuum (UHV) chamber without any pre-cleaning. The UHV chamber was evacuated to 5×10^{-6} Pa, and then the wafers were irradiated by the Si-containing Ar beam, by using a modified Si-walled atom beam source, at a power of 1.0 kV × 100 mA. For comparision, conventional Ar beam irradiation was also used, at the same power. After surface irradiation, the wafers were transported into the load-lock chamber and exposed to humid N_2 before wafer unloading from the lock-load chamber. The humid N_2 was purged into the chamber through a bubbling system at N_2 flow rate of 100 ccm. The wafers were then exposed to air at relative humidity (RH) of around 40% and optionally cleaned by pure

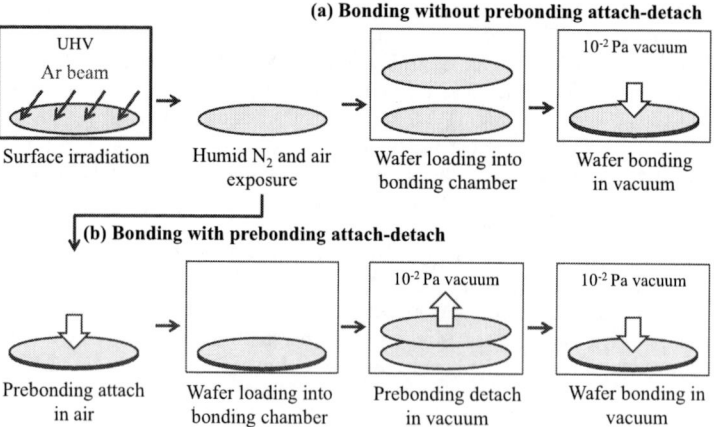

Figure 1. Process flow of the combined SAB with process (a) without prebonding attach-detach and (b) with prebonding attach-detach.

water and dried by spin-drying and N_2 gas. Subsequently, wafer bonding was conducted in two ways termed as process (a) and (b), as shwon in Figure 1. In process (a), the wafers were loaded into a bonding chamber that was subsequently evacuated to ~10^{-2} Pa, which took several minutes, followed by bringing the wafers into contact at room temperature (RT) and under a compression pressure of 2.5 MPa for 5 min. In process (b), the wafers were prebonding-attached, i.e., brought into contact in air, at 40% RH without external compression. After storage in air for more than 10 min, the attached wafers were loaded onto the lower electrostatic bonding chuck inside a bonding chamber. The wafers were prebonding-detached in vacuum of 10^{-2} Pa by the electrostatic forces of the bonding chucks, and then bonded in vacuum at RT under 2.5 MPa for 5 min. The time between the prebonding detach and bonding steps was about 1-2 minutes. In all experiments, postbonding annealing at 200°C was finally carried out in batch at atmospheric pressure.

For Cu-Cu bonding, only process (b) was used in the experiments. It was found that the Cu-Cu pairs as-bonded at RT has large non-bonded area at the center, probably because of the non-flatness of the wafers. Therefore, thermo-compression at 200°C under 2.5 MPa for 30min was employed during the bonding of Cu-Cu pairs. SiO_2-SiO_2 and SiO_2-SiN_x boonding was also conducted with the same process.

Table I lists the bonding experiments with different conditions. Wafer bonding experiments of SiO_2-SiO_2, SiO_2-SiN_x, and Cu-Cu were carried out by using blanket wafers covered by various films. Thermally grown SiO_2 film with thickness of 100 nm was used in the SiO_2-SiO_2 and SiO_2-SiN_x bonding. For the SiO_2-SiN_x bonding, the SiN_x film was deposited on the SiO_2 by using plasma-enhanced chemical vapor deposition (PECVD) and has a thickness of 100 nm. The Cu film was sputtering deposited on a 100-mm p-type Si wafer. A 50-nm thick Ti adhesion layer was inserted between the Cu film and Si substrate. After a standard CMP step, the Cu film has a thickness of 300 nm and surface roughness 0.57 nm.

TABLE I. Bonding experiments carried out in this work. The bonding was conducted in 10^{-2}Pa vacuum and under a compression pressure of 2.5 MPa; the annealing was conducted at atmospheric pressure without compression.

Beam type	Process	Bonding pair	Bonding	Annealing
Conventional Ar beam	Process (a)	SiO_2-SiO_2	RT 5min	200°C 7h
Si-containing Ar beam	Process (a)	SiO_2-SiO_2	RT 5min	200°C 7h
Si-containing Ar beam	Process (b)	SiO_2-SiO_2	RT 5min	200°C 7h
Si-containing Ar beam	Process (b)	SiO_2-SiO_2	200°C 30min	200°C 2h
Si-containing Ar beam	Process (b)	SiO_2-SiN_x	200°C 30min	200°C 2h
Si-containing Ar beam	Process (b)	Cu-Cu	200°C 30min	200°C 2h

The crack opening method was used to evaluate the bonding strength (13). The blade insertion was conducted manually in air at 40% RH. Bonding energy in terms of surface energy (γ) was calculated. Infrared (IR) imaging was used to measure the crack length in the SiO_2-SiO_2 and SiO_2-SiN_x strength tests, after the blade insertion within 1 min. Because it was impossible to insert the blade into the edges of the Cu-Cu bonded wafers without breaking the Si substrates, crack length measurements were not necessary. For Cu-Cu bonding interface observations, bonded specimens were thinned by an ion slicer (JEOL EM-09100IS) and inspected by TEM (JEOL JEM-ARM200F) at an acceleration voltage of 200 kV. EDS analysis was performed by using the JEOL JED-2300T analyzer equipped with a silicon drift detector with a detection area of 100 mm^2.

Results and Discussion

Bonding Results

Our experiments show that the combination of Si-containing Ar beam irradiation and process (b) (i.e. with the prebonding attach-detach procedure) yields the highest bonding energy of the SiO_2-SiO_2 pairs. The crack opening test results of the SiO_2-SiO_2 pairs bonded at RT followed by 200°C annealing are listed in Table II. By using the conventional Ar beam irradiation and process (a), the surface energy was around 0.6 J/m^2 (10). The surface energy was improved to 1.3 J/m^2 by using Si-containing Ar beam instead of the conventional Ar beam. This implies the Si-containing Ar beam is more effective for the SiO_2 surface activation. By using the Si-containing Ar beam combined with the process (b), the surface energy was further improved. The SiO_2-SiO_2 pairs fractured in bulk at the edges during the crack opening tests, which means the surface energy is close to the Si fracture energy of 2.5 J/m^2.

TABLE II. Results of crack opening test of SiO_2-SiO_2 pairs bonded at RT followed by annealing at 200°C for 7h.

Beam type	Process	Bonding pairs
Conventional Ar beam	Process (a)	$\gamma \sim$0.6 J/m^2
Si-containing Ar beam	Process (a)	$\gamma \sim$1.3 J/m^2
Si-containing Ar beam	Process (b)	Wafer bulk fracture occurred ($\gamma \sim$2.5 J/m^2)

TABLE III. Results of crack opening test of SiO_2-SiO_2, SiO_2-SiN_x, and Cu-Cu pairs bonded at 200°C by using Si-containing Ar beam irradiation combined with process (b).

Bonding pair	As-bonded (200°C 30min)	After Annealing (200°C 2h)
SiO_2-SiO_2	$\gamma \sim$0.5 J/m^2	$\gamma \sim$2.25 J/m^2
SiO_2-SiN_x	Not measured	Wafer bulk fracture occurred
Cu-Cu	Wafer bulk fracture occurred	Wafer bulk fracture occurred

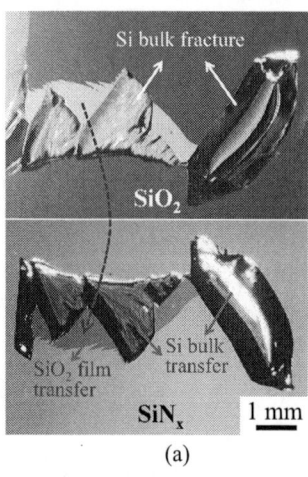 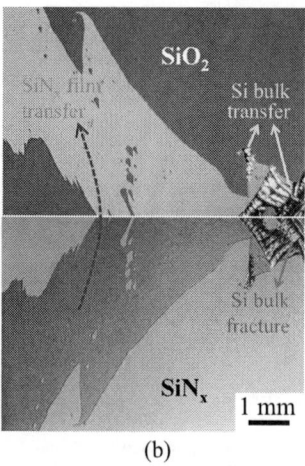

Figure 2. Fracture surfaces of SiO_2-SiN_x bonded samples with postbonding annealing at 200°C for 2h, showing Si bulk fracture and transfer of either SiO_2 or SiN_x films in (a) and (b), respectively.

By using the Si-containing Ar beam irradiation combined with the process (b), wafers were also bonded at 200°C under 2.5 MPa for 30min and annealed at 200°C for 2h. The crack opening test results are listed in Table III. The SiO_2-SiO_2 pairs exibit surface energy of 0.5 J/m^2 before the postbonding annealing and surface enrgy of 2.25 J/m^2 after the postbonding annealing, respectively. The SiO_2-SiN_x pairs were also very strong after the postbonding annealing, so that the wafers fractured in bulk during the crack opening test. Figure 2 shows the fracture surfaces of the SiO_2-SiN_x samples, showing Si bulk fracture and transfer of either SiO_2 or SiN_x films from one side to the other. In Cu-Cu bonding, all the as-bonded pairs and annealed pairs were so strong that bulk fracture occurred during the tests. Therefore, we conclude that the combined process is effective for all the SiO_2-SiO_2, SiO_2-SiN_x, and Cu-Cu bonding at temperature as low as 200°C.

TEM and EDS measurements were employed to investigate the Cu-Cu interface obtained after the postbonding annealing. Figure 3 shows the TEM image of the Cu-Cu bonding interface after the postbonding annealing. It can be seen that the interface mainly consists of portions without interlayers, with a thin CuO_x interlayer, and with some voids (in bright). Figure 4(a) and 4(b) shows the high-resolution TEM images of the low-O and CuO_x interfacial portions, respectively. It can be seen that the CuO_x at the interface is amorphous and around 6-nm thick.

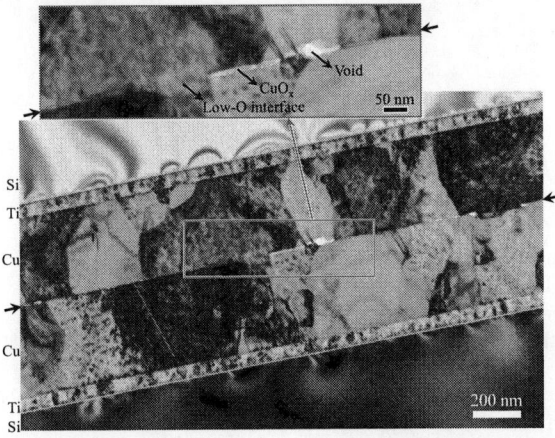

Figure 3. TEM image of the Cu-Cu hydrophilic bonded interface.

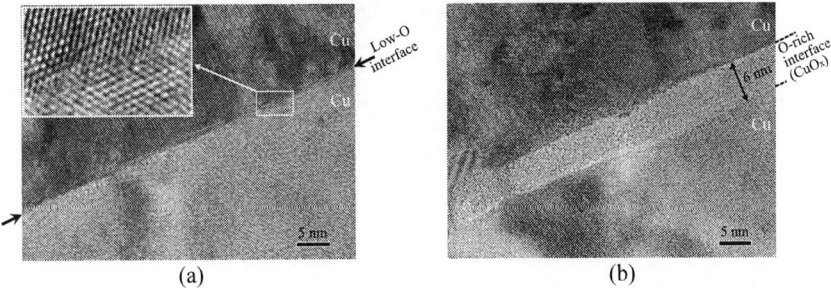

(a)   (b)

Figure 4. TEM images of the (a) O-less and (b) O-rich interfacial protions.

Figure 5 shows the EDS results of chemical compositions of the different positions of a bonded sample. It confirms the presence of both low-O and CuO_x interfacial portions at the bonding interface. We estimated that around 75% area of the bonding interface is low-O by the TEM observations. The low-O interface contains even less O than in the Cu bulk, which is considered to be owing to some contaminants adsorbed on the Cu bulk area during transportation of the specimen from the ion slicer to TEM/EDS. The ratio of the numbers of Cu and O atoms at the CuO_x interfacial region is around 1:0.78. The amount of Si atoms (Si/Cu ratio ~0.03/1) is at the contaminant level at all the EDS analyzed points. The formation of low-O interface is owing to diffusion of the interfacial Cu and O species, which is consistent with previous studies (1,14).

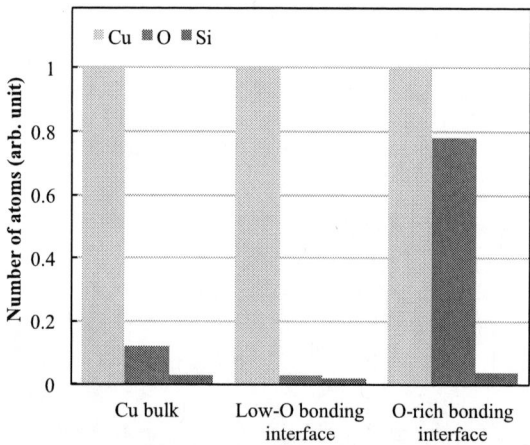

Figure 5. Elemental analysis results of different positions (indicated by the inset) of the Cu-Cu hydrophilic bonded structure.

Effect of Si-containing Ar beam irradiation

The experiment results showed that the Si-containing Ar beam is more effective for SiO_2 surface activation. We suggest the improvement can be attributed to modification of SiO_2 surface chemistry by the Si atoms added in the Ar beam. SRIM simulations were used to evaluate the subplantation depth of the Si atoms into SiO_2 surface. The simulation results in Figure 6 show that Si atoms can be subplanted into 0.5-1.5 nm depth range, as shown by the blue bars, on an amorphous SiO_2 surface even at very low atomic energies of 20-100 eV. This behavior implie that The Si-containing Ar beam can modify the SiO_2 surface chemistry to be Si-enriched, comparing to the conventional Ar beam that mainly remove the contaminants by physical surface sputterring. The Si subplantation and surface sputtering is simultaneous during the Si-containing Ar beam irradiation, as shown in Figure 7. This results in a contaminant-free oxide surface that has Si-enriched chemistry.

A Si-enriched surface contains more Si-Si and dangling Si- bonds, or called O-vacancy defects, which were suggested to be very reactive for OH adsorption on oxide surfaces (15). As a consequence, the number of Si-OH bonding sites is increased and hence the bonding energy is improved. On the other hand, an O-rich surface is less reactive for OH adsorption, which can be indicated by the decreased bonding energy of Si-SiO$_2$ by using oxygen plasma activation compared to nitrogen plasma (3,16).

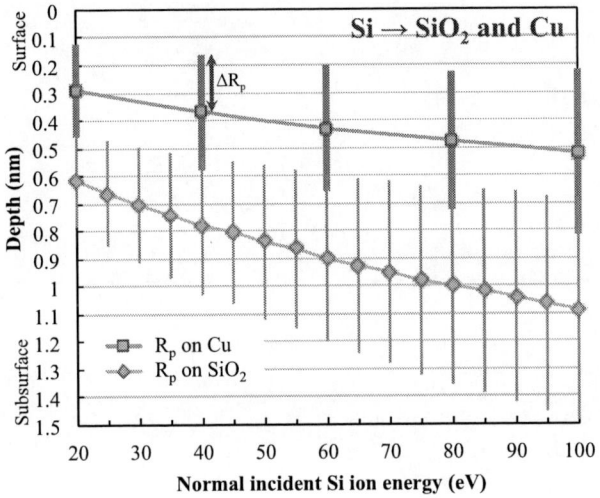

Figure 6. SRIM simulation results of projgeted range (R$_p$) and straggling (ΔR$_p$) of energitic Si ions at 20-100 eV energy into SiO$_2$ and Cu surfaces.

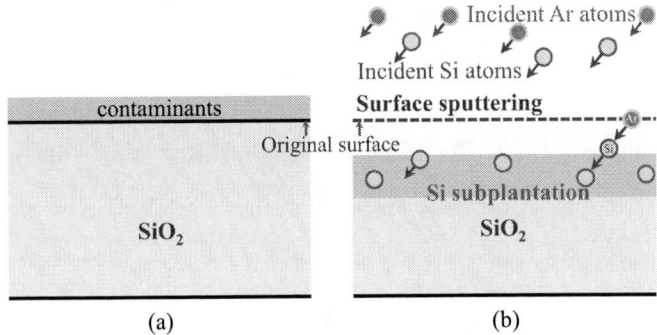

Figure 7. Schematic representation of the Si-containing Ar beam irradiation on SiO$_2$ surface: (a) before irradiation and (b) after irradiation with simultaneous surface sputtering (contaminants removal) and Si subplantation.

The Si atoms can be also subplanted into the Cu surface. However, the depth is much more shallow on Cu compared to that on SiO$_2$, as shown in Figure 6. This means the Si atoms in the Ar beam are difficult to be subplanted onto Cu. Re-sputtering of the Si from the Cu surface can happen during the simultaneous Si subplantation and surface sputtering, as illustrated in Figure 8. Moreover, the re-sputtering of Si can be further enhanced by the preferential sputtering effect (17). This explains the absence of Si species at the Cu-Cu bonding interface confirmed by the small Si peaks at contaminant level in the EDS results.

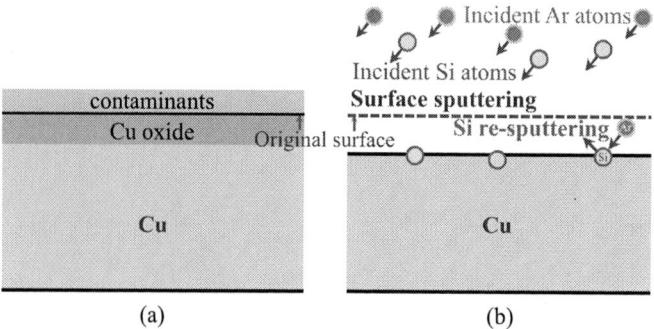

(a)   (b)

Figure 8. Schematic representation of the Si-containing Ar beam irradiation on Cu surface: (a) before irradiation and (b) after irradiation with surface sputtering (contaminants and oxide removal) and Si re-sputtering.

<u>Prebonding attach-detach process</u>

<u>SiO_2-SiO_2 Bonding</u>. The mechanism of the SiO_2-SiO_2 bonding combining the prebonding attach-detach process is shown in Figure 9. After surface irradiation and exposure to humid N_2 and air, the SiO_2 surface is hydrophilic, but still covered by insufficient OH sites and by a H_2O film adsorbed from air, as shown in Figure 9(a). During the prebonding attach in air, the adsorbed H_2O is trapped at the interface and reacts with the attached surfaces, as shown in Figure 9(b). The trapped H_2O is dissociatively adsorbed (as OH sites) on the attached surfaces, through the reaction [1] between H_2O and Si-O-Si bonds on the two attached surfaces.

$$Si\text{-}O\text{-}Si + xH_2O + Si\text{-}O\text{-}Si \rightarrow 2(Si\text{-}OH) + (x-2)H_2O + 2(HO\text{-}Si) \qquad [1]$$

The possible reasons of the enhancement of the reaction [1] may be related to the shortened distance between the Si-O-Si and H_2O and the modified property of the trapped H_2O. S. H. Garofalini et al. suggested that, when H_2O is confined in SiO_2 pores, the density and structure of water near the SiO_2-H_2O interface is similar to that of bulk H_2O at a high pressure of 750 atm (75 MPa) based on their results of molecular dynamics simulations (18). Their simulations also indicate the dissociative chemisorption and penetration of OH/H_2O extend about 0.7 nm below the outermost surface of SiO_2. This implies a high coverage degree and a great number of OH sites on the prebonding-attached SiO_2 surfaces, as shown in Figure 9(c). The high bonding energy close to 2.5 J/m^2 indicates the number of the Si-OH groups can be estimated to be around 6.87 per square nanometer, which is even greater than the saturated value of 4.6 per square nanometer on a SiO_2 surface (19). Subsequently, the excess undissociated H_2O molecules (H_2O in the right side of reaction [1]) are removed by the prebonding detach in vacuum, thanks to a rapid H_2O desorption from the wafer surfaces exposed to vacuum, as shown in Figure 9(d). Then, bonding with less interfacial H_2O can be realized by conducting the bonding in vacuum. The Si-OH bonding sites on the opposing surfaces can be connected via a very short distance, without seperation by H_2O films, as shown in Figure 9(e).

Finally, the Si-OH pairs transform into strong covalent Si-O-Si bonds during postbonding annealing through the reaction [2]:

$$Si\text{-}OH + HO\text{-}Si \rightarrow Si\text{-}O\text{-}Si + H_2O \qquad [2]$$

As a result, the bonding interface closes during annealing at a low temperature no more than 200°C, as shown in Figure 9(f). The transformation of Si-OH into Si-O-Si is more prone to occur because of the H_2O management. Furthermore, it is also possible that residual H_2O, if any, and generated H_2O by-product can be transformed into H_2, which can dissolve in SiO_2, through oxidation of residual Si-Si bonds in the Si-enriched subsurfaces.

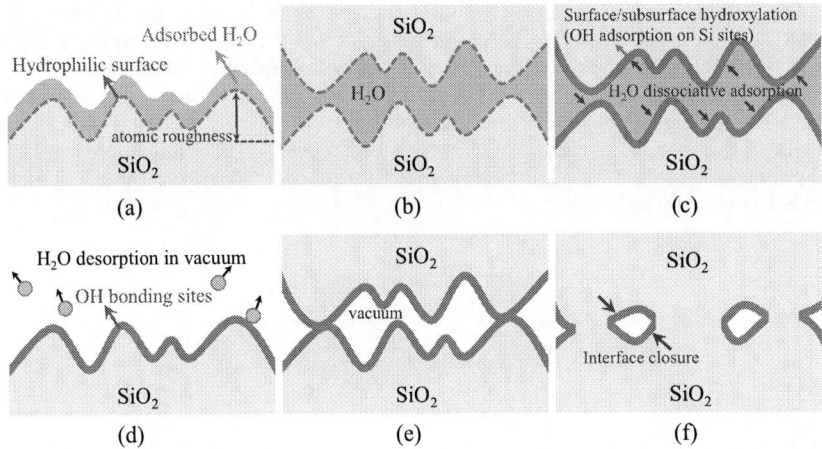

(a)   (b)   (c)

(d)   (e)   (f)

Figure 9. Mechanism of the hydrophilic SiO_2-SiO_2 bonding (11): (a) wafer covered by OH and H_2O when exposed to air, (b) prebonding attach in air with H_2O trapped, (c) surface hydroxylation by trapped H_2O, (d) prebonding detach in vacuum, H_2O removal, leaving OH sites on the surfaces, (e) wafer bonding in vacuum without H_2O trapping, and (f) covalent Si-O-Si bonds formation and gap closure during annealing.

The above meachnism indicates the interfacial H_2O can play a positive role for Si-OH formation at the early stage of hydrophilic bonding. This explains the decreased SiO_2-SiO_2 bonding strength obtained in vacuum than in air by existing bonding methods. In hydrophilic bonding in air, the trapped H_2O increases the Si-OH bonding sites, as shown in Figure 9(c), although the excess H_2O reduces the number of interfacial Si-O-Si bonds by H_2O-filled defects and prevents complete interface closure at low temperatures. Thus, a moderate bonding energy ($\gamma \sim 1.5$ J/m^2) (3) can be achieved after postbonding annealing. However, in conventional bonding in vacuum such as SAB and in situ plasma activated wafer bonding, the numbers of Si-OH bonding sites on SiO_2 surfaces can be far from sufficient for a strong bonding formation, thus it yields a significant lower bonding strength at a low temperature compared to SiO_2-SiO_2 bonding in air (3,7–9). Furthermore, the interfacial H_2O has to be removed to improve bonding formation and interface closure. Bonding in vacuum is an effective way for the prebonding removal of excess trapped H_2O.

Cu-Cu Bonding. In the Cu-Cu bonding, the as-irradiated Cu surface is re-oxidized and covered by adsorbed OH/H_2O during exposure to humid N_2/air and prebonding attach, as illustrated in Figure 10(a) and (b), respectively. Ideally, this should be avoided so that room temperature Cu-Cu bonding can be realized between clean Cu surfaces, like in the conventional SAB. However, to enable both Cu-Cu and SiO_2-SiO_2 bonding in the Cu/SiO_2 hybrid bonding, the re-oxidation and OH adsorption is unavoidable because it is impossible to terminate the SiO_2 surface with OH and keep the Cu surface clean on the same wafer at the same time. Because the reactions between H_2O and Cu-O-Cu and Si-O-Si bonds are similar, we suggest similar reaction happens during the prebonding attach of Cu-Cu and SiO_2-SiO_2 pairs. By prebonding detach in vacuum, the trapped H_2O molecules between the prebonding-attached Cu surfaces are also released, as shown in Figure 10(c). Cu-Cu bonding is then conducted in vacuum, without H_2O and gases trapped at the bonding interface, as shown in Figure 10(d). During heating and annealing at 200°C, the Cu-OH sites are converted into Cu-O-Cu bonds across the interface forming a CuO_x interlayer (see Figure 10(e)), and then interfacial Cu/O diffusion happens, resulting in a interface containing inhomogeneous low-O and ultrathin CuO_x portions (refer to Figure 10(f)). Comparing to the result of Cu-Cu bonding with a 15-nm-thick O-containing bridging layer having a electrical resistivity of around 4×10^{-8} Ωm reported by A. Shigetou and T. Suga (19,20), a low-resistance Cu-Cu bonding can be expected by using the present bonding technique.

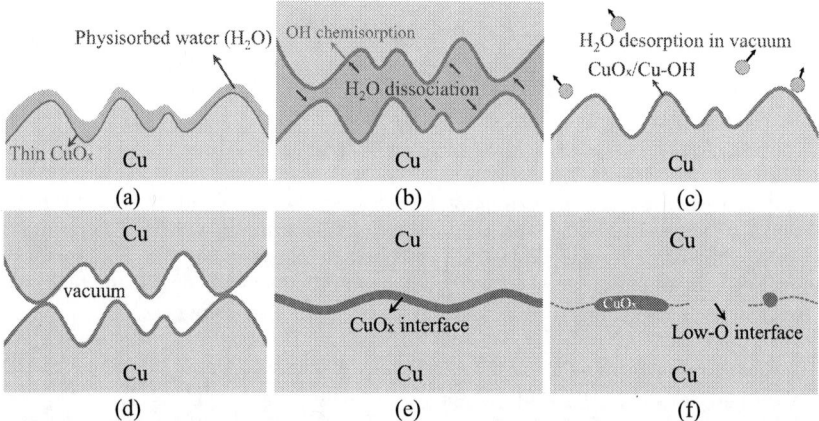

Figure 10. Mechanism of hydrophilic Cu–Cu bonding: (a) Hydrophilic surface exposed to air, (b) prebonding attach in air and wafer surface/subsurface hydroxylation by interfacial H_2O, (c) prebonding detach and H_2O removal in vacuum, leaving OH sites on the surface, (d) wafer bonding in vacuum under a bonding force, and (e) Cu–Cu bonding with CuO_x interface, and (f) Cu/O interdiffusion during 200 °C annealing.

Combined SAB for Cu/SiO_2 and Cu/SiO_2/SiN_x Hybrid Bonding

The present combined SAB is feasible for low-temperature Cu/SiO_2 and Cu/SiO_2/SiN_x hybrid bonding, since it has been demonstrated to be effective for all the SiO_2-SiO_2, SiO_2-SiN_x, and Cu-Cu bonding involved in the hybrid bonding. Figure 11 illustrated the hybrid bonding process by using the combined SAB. By using Si-containing Ar beam irradiation, the contaminants and Cu oxides are removed from the wafer surfaces. The wafers are then covered by OH sites (Si-OH and Cu-OH) after the prebonding attach-detach process, as shown in Figure 11(a). Afterwards, the wafers are brought into contact in vacuum, as shown in Figure 11(b), without trapping of excess H_2O, O_2, and N_2 molecules, which may occur especially in the gaps between the CMP-dished Cu surfaces if the bonding is conducted in air. During heating, an external compression can be applied to overcome the wafer non-flatness and to prevent bonding defects induced by thermal stress generated by Cu thermal expansion, as shown in Figure 11(c). The Cu-Cu is obtained with help of the external compression and the internal compression generated by thermal expansion of Cu. Finally, the bonded wafers are postbonding annealed to enhance the SiO_2-SiO_2 or SiO_2-SiN_x bonding and Cu/O diffusion at the Cu-Cu bonding interface, as shown in Figure 11(d). Based on the results presented in this paper, strong and low-resistance bonding can be expected by using the combined SAB technique at temperature as low as 200°C.

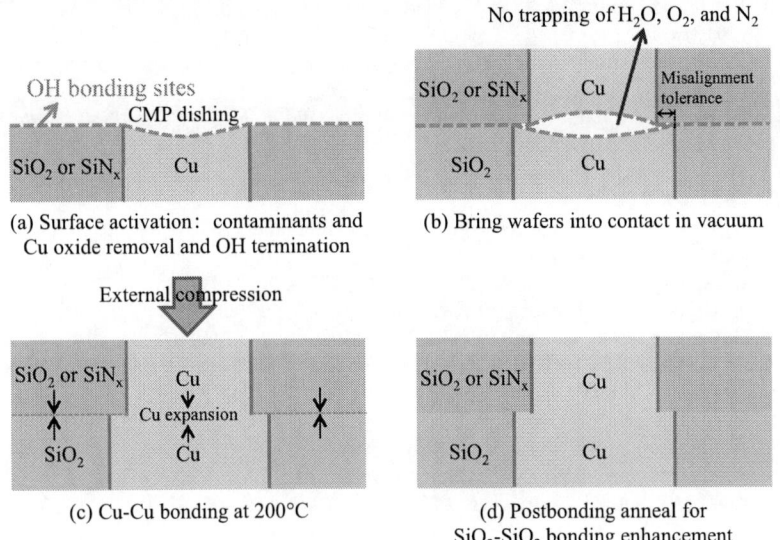

Figure 11. Schematic representation of Cu/SiO_2 and Cu/SiO_2/SiN_x hybrid bonding by using the present combined SAB method.

Conclusion

A combined SAB technique using Si-containing Ar beam irradiation and prebonding attach-detach was investigated. The effectiveness of this technique for bonding of SiO_2-SiO_2, and SiO_2-SiN_x bonding has been shown. It provides enhanced SiO_2-SiO_2 and SiO_2-SiN_x bonding and achieves excellent Cu-Cu bonding containing low-O and ultrathin O-rich interfacial portions at temperature as low as 200°C. Therefore, the present combined SAB holds promise for low-temperature bonding with high bonding quality for 3D integration applications.

References

1. L. D. Cioccio et al., *J. Electrochem. Soc.*, **158**, P81–P86 (2011).
2. H. Moriceau et al., *Microelectron. Reliab.*, **52**, 331–341 (2012).
3. T. Suni, K. Henttinen, I. Suni, and J. Mäkinen, *J. Electrochem. Soc.*, **149**, G348–G351 (2002).
4. C. Rauer et al., *ECS J. Solid State Sci. Technol.*, **2**, Q147–Q150 (2013).
5. C. Ventosa et al., *Electrochem. Solid-State Lett.*, **12**, H373–H375 (2009).
6. C. Sabbione, L. D. Cioccio, L. Vandroux, J.-P. Nieto, and F. Rieutord, *J. Appl. Phys.*, **112**, 063501 (2012).
7. F. Fournel et al., *ECS J. Solid State Sci. Technol.*, **4**, P124–P130 (2015).
8. H. Takagi, J. Utsumi, M. Takahashi, and R. Maeda, *ECS Trans.*, **16**, 531–537 (2008).
9. Q.-Y. Tong and U. M. Gösele, *Adv. Mater.*, **11**, 1409–1425 (1999).
10. R. He, M. Fujino, A. Yamauchi, and T. Suga, *Jpn. J. Appl. Phys.*, **54**, 030218 (2015).
11. R. He, M. Fujino, A. Yamauchi, and T. Suga, *Jpn. J. Appl. Phys.*, **55**, 04EC02 (2016).
12. R. He, M. Fujino, A. Yamauchi, Y. Wang, and T. Suga, *ECS J. Solid State Sci. Technol.*, **5**, P419–P424 (2016).
13. W. P. Maszara, G. Goetz, A. Caviglia, and J. B. McKitterick, *J. Appl. Phys.*, **64**, 4943–4950 (1988).
14. S. Y. Lee, N. Mettlach, N. Nguyen, Y. M. Sun, and J. M. White, *Appl. Surf. Sci.*, **206**, 102–109 (2003).
15. M. Salmeron et al., *Faraday Discuss*, **141**, 221–229 (2009).
16. V. Dragoi and P. Lindner, *ECS Trans.*, **3**, 147–154 (2006).
17. S. Berg and I. V. Katardjiev, *J. Vac. Sci. Technol. A*, **17**, 1916–1925 (1999).
18. S. H. Garofalini, T. S. Mahadevan, S. Xu, and G. W. Scherer, *ChemPhysChem*, **9**, 1997–2001 (2008).
19. Q.-Y. Tong and U. Gösele, *J. Electrochem. Soc.*, **143**, 1773–1779 (1996).
20. A. Shigetou and T. Suga, *Appl. Phys. Express*, **2**, 056501 (2009).
21. A. Shigetou and T. Suga, *J. Electron. Mater.*, **41**, 2274–2280 (2012).

Impact Of Water Edge Absorption On Silicon Oxide Direct Bonding Energy.

F. Fournel[a], M. Tedjini[a], V. Larrey[a], F. Rieutord[b], C. Morales[a], C. Bridoux[a], H. Moriceau[a]

a- CEA, Leti, Minatec Campus, 17 rue des Martyrs, 38054 Grenoble Cedex 9, France
b- CEA, INAC, Minatec Campus, 17 rue des Martyrs, 38054 Grenoble Cedex 9, France

Direct bonding energy is one of the key parameter for wafer direct bonding. Since few years the impact of the water stress corrosion on the measurement has been pointed out. It has been also proven that water can diffuse into the direct bonding interface from the sample edges. This study is then based on the impact of the edge water diffusion on the direct bonding energy. A positive and a negative impact will be shown depending if the diffusion occurs before or after the bonding annealing. A mechanism will be proposed to explain these results and will mainly be based on water stress corrosion influence.

Introduction

Direct bonding of silicon and silicon dioxide thin films is now widely used in industry for SOI (Silicon on insulator) elaboration for instance or in some backside imager manufacturing process which is now in mass production process. Silicon dioxide layer is very often used as a convenient bonding layer allowing easy topography planarization with classical CMP (Chemical Mechanical polishing) process. Silicon dioxide//Silicon dioxide direct bonding interface deserves to be deeply studied. The direct bonding energy evolution is one of its key parameter and has been widely studied since many years [1,2, 3]. However, only recently, water edge penetration at the bonding interface after the direct bonding has been reported [5, 6]. Thereby, this work is focused on the impact of the water edge penetration on silicon dioxide direct bonding energy.

Experiment

200 mm <001> silicon wafers are thermally oxidized at 950°C under steam atmosphere in order to growth 145 nm of silicon dioxide thin film. After classical RCA cleaning, these wafers are bonded at room temperature and ambient pressure. Just after the bonding, they are diced into 2 cm wide beams for DCB (double cantilever beam) anhydrous direct bonding energy measurement [3]. For this measurement a blade is inserted at the bonding interface inside a glove box filled with anhydrous nitrogen (water concentration is below 1ppm). The induced debonding length is measured by infrared camera and used to calculate the bonding energy with the El-Zein [4] equations. The bonding energy measurement is done after a 300°C annealing for all beams. But after the dicing, the atmospheric water can start to penetrate the direct bonding interface especially from the diced edges. Two times are then considered: t_1 is the time after dicing and before annealing; t2 is the time after annealing and before bonding energy measurement. The storage are mainly done in humid atmosphere of 40% relative humidity (RH) The bonding energy dependence regarding t1 and t2 will then be studied.

As shown on figure 1, the impact of t_2 in humid atmosphere is clearly detrimental for the oxide/oxide direct bonding energy. In 6 days, the bonding energy is decreased by more than 50%.

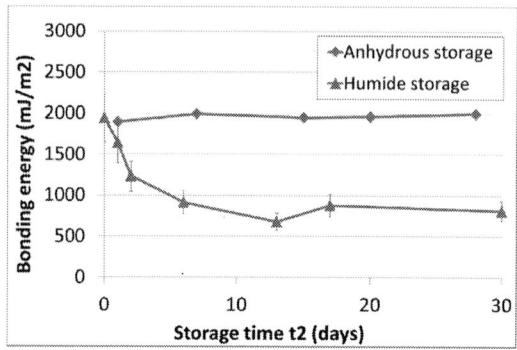

figure 1. t2 storage time impact on the direct bonding energy of silicon oxide bonding. The storage takes place between the annealing and the bonding energy measurement in humid (40%RH) or anhydrous atmosphere.

As recently shown by M. Tedjini et al. water is able to diffuse at a not completely closed direct bonding interface. This penetration and its negative impact on the direct bonding energy are obviously shown on figure 2.

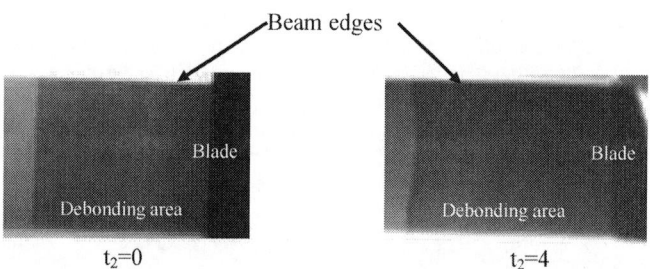

figure 2. Infra-red images of debonding front during the DCB measurement for two t2 (0 and 4 days). A straight front is observed at t2=0 days and a wavy front is observed at t2=4 days

Indeed these infra-red images show the debonding front during the direct bonding energy measurement before and after a storage time of 4 days ($t_2=0$ and $t_2=4$). A straight debonding front is observed at the storage beginning whereas a wavy debonding front appears after a storage time of 4 days. A more important debonding length is observed at both beam edges of this last sample. The bond strength decrease starts then from the edges and progresses toward the beam center. This weakening effect is clearly due to the water penetration as the same storage times in anhydrous atmosphere do not show any bonding energy decrease as shown on figure 1.

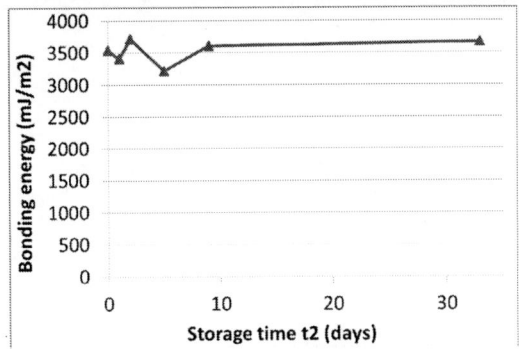

figure 3. Sketch showing the water stress corrosion at room temperature of the stressed bonded asperities.

The proposed mechanism is the water stress corrosion of the asperity contact areas which are under stress due to the bonding evolution [7]. As the water reaches the oxide asperity at room temperature, the water can not diffuse into the oxide in order to soften the asperity and to facilitate the asperity expansion and the bonding interface closure. At room temperature, the water can just reach the asperity surface and especially the bonding interface between two asperities where all the stress is concentrated as shown on figure 3. As the water stress corrosion is induced and enhanced by stress, water can then break the siloxane bonds (Si-O-Si) at this point which then opens the bonding interface. The resulting macroscopic adhesion value is thus reduced. Obviously, this mechanism will stop when the system will have released enough stress. The water stress corrosion will stop and a constant bonding energy value will appear regarding t_2. In this experiment this value arrives after a t_2 of 10 days.

However, this t_2 impact on bonding energy can only append if the bonding interface is open enough to allow water diffusion. Indeed, as shown on figure 4, if the bonding energy at $t_2=0$ (just after the annealing) is high enough, almost no bonding energy decrease is observed. These high bonding energies values are obtained thanks to an optimised cleaning sequence derived from the classical RCA cleaning.

figure 4. t2 storage time impact on the direct bonding energy of silicon oxide bonding. No bonding energy decrease is observed thanks to the high bonding strength right after the annealing t2=0.

Indeed, as previously reported it is quite easy to vary the oxide/oxide bonding energy by changing slightly the oxide surface rinsing and drying type just before the bonding [2]. After the 300°C annealing, a bonding energy difference as high as 3 J/m² can be observed depending of the surface preparation (cf. figure 5). In addition to a high bonding strength, the use of the good oxide surface preparation is then also interesting to have a stable bonding energy regarding the storage time after the annealing.

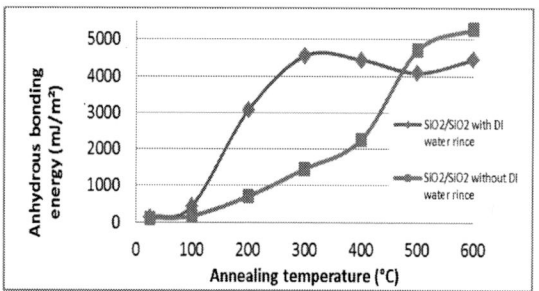

figure 5. Oxide/oxide bonding energy evolution regarding a "dry" or wet oxide surface preparation before bonding [5].

Despite t_2, t_1 seems not to have any impact on the bonding energy measured just after the annealing. Indeed, as shown on figure 6, t1 storage times in wet atmosphere, in water or in anhydrous atmosphere induce the same bonding energy just after the 300°C annealing ($t_2=0$).

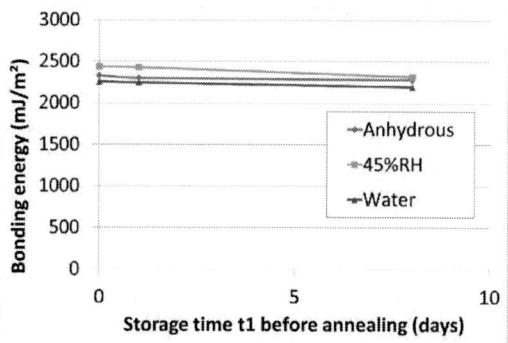

figure 6. t1 storage time impact on the direct bonding energy of silicon oxide bonding. The storage takes place after dicing and before annealing (obviously t2=0).

All the annealing of the previous experiments have been realized in a non-controlled atmosphere (clean room air) with 45% RH at the beginning of the annealing. Two others atmospheres have been tried: N_2 atmosphere or "wet" atmosphere (putting a glass of water inside the oven). As shown on table 1, the dry atmosphere (N_2) reduces the bonding energy obtained after the annealing compare to the non-controlled atmosphere. In contrary, the wet atmosphere increases the bonding energy.

Annealing atmosphere	N$_2$	Air (non controlled)	Wet
Bonding energy after 300°C, 2h (mJ/m^2)	1800	2300	2800

table 1 Impact of the annealing atmosphere on the bonding energy of silicon oxide bonding. The annealing is done at t1=0 and the bonding energy at t2=0 after 300°C during 2 hours.

The impact of the annealing atmosphere on the bonding energy explain the result obtained with the different t_1 storage time under different atmospheres. Indeed whatever the moisture level of the bonding interface before the annealing, the interface is reset during the 2cm wide beams annealing. This reset might be done during the first time of the annealing. T_1 storage time has then no impact on the future bonding energy for oxide/oxide bonding. Only the annealing atmosphere type is important and a wet ambiance leads to higher bonding strength.

Obviously using a full wafer, the whole bonding interface might not be reset during the annealing but a non-homogenous bonding energy along the diameter might be foreseen. Moreover, with a full wafer, the t_1 storage time and atmosphere might have an impact in order for instance to increase the amount of water at the bonding interface. A positive impact of wet atmosphere during t_1 and annealing is then foreseen for silicon oxide bonding. But, after the annealing, a dry atmosphere is then recommended in order to avoid any weakening effect due to the water stress corrosion, to keep a homogenous behavior of the bonding interface and to avoid any "edge effects".

Conclusion

Antagonistic water effect has been pointed out in this study for silicon dioxide to silicon dioxide bonding. Before annealing, a wet interface and a wet annealing atmosphere leads to better bonding energy than a dry bonding or annealing atmosphere. Water is needed for bonding interface strengthening thanks to the water stress corrosion during the annealing. But the annealed interface can then be corroded again by the same water at room temperature as shown by the t_2 storage time impact. These two different behaviours of the water are probably mainly depending on the location of the corrosion: only at the asperity surface at room temperature and inside the whole asperity during the annealing. Water management during storage and annealing is then a key parameter for silicon dioxide to silicon dioxide bonding in order to obtain and to maintain a good bonding energy with low annealing temperature under 400°C for instance.

Acknowledgments

The authors would like to thank SOITEC S. A. for financial support.

Reference

[1] G. Kräuter, A. Schumacher, U. Gösele, Sensors and Actuators A: Physical, Vol. 70 (3), 271–275 (1998).

[2] F. Fournel, C. Martin-Cocher, V. Larrey, H. Moriceau, F. Rieutord, WaferBond international conference (2015)

[3] F. Fournel, L. Continni, C. Morales, J. Da Fonseca, H. Moriceau, F. Rieutord, A. Barthelemy, I. Radu, J. Apl. Phys., 111, 104907 (2012)

[4] M. S. El-Zein and K. L. Reijsnider, J. Of compo. Tech. & Research 10(4) pp.151-155 (1988).

[5] M. Tedjini, F. Fournel, V. Larrey, F. Rieutord, WaferBond international conference (2015)

[6] M. Tedjini, F. Fournel, H. Moriceau, V. Larrey, D. Landru, O. Kononchuk, S. Tardif, F. Rieutord, Submitted to Journal of Applied Physics.

[7] F. Fournel, C. Martin-Cocher, D. Radisson, V. Larrey, E. Beche, C. Morales, P. A. Delean, F. Rieutord, and H. Moriceau, ECS Journal of Solid State Science and Technology, 4 (5) P1-P7 (2015)

Chapter 4

Bonding Mechanism & Photonic Integration

Control of Direct Bonding Behavior by Interlayers

M. Eichler[a], H. Dillmann[b], K. Nagel[a], and C.-P. Klages[a,b]

[a] Department of Atmospheric Pressure Plasma, Fraunhofer Institute for Surface Engineering and Thin Films IST, 38118 Braunschweig, Germany
[b] Institut für Oberflächentechnik IOT, Technische Universität Braunschweig, Braunschweig 38108, Germany

Atmospheric pressure plasma treatments were used to control free surface energy of different areas on silicon wafers before bonding. Surface energy measurements in situ during annealing for this different areas are presented for SF_6 etching as well as acetylene, glycidyl methacrylate, tetramethylsilane and C4F8 and coatings. The bonding energy can be permanently reduced by appropriate coatings or surface roughness. The results revel important aspects for the choice of precursors and parameters to obtain high contrast between the treated and untreated areas.

Introduction

For MEMS manufacturing as well as MEMS operation the surface contact behavior for substrates and integrated components plays an important role. In the area of the bonding frame high bond strengths are required. In MEMS with membranes, gyrating masses, and valves, on the other hand, bond frames are frequently located in close vicinity to areas where anti-sticking properties are needed. The deposition of thin layers allows to adjust adhesion by surface energy or roughness. Using the atmospheric-pressure plasma tool SELECT surfaces can be subjected to a patterned treatment, for example they can be coated locally by PECVD or a specific topography can be generated by plasma etching. For the characterization of the bonding and wetting behaviors minimum surface areas are needed. Therefore relatively large areas were homogeneously treated abutting to untreated zones.

Experimental

Experiments were carried out with single-side polished, B-doped (1–10 Ωcm), 100 mm diameter Czochralski-grown prime-grade silicon [100] wafers with a thickness of 525±25 μm and a total thickness variation across the wafer of less than 5 μm. Before PECVD coating or etching, the silicon wafers were surface-activated for 30 s by an atmospheric-pressure plasma in oxygen. Therefor the plasma tooling SELECT of the SUSS MicroTec aligner was utilized and an electrode gap distance of 300 μm and a pulse pause ratio of 4 ms : 4 ms was used [1].

A particular challenge is the characterization of the bonding behavior of PECVD layers during and after annealing. Therefore an electrode is used which limits plasma treatment to a 46 mm wide strip with a gap distance (electrode to substrate) of 70 μm over the full length in the middle of a 100 mm wafer (Figure 1). The 46 mm wide plasma-modified strip will be used to characterize the bonding behavior in comparison to non-modified areas beside the strip. After plasma treatment the wafer were cleaned using a megasonic DI-water nozzle in a spin dryer in order to remove particles caused from wafer handling. Subsequent the coated wafer were bonded to a non-coated flat silicon wafer.

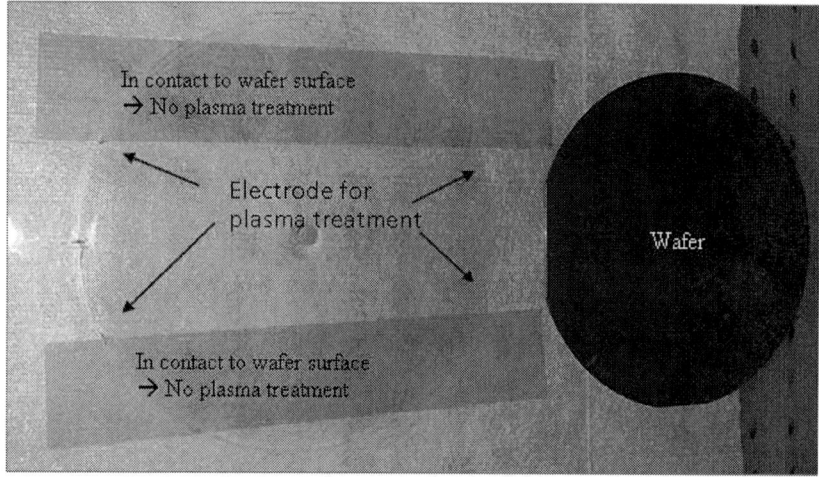

Figure 1. Transparent electrode for SELECT plasma tool covered by two stripes which limit the plasma treatment. For comparison, a 100 mm wafer on the right side is shown.

The bond strength (surface energy) can be determined by crack opening test *in situ* during annealing for increasing temperature [2].

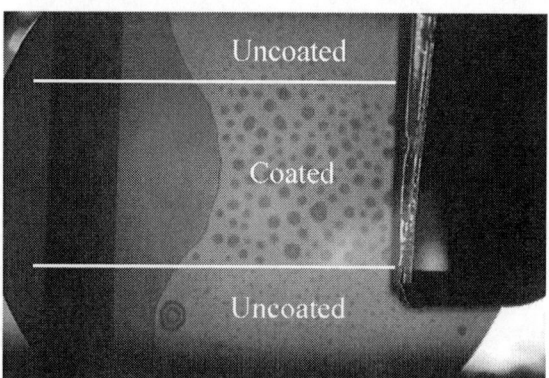

Figure 2. Parallel surface energy measurement for coated and uncoated surface areas. From left the blade cracks the wafer pair and causes a "crack line" which propagate depend from surface energy away from the blade.

Based on the different crack lengths in the coated and uncoated area the different strengths were calculated (Figure 2). In addition to the different bond strengths it is also evident that the formation of bubbles in the interface is affected by the coating.

Results and discussion

For coating and etching of the wafer different process gases were used. After bonding the wafers were heat-treated up to 200 ° C and the surface energy *in situ* characterized.

Acetylene coating

For this experiment an organic plasma coating was deposited from a process gas mixture of 4.95 slm He and 0.05 slm acetylene. After 1 s plasma treatment the plasma process was paused for 60 s in order to facilitate the process gas exchange and renewal in the treated area. This was repeated 10 times (1 s x 10 = 10 s treatment time). During plasma treatment a pulse pause ratio of 1 ms : 1 ms was used. The surface energy measurement is shown in Figure 3.

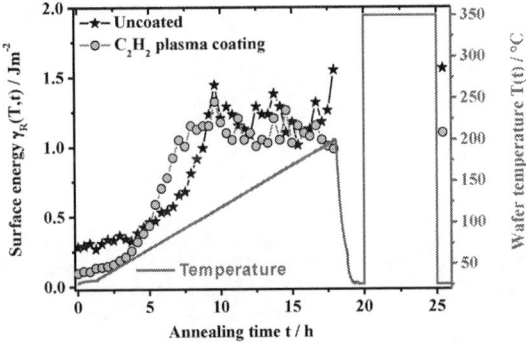

Figure 3. Parallel surface energy measurements for acetylene coated and uncoated surface areas. The acetylene coating changes the surface energy up to 100 °C compare to the uncoated area. The post-annealing at 350 °C causes no significant changes.

After the first annealing up to 200 °C a second annealing was done at 350 °C. The bond strength varies slightly between the coated and the uncoated area, depending on the temperature. But even after the 350 °C anneal a relatively high bond strength was measured in the coated area.

Glycidyl methacrylate coating

A further deposition experiments were done with the organic precursor glycidyl methacrylate (GMA). The coating was deposited from 5 slm He bubbled in liquid GMA. After 1 s plasma treatment the plasma process was paused for 60 s in order to facilitate the process gas exchange and renewal in the treated area. This was repeated 10 times

(1 s x 10 = 10 s treatment time). During plasma treatment a pulse pause ratio of 1 ms : 1 ms was used. The surface energy measurement is shown in Figure 4.

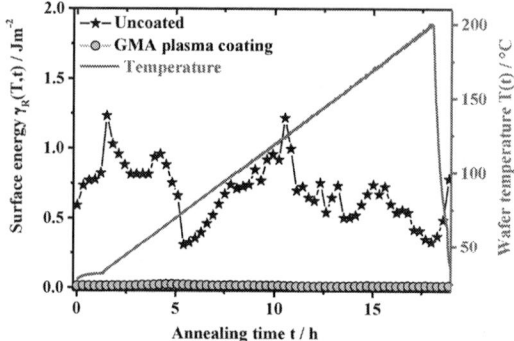

Figure 4. Parallel surface energy measurements for GMA coated and uncoated surface areas. The coating reduces the bond strength close to zero.

The GMA coating reduces the bonding energy close to zero. But there is also a change in bonding behavior for the uncoated area compare to the uncoated area in Figure 3. This might indicates unwanted contaminations on the uncoated area from the precursor.

Tetramethylsilane coating

A further deposition experiment was done with the silicon organic precursor tetramethylsilane (TMS). The coating was deposited from 4.95 slm He and 0.05 slm He bubbled in liquid TMA. After 1 s plasma treatment the plasma process was paused for 60 s in order to facilitate the process gas exchange and renewal in the treated area. This was repeated 10 times (1 s x 10 = 10 s treatment time). During plasma treatment a pulse pause ratio of 1 ms : 1 ms was used. The surface energy measurement is shown in Figure 5.

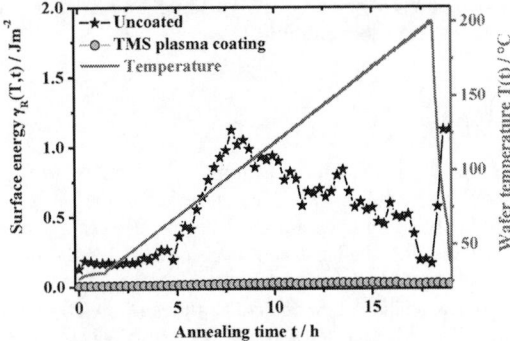

Figure 5. Parallel surface energy measurements for TMA coated and uncoated surface areas. The coating reduces the bond strength close to zero.

Also the TMA coating reduces the bonding energy close to zero and the change in bonding behavior for the uncoated compare to the uncoated area in Figure 3 for temperatures higher than 100 °C indicates unwanted contaminations from the precursor.

Octafluorocyclobutane coating

A further deposition experiment was done with the precursor octafluorocyclobutane (C_4F_8). The coating was deposited from a process gas mixture of 4.95 slm He and 0.05 slm C_4F_8. After 8 s plasma treatment the plasma process was paused for 60 s in order to facilitate the process gas exchange and renewal in the treated area. This was repeated 10 times (8 s x 10 = 80 s treatment time). During plasma treatment a pulse pause ratio of 1 ms : 1 ms was used. The surface energy measurement is shown in Figure 6.

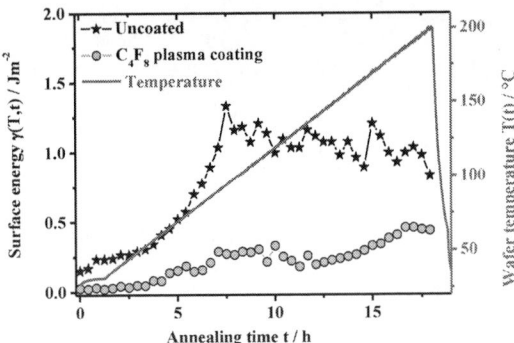

Figure 6. Parallel surface energy measurements for C_4F_8 coated and uncoated surface areas. The C_4F_8 coating reduces the surface energy.

As expected the C_4F_8 coating reduces the bonding energy. But the effect is not this strong like for the precursors TMS and GMA. This might be an also an effect of different coating thicknesses, which was not determined.

Sulfur hexafluoride etching

In addition to the above-described coating experiments the etching by sulfur hexafluoride (SF_6) was investigated. The idea is to roughen the surface by etching and to influence in this way the bonding behavior. For this experiment a process gas mixture of 4.58 slm He, 0.28 slm He and 0.14 slm SF_6 was used. After 8 s plasma treatment the plasma process was paused for 60 s in order to facilitate the process gas exchange and renewal in the treated area. This was repeated 10 times (8 s x 10 = 80 s treatment time). During plasma treatment a pulse pause ratio of 1 ms : 1 ms was used. The surface energy measurement is shown in Figure 7.

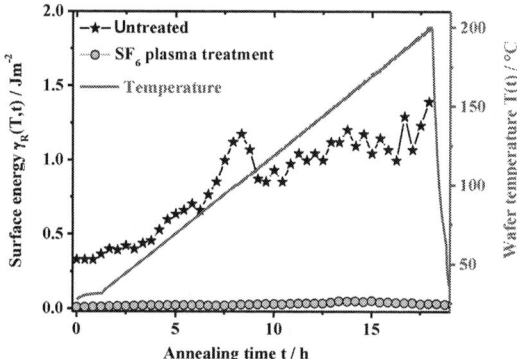

Figure 7. Parallel surface energy measurements for SF_6 etched and unetched surface areas. The SF_6 etching reduces the surface energy.

The etching process reduces the bond strength clearly and prevents contamination in the untreated area.

Patterned atmospheric pressure plasma coating

Initial studies for the patterned deposition were performed. Relatively large structures were used in order to measure the water contact angles and to characterize the wetting properties. As mask a 150 μm polyethylene terephthalate film was used. For simplification the mask was not fixed to the electrode and just placed between the electrode and substrate. For the coating processes electrode, mask and silicon wafer were pressed together and the plasma ignited. In Figure 8 on the left side a patterned coated wafers is shown.

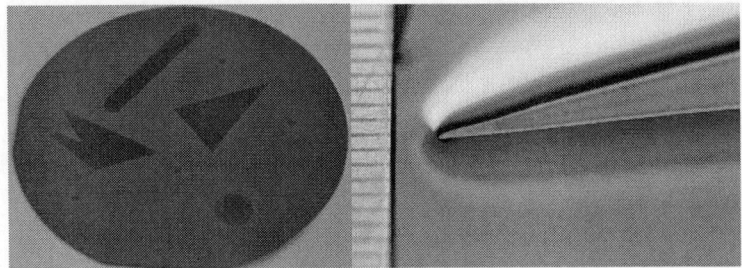

Figure 8. Left: Silicon wafer patterned coated by C_4F_8. Right: Wetting behavior on the border between C_4F_8-coated and uncoated area.

This patterned deposition experiment were done with the precursor C_4F_8. The coating was deposited from a process gas mixture of 4.75 slm He and 0.25 slm C_4F_8. After 1 s plasma treatment the plasma process was paused and the electrode 250 µm evaluated for 10 s in order to facilitate the process gas exchange and renewal in the treated area. This was repeated 20 times (1 s x 20 = 20 s treatment time). During plasma treatment a pulse pause ratio of 1 ms : 4 ms was used.

Threes water contact angle measurements were done each in the coated and the uncoated surface area:
C_4F_8-coating areas: 125.4° +/- 2.8°
Uncoated area: 6.9° +/- 0.3°
The measurements have shown that the water contact angle over the entire wafer are almost equal. Moreover, for these process parameters there are obviously no contamination of the untreated areas, since the contact angle has not increased. The wetting experiment in Figure 8 (left) illustrate the contrast between the coated and uncoated area.

This same experiments were done for the precursor TMS. The coating was deposited from a process gas mixture of 4.95 slm He and 0.05 slm He bubbled in liquid TMA. After 1 s plasma treatment the plasma process was paused for 30 s and the electrode 250 µm evaluated in order to facilitate the process gas exchange and renewal in the treated area. This was repeated 10 times (1 s x 10 = 10 s treatment time). During plasma treatment a pulse pause ratio of 1 ms : 1 ms was used.

Threes water contact angle measurements were done each in the coated and the uncoated surface area:
TMS-coating areas: 91.6° +/- 3.9°
Uncoated area: 62.7° +/- 32.6°
The measurements have shown that the water contact angle over the entire wafer are not equal, especial in the uncoated area. There are obviously contaminations of the untreated areas, since the contact angle has strongly increased.

Conclusions

Investigations have shown that PECVD coatings of fluorocarbons, silicon organics, and organics as well as an etching of the surface with fluorine-containing etching gas enables non-sticking properties. For a good contrast between treated and untreated areas the process gas must not contaminate the untreated areas by overflow. Therefore, it is important to optimize the process parameters sufficiently.

Acknowledgments

The authors thanks to C. L. Reim for his support in the atmospheric pressure plasma lab.

References

1. M. Eichler and M. Gabriel, SUSS report (www.suss.com), issue April 2010, 8 (2010).
2. M. Eichler, B. Michel, P. Hennecke, and C.-P. Klages, J. Electrochem. Soc., 156, H786 – H793 (2009)

Adhesion energy and bonding wave velocity measurements

V. Larrey[a,c], G. Mauguen[a,c], F. Fournel[a,c], D. Radisson[a,c], F. Rieutord[b,c], C. Morales[a,c], C. Bridoux[a,c], H. Moriceau[a,c]

a- CEA, Leti, Minatec Campus, 17 rue des Martyrs, F-38054 Grenoble, France
b- CEA, INAC, Minatec Campus, 17 rue des Martyrs, F-38054 Grenoble, France
c- Univ. Grenoble Alpes, F-38000 Grenoble, France

Adhesion energy and bonding wave velocity are key parameters that should be controlled in an industrial direct bonding process because it reports what exactly occurs at the time of the bonding. While dynamics of the bonding front has been analytically studied and modelled [1]-[4], the literature gives not much information about its characterization. This study is then focused on adhesion energy (Ea) and bonding wave velocity (Vo) measurement. After considerations about immediate and stabilized adhesion energy, we will show the dependence between Ea and Vo and we will confirm Rieutord's model [1] for hydrophilic direct bonding. Roughness dependence, plasma pre-treatment, hydrophobic direct bonding and re-bonding will also be discussed.

Introduction

Direct bonding is now widely used in microelectronics for SOI elaboration or 3D integration. One of its main characterizations is the adherence energy, also called bonding energy, which estimates the strength of the bonded structure and its evolution with the different process parameters (annealing temperature, plasma activation...). The double cantilever beam (DCB) technique described by Maszara et al. [5] is the most common technic to measure the adherence energy. During the bonding another key characterization, the adhesion energy, deserves to be measured. This quantity gives information about the work of forces involved when contacting the two surfaces and helps to understand the bonding wave velocity. Rieutord et al. [1] suggested an analytical model based on air flow dynamics between the two wafers and derived an analytical relation between the adhesion energy and the bonding wave velocity. This model predicts a variation of the bonding wave velocity as the power 5/4 of the adhesion energy. Moreover, Rieutord et al. introduced a lower fluidic cut off distance z0 to avoid a divergence in their calculation near the crack tip; they considered that z0 should be equal to the molecular mean free path (50nm for air at standard temperature and pressure). This study is then focused on the development of methods to measure these quantities and the comparison of the experimental data with the analytical model.

Experiment

In this work, 200 mm <001> silicon (Si) wafers are bonded at room temperature and ambient pressure after cleaning and surface preparation. Some of them are previously thermally oxidized at 950°C under steam atmosphere in order to growth 145 nm of silicon dioxide (SiO_2) thin film. We can then achieve Si to Si, Si to SiO_2 and SiO_2 to SiO_2 direct bondings. As cleaning and surface preparation, we use a standard chemical RCA sequence, sometimes followed by a plasma treatment in nitrogen atmosphere. Bonding can occur just after the chemical preparation (referenced as chemical bondings) or after the plasma treatment (referenced as plasma bondings). Note that these two kind of preparations result in hydrophilic direct bondings.

Prior to contacting the surfaces, we put a 395μm thick blade on the lower substrate, at the opposite side of the bonding initiation. During the bonding wave propagation, we record multiple infra-red pictures of the bonding front delayed by an adjustable duration time. By an automatic contrast detection, the bonding front position is determined at each time. One can then calculate the bonding wave velocity. Because of the presence of the blade, the bonding wave then stops and the remaining "non-bonded" length between the blade and the bonding front is measured. Using Maszara [5] or El Zein [6] formalism, the associated energy corresponding to the adhesion energy can be estimated.

Results and discussion

Hydrophilic Bondings

Measuring the adhesion energy time dependence for hydrophilic bondings, one can note a strong evolution occurring during a time range of about one hour and a half (figure 1). After the bonding wave stop, a first plateau is observed: the adhesion energy slowly increases during about 100 seconds; we can then determine a first value named immediate adhesion energy. Then the adhesion energy strongly increases and reaches a second plateau whose value is named stabilized adhesion energy. Stabilized adhesion energy value can be twice immediate adhesion energy one.

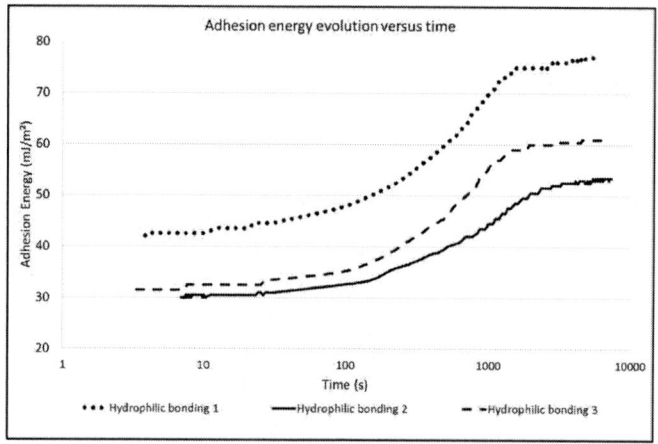

Figure 1. Adhesion energy evolution versus time for various hydrophilic bondings

We propose to explain this long time evolution with the formation of capillary bridges between the two surfaces as shown in figure 2. Indeed, the Kelvin radius can be seen as the maximal curvature radius of the liquid bridge between asperities coming from the two separated surfaces. Calculating the Kelvin radius from [7], we note that it is of the order of magnitude (nanometer scale) of the typical peak to valley roughness of our silicon surfaces (PV~2nm). Because of the presence of the blade, the bonding wave propagation ends as a crack front and then capillary bridges can be formed, slowly bringing closer the two surfaces and establishing new contacts between the asperities of the two surfaces [8]; the debonded length then decreases and the adhesion energy increases. It is really important to note that this adhesion energy increase cannot be due to any adherence energy (also called bonding energy) enhancement. Indeed, as proven by Ventosa et al. [9], the two surfaces do not get closer when the adherence energy increases and therefore no information about this increase can reach the bonding front.

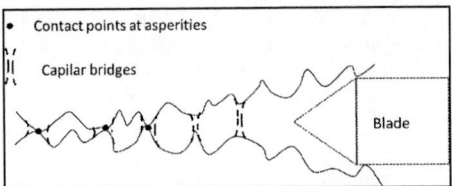

Figure 2. Adhesion energy evolution versus time for various hydrophilic bondings

We have now to choose which adhesion energy is pertinent. We first consider that the propagation of the bonding wave takes into account all the phenomena that can bring closer the two surfaces; the stabilized adhesion energy seems to realize that consideration and we will further present these stabilized values.

Considering the bonding wave velocity, one can observe that a plateau is rapidly reached after the bonding initiation. White lines in figure 3.a represents the successive positions of the bonding front leading to the graph presented in figure 3.b showing the bonding wave velocity at various point of the bonding front. This plateau value is the stabilized bonding wave velocity; because of the blade presence, the velocity then decreases and the propagation stops.

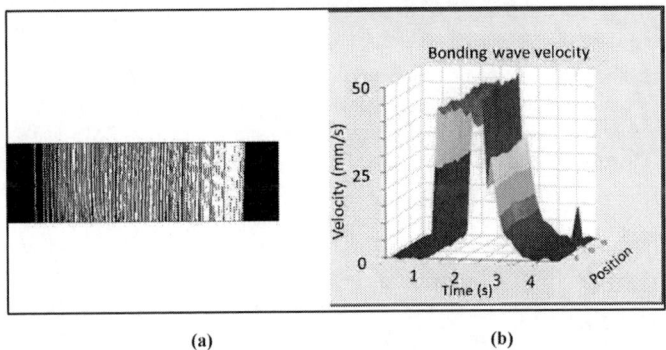

(a)   (b)

Figure 3. (a) detection of successive positions of the bonding front – (b) Bonding wave velocity calculation of the bonding front

Hydrophilic measurement

Varying nature (Si or SiO2) and preparation (chemical or plasma) of the surfaces to be bonded, a large scale of stabilized adhesion energies and bonding wave velocities are obtained. As shown in figure 4, the dependence of the bonding wave velocity to the adhesion energy is plotted for all the conditions and the measured values are compared to the Rieutord's analytical model [1]. In this work, we assume the Rieutord's dependence of equation 1

$$v = A \frac{Ea^{5/4} z0^{1/2}}{\eta D^{1/4}}$$

Eq. 1

where v [m/s] is the bonding wave velocity, Ea [J/m²] is the adhesion energy, z0 [m] the cut off distance, η [=18.6x10⁻⁶ Pa.s] the air viscosity and D [=5.39J] the substrate rigidity (linked to substrate thickness, Silicon Young modulus and Poisson ratio). A is a constant equal to 0,057. With a standard linear regression of our experimental values, a z0 value equal to 55nm is obtained which is very close to the molecular mean free path considered by Rieutord et al. A very good agreement between the theoretical model and the measured values confirms the relation between the two measured quantities. Noteworthy, an additional point measured for direct bonding in anhydrous atmosphere (0.1ppm of water) is added with very low adhesion energy and bonding wave velocity (12.5 mJ/m^2 and 1.2mm/s). A very efficient surface drying with a certain consumption of silanol groups results in a very slow wave propagation.

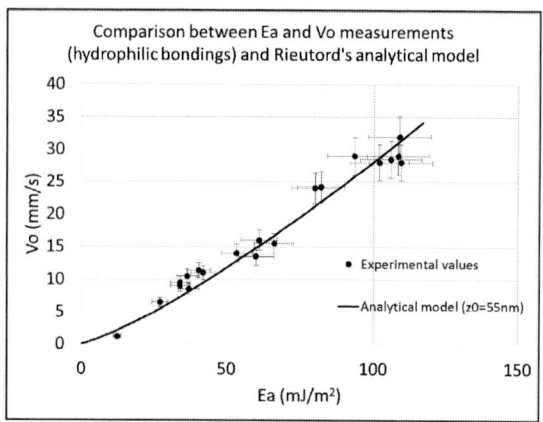

Figure 4. Comparison between experimental data (hydrophilic bondings) and Rieutord's analytical model

Among SiO$_2$ to SiO$_2$ bondings, we present more in details a study concerning microroughness: after thermal oxidation, some substrates are isotropically etched in a 10% fluorhydric acid solution for various times, resulting in various microrougnesses in the range of 2.2-5.1Å RMS characterized using Atomic Force Microscopy (AFM) with 1*1µm^2 scans. Hydrophilic chemical bondings contacting equivalent rough surfaces are then performed. The dependence of adhesion energy and bonding wave velocity on microroughness is plotted in figure 5 and a linear dependence is clearly showed. One can extrapolate these data to obtain the microroughness limit at which the direct bonding wave cannot propagate (Vo=0mm/s): this value of 6.5Å RMS is in good agreement with previous works [10].

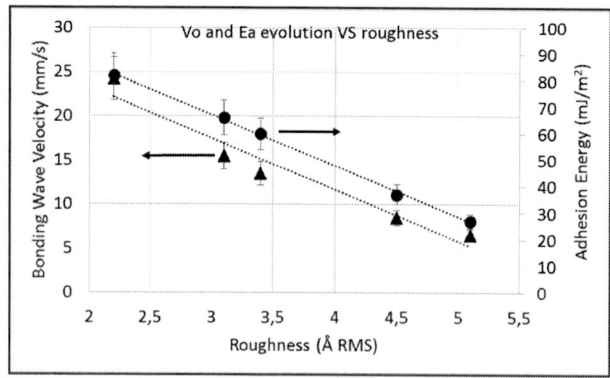

Figure 5. Bonding wave velocity and Adhesion energy evolution versus roughness

The results of figure 4 can be also analysed regarding each hydrophilic bonding type (table 1). Chemical bondings show lower adhesion energy and bonding wave velocity each time a silicon surface is involved. Actually, a 1nm thin native oxide layer is generated at the silicon surface during

148

hydrophilic chemical treatment. As surface microroughness is lower on native oxide (1.8Å RMS) than on 145nm thick thermal oxide (2.2Å RMS), this behaviour cannot be correlated with roughness. These first results thereby suggest a smaller adhesion on native oxide.

After plasma treatments, adhesion energies and bonding wave velocities increase for all types of bondings: surfaces have been modified so as the air flow at its neighbourhood is favoured. This increase is especially important when a native oxide surface is involved, showing that the surface transformation by RIE Nitrogen plasma is more sensitive on silicon. Surface microroughness is reduced [11] and surface properties are changed by a plasma treatment: Moriceau et al [11] showed by interfacial X-Ray reflectivity that density gap depths of RIE plasma pre-treated structures are strongly reduced and are correlated with high surface energies. The benefit of plasma treatment comes from both chemical and/or mechanical evolution of the subsurface and the exact role of each contribution is still under discussion. Nevertheless its impact on silicon surface seems more important than on oxide surface. Considering a pure mechanical point of view, the plasma treated silicon asperities may be even softer than the oxide ones.

Bonding type	Chemical bonding		Plasma Bonding	
	$Ea(mJ.m^{-2})$	$Vo(mm.s^{-1})$	$Ea(mJ.m^{-2})$	$Vo(mm.s^{-1})$
SiO2/SiO2	61-82	14-24	80-102	24-28
SiO2/Si	34-36,5	9-10,5	106-108,5	28,5-29
Si/Si	40,5-42	11-11,4	93,5-109,5	28-32

Table 1. Chemical and plasma bonding data

Hydrophobic measurement

After hydrophilic bonding, hydrophobic bonding are also studied in this work. To achieve high bonding wave velocities, reconstructed surfaces are prepared by a multiple steps sequence including annealings under H2 at high temperature (950-1110°C) and silicon epitaxy [12]. Such surfaces are known to have atomically flat terraces separated by monoatomic steps and to exhibit then very low surface roughness. Two pairs of such Silicon reconstructed surfaces are prepared and one of these have been slightly degraded using a 50% HF solution. Bonding wave propagation and adhesion energy is then measured contacting equivalent surfaces and two additional couples of values are reported on figure 6. Noteworthy, during the adhesion energy measurement, no energy evolution over the time could be seen for hydrophobic bonding. Immediate and stabilized adhesion energy are the same values. Moreover, we can clearly note that hydrophobic data do not follow the same trend as hydrophilic ones: analytical model with $z_0=55nm$ does not depict the bonding wave propagation with hydrophobic surfaces.

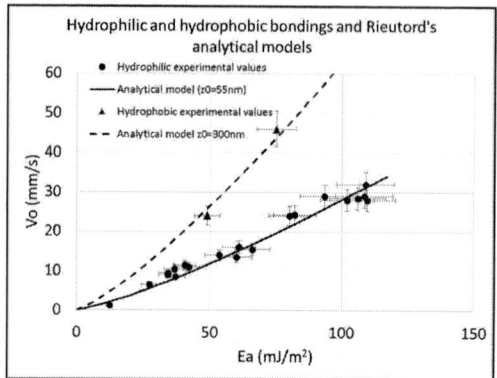

Figure 6. Comparison between experimental data (hydrophilic and hydrophobic bondings) and Rieutord's analytical model (z_0=55nm and z_0=300nm)

It is pre- supposed in Rieutord's study that there is no slip effect at the surface, which results in a distribution of amplitudes $V_x(y)$ as a parabolic Poiseuille profile with $Vx(\pm h)=0$ at the walls, as shown in Figure 7. This assumption is realized on a hydrophilic surface. However, on a hydrophobic surface, it has been shown [13] that the fluid flow of water or vapor atmosphere containing water in the vicinity of a hydrophobic surface could have a different behavior. A slip effect at the surface vicinity could then be considered. This results in a non-zero velocity of the fluid flow on the hydrophobic surface (distribution $V_x(y)$ in dashed lines in Figure 7), and thus the introduction of a slipping length b. Such new considerations deserve a complete analytical development. But, as a first approach, one can suggest that the introduction of a slipping length influences the fluidic cut off distance z0. A new z0=300nm well depicts the hydrophobic data (dashed curve in figure 6).

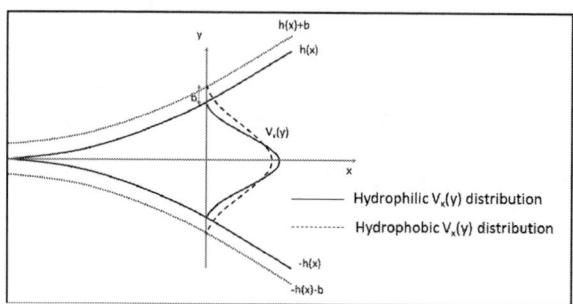

Figure 7. Hydrophilic $V_x(y)$ distribution (no slip boundary) and Hydrophilic $V_x(y)$ distribution (slip boundary at surface)

Re-bonding

In order to easily study reproducibility and homogeneity within the wafer, it should be useful to have a method that allows beams measurement. In this work, a simple measurement apparatus derived from the DCB technique has been developed (figure 8). After an initial wafer direct bonding, beams cut in <001> bonded silicon wafers are opened at the bonding interface by a very thick blade insertion; then the blade is suddenly partially removed allowing the structure to be bonded again. Recording multiple infra-red pictures one can follow the rebonding wave propagation and measure its velocity. The thick blade allows an easy and long debonded area in order to be sure to detect an enough long plateau in rebonding wave speed. By putting a thin classical blade just in front of the thick blade and taking care to leave this thin blade in between the two silicon beams, an adhesion bonding energy measurement can be performed at the same time. In one bonding, the wave velocity and the adhesion energy can be evaluated.

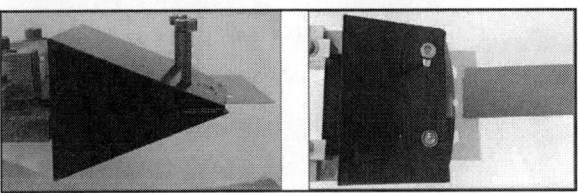

Figure 8. Apparatus for re-bonding beams measurement

In this section, this technic reliability is discussed because it requires a bonding with a potential degraded surface due to a first debonding and the additional dicing step influence should also be estimated. The first results obtained with this technic are plotted in figure 9. Only beams derived from chemical hydrophilic bondings are considered. Globally, experimental data still follows analytical model with z0=55nm. However slight differences have to be pointed out. First, Si to Si samples exhibit higher bonding wave velocities (13-15 mm/s) than those reported in the previous section (initial bonding measurement) whereas adhesion energy are equivalent (40-44 mJ/m²). Secondly, Si to SiO₂ samples show disparate adhesion energy results (58-80mJ/m²). Finally, SiO₂ to SiO₂ samples present a good reproducibility and a good agreement with previous measurements. Obviously, the beam edge could influence the viscous flow dynamics as shown by Radisson [14]. But even considering this effect, the different behaviour cannot be explained. Thereby, re-bonding technic and maybe dicing steps seem to exhibit a non-negligible influence on adhesion energy and bonding wave velocities values. These first results have to be confirmed and other experiments and characterizations have to be performed.

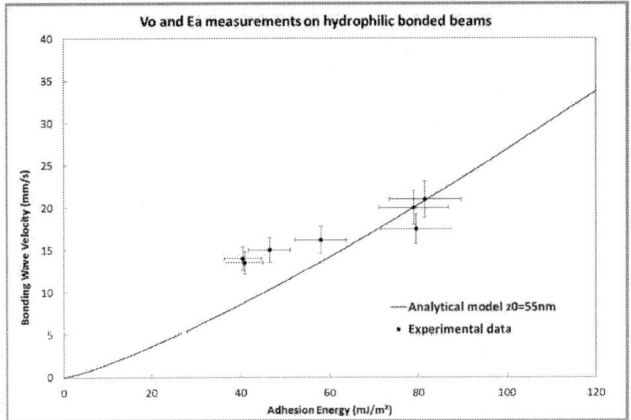

Figure 9. Comparison between experimental data (hydrophilic re-bondings on beams) and Rieutord's analytical model

Conclusion

A method to accurately measure bonding wave velocity and adhesion energy has been developed. The Rieutord's theoretical relation between these two quantities has been confirmed by our experimental data. Following this analytical model, a fluidic cut-off distance of 55nm, very close to the molecular mean free path, is found to well depicts the bonding wave propagation. Experimental data dependence on surface roughness has been showed and the effect of a nitrogen plasma pretreatment has been discussed. Finally, re-bonding experiments have been performed thanks to a special apparatus and the first results of these technic have been presented.

REFERENCES

[1] F. Rieutord, B. Bataillou, and H. Moriceau, Physical Review Letters, vol. 94, 236101, (2005)
[2] E. Navarro, Y. Bréchet, R. Moreau, T. Pardoen, J.-P. Raskin, A. Barthelemy, and I. Radu , Applied Physics Letters 103, 034104 (2013)
[3] Guanglan Liao, Yuping Shi, Tielin Shi, and Lei Nie, IEEE Transactions on advanced Packaging, Vol.33, n°2 (2010)
[4] Damien Radisson, Frank Fournel, Elisabeth Charlaix, Microsyst Technol 21:969–971 (2015)
[5] W. P. Maszara, G. Goetz, A. Caviglia and J. B. McKitterick, J. Appl. Phys., 64 p 4943 (1988)
[6] M. S. El-Zein and K. L. Reijsnider, Journal of Compo. Tech. & Research, 10(4), pp.151-155, (1988)
[7] L. Bocquet, E. Charlaix, S. Ciliberto, and J. Crassous, Nature 396 pp.735-737 (December 1998)
[8] F. Fournel, C. Martin-Cocher, D. Radisson, V. Larrey, E. Beche, C. Morales, P. A. Delean, F. Rieutord, and H. Moriceau, ECS Journal of Solid State Science and Technology, 4 (5) P124-P130 (2015)
[9] C. Ventosa, F. Rieutord, L. Libralesso, C. Morales ,F. Fournel , H. Moriceau, Journal of Applied Physics Volume 104, Issue 12, 2008
[10] O. Rayssac, H. Moriceau, B. Aspar, A. M. Cartier, A. Soubie and M. Bruel, Conference on Materials for Microelectronics, IOM Communications, 183, (1998) 2.
[11] H. Moriceau, F. Rieutord, C. Morales, S. Sartori, A.M. Charvet, Electrochemical Society Proceedings, Vol 2005-6, pp 34-49
[12] C. Rauer, F. Rieutord, J. M. Hartmann, A.-M. Charvet, F. Fournel, D. Mariolle, C. Morales, H. Moriceau, Microsyst Technol pp. 675–679 (19-2013)
[13] C. Cottin-Bizonne, B. Cross, A. Steinberger, and E. Charlaix, Physical Review Letters, PRL 94, 056102 (2005)
[14] D. Radisson, PhD Thesis (2013)

A Study of Void Formation in Fluorine Containing Plasma Activated Wafer Bonding

Chenxi Wang[a], Yannan Liu[a], and Tadatomo Suga[b]

[a] State Key Laboratory of Advanced Welding and Joining, Harbin Institute of Technology, Harbin 150001, China
[b] Department of Precision Engineering, The University of Tokyo, Tokyo 113-8656, Japan

A quantitative evaluation of void formation in the fluorine containing plasma activated wafer bonding is demonstrated. Results shows adding a small amount of carbon tetrafluoride (CF_4) into oxygen plasma could effectively mitigate void formation in subsequent annealing process. Moreover, long plasma treatments (e.g. >60 s) may produce more porous surfaces and introduce undetected microdefects. Therefore, a short fluorine containing plasma treatment (e.g. ~10 s) is able to remove organic contaminants with fewer defects and the bonding interfaces are nearly void-free during heating from 200 to 800°C. A void formation model is proposed to gain insight the mechanism of the fluorine containing plasma activated bonding.

Introduction

Plasma activated bonding is one of promising candidates in wafer bonding field (1). In the past decade, much progress in terms of plasma generation, plasma treatment parameters, and annealing conditions has been achieved to allow high strength bonding at low temperatures with minimal defects (2). Since wafer bonding is often used as a middle process during device manufacturing, subsequent heated (or annealing) processes are inevitable for bonded wafers (3). However, the void generation in annealing process (known as "annealing voids") is a serious problem in plasma activated bonding, especially for Si/Si wafer pairs. In particular, voids more frequently occur and are significantly severer as the wafer sizes further increase. Recently, our group has developed a fluorine activated bonding process enabling strong bonding of Si/Si bonded pairs (~2.4 J/m^2 in surface energy) even at room temperature (~25°C) without requiring annealing by introducing a small amount of carbon tetrafluoride (CF_4) into oxygen plasma treatments (4, 5). It seems that the fluorinated oxide allows water and hydrogen to diffuse out more easily from the interface at room temperature. But we have no idea how voids would be generated in elevated temperatures during subsequent annealing process. In this work, we demonstrate a quantitative evaluation of the void formation. As a comparison, void generation in the samples prepared by O_2 plasma activated bonding is also systemically investigated. A void formation model is proposed to gain insight the mechanism of the fluorine containing plasma activated bonding.

Experimental

8-inch (200-mm diameter), p-type, (100)-oriented, one-side-polished silicon wafers were used in the experiments. The thickness of all the wafers was 725 μm. To inspect void formation, a couple of bonded wafer pairs were prepared with different plasma treatment conditions. The plasma treatments for wafer surfaces were accomplished using a reactive ion etching (RIE) plasma system with a discharge power of 200 W and a radio frequency of 13.56 MHz. For the fluorine containing plasma, a mixture gas of oxygen and CF_4 was kept at a pressure of 50 Pa during the RIE plasma treatment, called shortly as "$(O_2 + CF_4)$ RIE" hereinafter. The gas flow of O_2 and CF_4 was fixed at 500 and 1.25 sccm. As we reported previously, a maximum surface energy of ~2.4 J/m^2 achieved at room temperature (~25°C) without heating (4).

All the wafers were bonded at room temperature and rolled under a pressure force of 75 kgf. The bonded wafer pairs were categorized into Groups A, B and C on basis of different plasma treatments as listed in TABLE I. All the bonded wafer pairs were heated at 200, 300, 400, 500, 600, 700, and 800°C for 8 h in ambient air. The heating steps do not aim to improve the bonding strength, but to investigate void formation at high temperatures. The IR imaging system was mainly used to characterize the voids in bonded pairs. To track the void emergence and growth, the digital image processing was employed to calculate void densities over the entire bonded wafer pairs. The water contact angles were measured by a contact angle meter CA-X (Kyowa Interface Science Co. Ltd., Japan) using distilled water to clarify the hydrophilicity of silicon surfaces prior to bonding.

TABLE I. The bonded pairs were prepared with different plasma treatments prior to bonding for the study of the void formation.

	Plasma treatment		
	Gas	Mode	Time
Group A	No plasma treatment		
Group B1	O_2	RIE plasma	60 s
B2	O_2	RIE plasma	10 s
Group C1	(O_2+CF_4)	RIE plasma	120 s
C2	(O_2+CF_4)	RIE plasma	60 s
C3	(O_2+CF_4)	RIE plasma	30 s
C4	(O_2+CF_4)	RIE plasma	10 s

Results and Discussion

The main origins of voids in wafer bonding have been classified in trapped voids, thermal voids and annealing voids. The void density can be defined as the ratio of the total area of detective voids to the total area of the bonded wafers.

$$\text{Void density (\%)} = \frac{\text{the total area of detective voids}}{\text{the total area of bonded wafers}} \qquad [1]$$

The bonded wafer pairs of Group A were prepared without the plasma treatment, as a reference sample. Only few voids appear in peripheral regions, as shown in Figure 1(a).

They attributed to trapped voids. When the bonded pairs were annealed at 200°C, many small thermal voids appeared in non-peripheral regions as shown in Figure 1(b). Because the wafers were never cleaned prior to bonding, organic contaminants, e.g. hydrocarbon, may be adsorbed on the surfaces. And they behave as thermal voids at 200°C annealing. These thermal voids develop in higher temperature from 300 to 700°C. However, upon to 800°C, many thermal voids suddenly disappeared as shown in Figure 1(h). It is reported that thermally unstable hydrocarbons on the wafer surfaces are vaporized and/or oxidized during the annealing at 800°C (6). Therefore, most of thermal voids vanished in IR image.

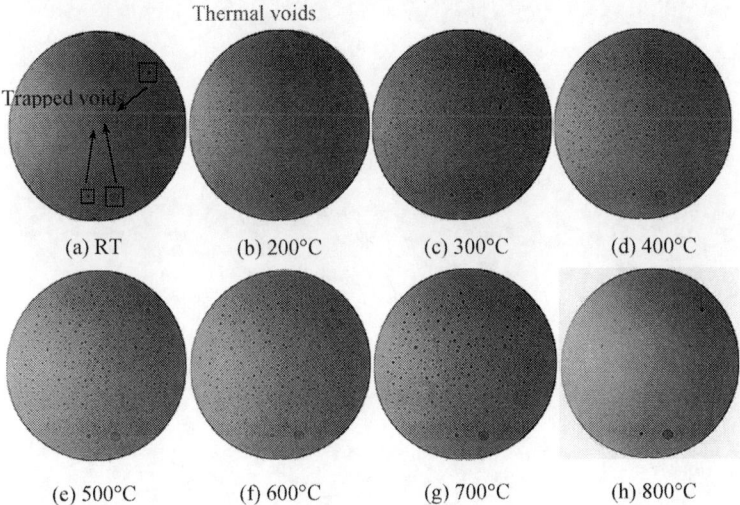

Figure 1. IR images of the 200-mm bonded pairs prepared without plasma treatment at RT and annealing from 200 to 800°C (Group A). The thermal voids dominate during annealing process and they vanish at 800°C.

The bonded wafer pairs of Group B1 were prepared with O_2 plasma treatment for 60 s. At 200°C, there was no thermal void in the IR image [see Figure 2(b)]. It means the organic contaminants may be removed by O_2 plasma treatment according to the reaction as below:

$$C_xH_y + (O^*, O) \rightarrow CO_2 + H_2O \qquad [2]$$

where O^* refers to the oxygen radicals generated during the oxygen plasma treatment. The annealing voids started arising in non-peripheral regions at temperature higher than 400°C. A large number of voids were observed at 700°C, as shown in Figure 2(g). Many voids still remained even at 800°C.

The bonded wafer pairs of Group C were prepared with $(O_2 + CF_4)$ RIE plasma treatment. Group C1, C2, C3 and C4 were treated by the plasma for 120, 60, 30 and 10 s, respectively. For long treatment of 120 s (Group C1), a number of annealing voids were generated from 400 to 700°C. For $(O_2 + CF_4)$ plasma treatment of 60 s (Group C2), the void densities were fewer at each temperature. The void density of Group C2 is 4.8% at

700°C. This is much fewer than that of Group B1 (12.6% at 700°C), although the plasma treatment time of Group C2 and B1 is identical. It reveals that adding CF_4 to O_2 plasma can mitigate void formation.

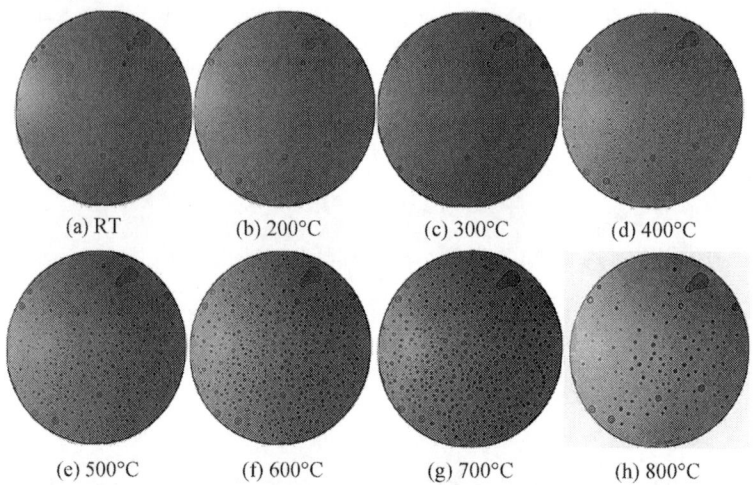

(a) RT　　　　(b) 200°C　　　　(c) 300°C　　　　(d) 400°C

(e) 500°C　　　　(f) 600°C　　　　(g) 700°C　　　　(h) 800°C

Figure 2. IR images of the 200-mm bonded pairs prepared with O_2 RIE plasma for 60 s (Group B1) and a large number of annealing voids were generated from 400 to 700°C.

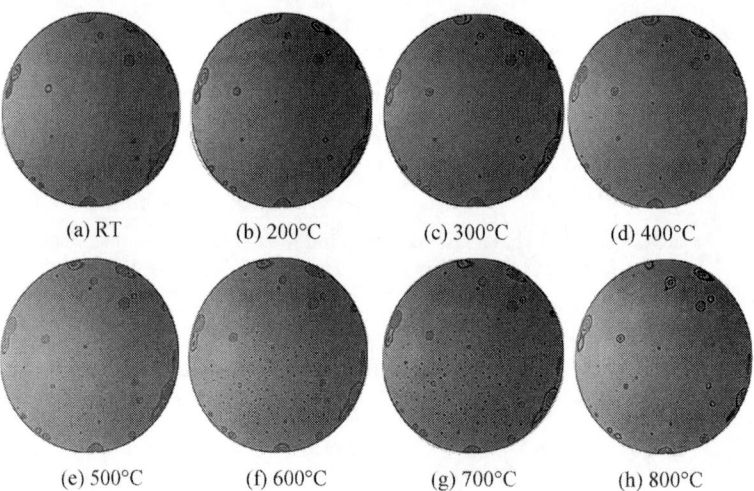

(a) RT　　　　(b) 200°C　　　　(c) 300°C　　　　(d) 400°C

(e) 500°C　　　　(f) 600°C　　　　(g) 700°C　　　　(h) 800°C

Figure 3. IR images of the 200-mm bonded pairs prepared with $(O_2 + CF_4)$ RIE plasma for 60 s (Group C2). Many annealing voids are generated from 400 to 700°C but they are fewer than those of Group B1.

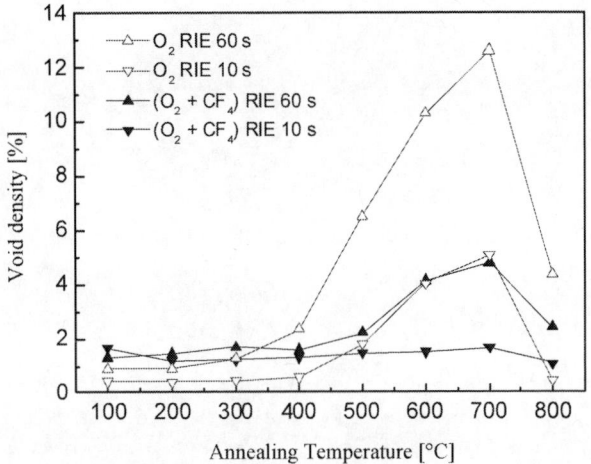

Figure 4. Void densities for the bonded pairs prepared with O_2 RIE plasma and (O_2 + CF_4) RIE plasma treatments for 60 s or 10 s (Group B1 vs. Group C2 and Group B2 vs. Group C4).

Figure 4 shows a comparison of void densities of Group B1, B2, C2 and C4. For Group B2, the bonded wafer pairs were prepared with O_2 plasma treatment for 10 s. Although there are still many voids at 700°C, the void density of Group B2 (5.0%) at 700°C was much fewer than that of Group B1 (12.6% at 700°C). It reveals that a shorter O2 plasma treatment can decrease the annealing voids at bonding interface.

The bonded wafer pairs of Group C4 were prepared with (O_2 + CF_4) RIE plasma treatment for 10 s. Voids of Group C4 from 200 to 800°C are fewer than any other bonded pairs. It may benefit from the CF_4 containing plasma and the short treatment time (~10 s). The void density at 700°C is only 1.7%. Since initial trapped voids at room temperature began to disappear at 800°C [see Figure 5(h)], the potential annealing voids would not be generated continuously at higher temperatures (7). To our knowledge, there is still no criterion for the void density of void-free bonding. For this case, the voids during the annealing steps are mainly initial trapped voids, caused by initial particle contamination in our class 10 K clean room. It is nearly void-free during whole annealing process. The void density may be minimized in a better clean room environment.

Figure 7 shows on the water contact angles on the silicon surfaces as a function of plasma treatment time. The generation conditions of O_2 RIE plasma and (O_2 + CF_4) RIE plasma were identical to previous conditions. CF_4 flow is 1.25 sccm. The water contact angle on silicon surface without plasma treatment is ~44.9°. After treated by O_2 RIE plasma for 10 s, the contact angle decreases to ~6° significantly. It means the surface is fairly hydrophilic. And longer O_2 RIE treatment time for 60 s, the contact angles did not degrease continuously and keep at ~4.8°. For the treatment time of 120 s, the contact angle increase to ~8.8°.On the other hand, the silicon surface treated by (O_2 + CF_4) RIE plasma shows a worse wetting character than that of O_2 RIE plasma. This means the adding of CF_4 to plasma renders the oxidized silicon surface less hydrophilic.

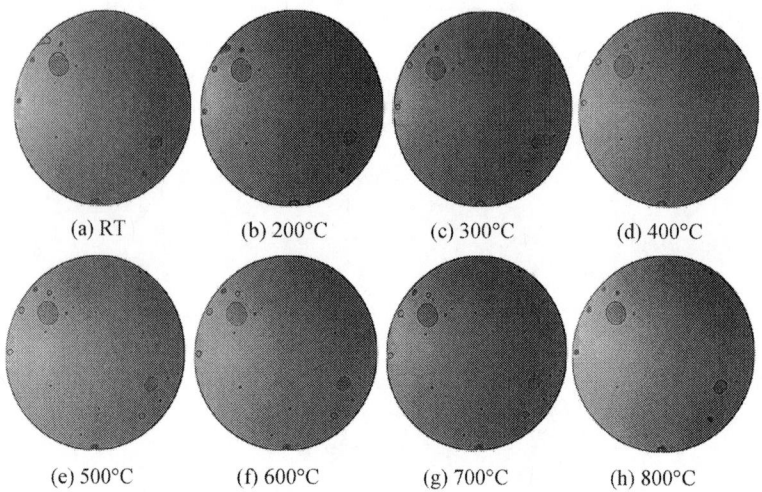

(a) RT   (b) 200°C   (c) 300°C   (d) 400°C

(e) 500°C   (f) 600°C   (g) 700°C   (h) 800°C

Figure 5. IR images of the 200-mm bonded pairs prepared with (O_2 + CF_4) RIE plasma for 10 s (Group C4). Very few small annealing voids are generated and most of voids are initial air trapped voids.

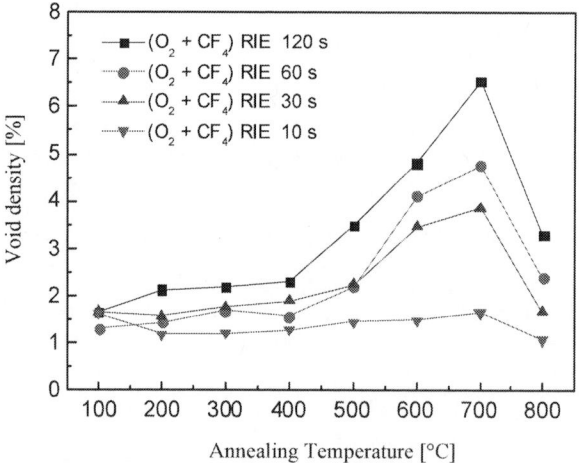

Figure 6. Void densities as a function of annealing temperature for the Si/Si bonded pairs prepared with (O_2 + CF_4) RIE plasma treatments (Group C1, C2, C3 and C4). With shorter plasma treatment time, fewer voids are generated.

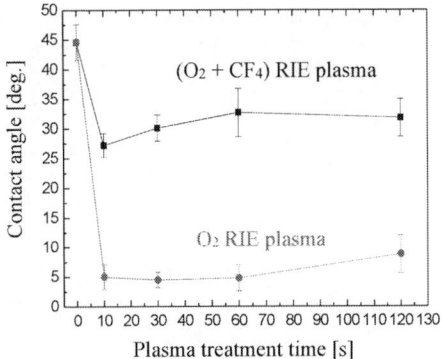

Figure 7. Water contact angles on the silicon surfaces as a function of plasma treatment time.

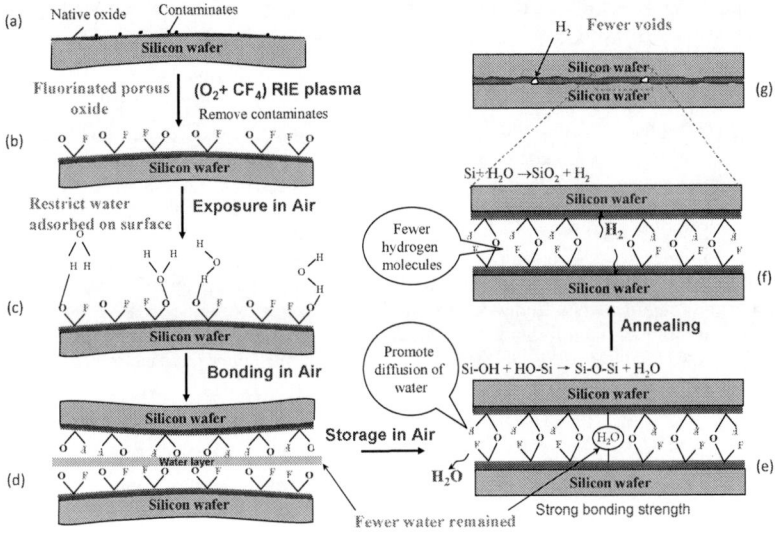

Figure 8. Schematic of the mechanism of fluorine containing plasma activated bonding and its void formation.

In our experiment, adding a small amount of CF4 to O2 plasma leads to increasing of bonding strength in our previous study (4). It is known that fluorine radicals may generated from the discharged CF4. These fluorine radicals are so active and very easily react with the native oxide layer to form a fluorinated oxide layer. Thus, adding CF_4 into O_2 plasma could restrict the amount of water molecules adsorbed on surface.

Figure 8 shows the proposed mechanism of fluorine containing plasma activated bonding and its void formation. Since the fluorinated oxide layers restricts the water molecules on the surfaces, fewer voids are generated in annealing process for the bonded wafer pairs by $(O_2 + CF_4)$ plasma than by O_2 plasma. During annealing process, hydrogen gas and remaining water molecules result in annealing voids at elevated temperatures. In our study, the RIE plasma treatment time is a vital parameter for achieving voidless bonding. Long RIE plasma treatment enables us to remove organic contaminants and active the surface sufficiently, but it may produce more porous surfaces and introduce undetected microdefects (7). These microdefects will act as the bubble nucleation centers at the interface. Hydrogen desorbed at a high temperature and remaining water molecules may aggregate around microdefects, and many annealing voids are generated. In addition, the bonded pairs are annealed at high temperatures immediately after bonding at room temperature. The rapid enhancement of the bonding strength blocks the diffusion of water molecules and gas along the bonding interface in a short time.

As a result, RIE plasma treatment time should be optimized carefully. In our case, O_2 RIE treatment for 10 s seems enough to remove the organic contaminants because nearly no thermal voids are generated at 200°C. Moreover, the short RIE treatment process should bring about little damage on surfaces, which is suitable for the bonding of electrically active substrates.

Conclusion

We found adding CF_4 into oxygen plasma could effectively mitigate void formation in subsequent annealing process. Moreover, the plasma treatment time is a vital parameter to optimize the bonding strength as well as the void generation. Long plasma treatment (e.g. >60 s) may produce more porous surfaces and introduce undetected microdefects. Consequently, a short fluorine containing plasma treatment (e.g. ~10 s) is suitable to remove organic contaminants, it is thus nearly void-free at the bonding interfaces during heating from 200 to 800°C. In practice, an optimum treatment time depends on the application requirements. Compared to the O_2 plasma treatment, the roles of the $(O_2 + CF_4)$ plasma in the void formation is also discussed. We believe that the voidless room-temperature bonding is feasible to be realized by the fluorine containing plasma activation.

Acknowledgments

The authors are grateful for the financial support from the National Science Foundation of China (Grant No. 51505106).

References

1. S. H. Christiansen, R. Singh, and U. Gösele, *Proc. IEEE*, **94**(12), 2060, (2006).
2. V. Masteika, J. Kowal, N. S. J. Braithwaite, and T. Rogers, *ECS J. Solid State Sci. Technol.*, **3**(4), Q42, (2014).
3. X. Ma, W. Liu, Z. Song, W. Li, and C. Lin, *J. Vac. Sci. Technol. B*, **25**(1), 229, (2007).

4. C. Wang and T. Suga, *J. Electrochem. Soc.*, **158**(5), H525, (2011).
5. C. Wang and T. Suga, *Microelectron. Reliab.*, **52**(2), 347, (2012).
6. Q.-Y. Tong and U. Gösele, *Semiconductor Wafer Bonding: Science and Technology*, Wiley, New York (1999).
7. X. X. Zhang and J. P. Raskin, *J. Microelectromech. Syst.*, **14**(2), 368, (2005).

Edge Water Penetration in Direct Bonding Interface

F. Rieutord[a,c], S. Tardif[a,c], D.Landru[d], O. Kononchuk[d], V. Larrey[b,c], H. Moriceau[b,c], M. Tedjini[b,c] and F.Fournel[b,c]

a- CEA, INAC, Minatec Campus, 17 rue des Martyrs, F-38054 Grenoble, France
b- CEA, Leti, Minatec Campus, 17 rue des Martyrs, F-38054 Grenoble, France
c- Univ. Grenoble Alpes, F-38000 Grenoble, France
d- SOITEC , Parc Technologique des Fontaines, F-38500 Bernin, France

We report here on measurements demonstrating that water is able to enter a hydrophilic wafer bonding assembly from the edge. The driving forces for water entry are capillary forces driven by the negative Laplace pressure. The characteristic time for water to penetrate the interface is consistent with standard values for water viscosity and gap width.

Introduction

Direct bonding of bulk materials always leaves a gap between the two solids due to the roughness of the contacting surfaces. The height of this gap is the result of a balance between attractive interaction between the two solids (usually van der Waals forces) and repulsive interactions due to the compression of the asperities. With standard surface preparations that result in a Gaussian distribution of asperity heights with a rms of 0.2nm typically, the equilibrium distance between the solids is of the order of 1nm(1). When performing hydrophilic bonding surface preparation, this gap is partly filled by water adsorbed on the hydroxyl terminated silicon surfaces(2). While some water may have some beneficial effects increasing the adherence energy at ambient temperature, water management upon annealing is also a key issue as water will eventually react with silicon producing hydrogen gas, whose pressure may lead to debonding and defects(3). It is thus essential to optimize the amount of water incorporated into the bonding and in any case avoid saturation. We are then left with a gap formed by two highly hydrophilic surfaces that is not saturated with water.

On the other hand, it is known that flow in micro- and nanochannels is considerably slowed down by viscosity effects (4). Hence the problem of water flow from outside to the gap is an issue.

Experimental observations

Observations of annealed direct silicon/silicon bondings after storage times ranging from minutes to several days have shown a distribution of defects along a rim whose extent increases toward the center of the assembly (5). It is clear from these data that water has flowed from the edge of the wafers to the center, contributing to an excess of water responsible for a large number of fine bubbles (Fig.1).

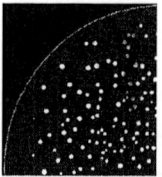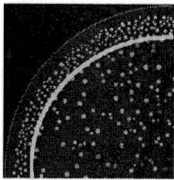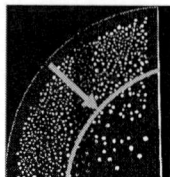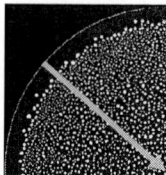

Figure 1. Evolution of the defect repartition for a Si/Si hydrophilic bonding annealed after different storage time in a wet atmosphere. Fom left to right the storage time is 0, 2, 13 and 73 days.

The fact that water is able to flow from the wafer bevel of the assembly to the center has been directly evidenced using interfacial X-ray reflectivity (6). Using hard X-rays, we could directly measure the reflection coefficient of the interface between the wafers. The reflectivity is a direct function of the electron density profile and thus varies when the water filling changes. Due to the short wavelength of X-rays, this allows one to measure directly the mean distance between the wafers (nanometric gap width) and also the filling of the gap (interference contrast). To simplify the experiment, a stripe with constant width has been cut in a 300mm bonding assembly. Sides of the stripe were protected using bee wax to avoid water intake from all sides but the bevel of the wafers.

Reflectivity results. At time t=0, a large drop of deionized water was placed against the assembly bevel to saturate the interface with water. Then reflectivity was periodically recorded at different distances from the edge. An example is shown on fig.2 where the reflectivity at 17mm from the edge is recorded over a 50000s time span. From these data, the arrival time for the imbibition front can be measured accurately.

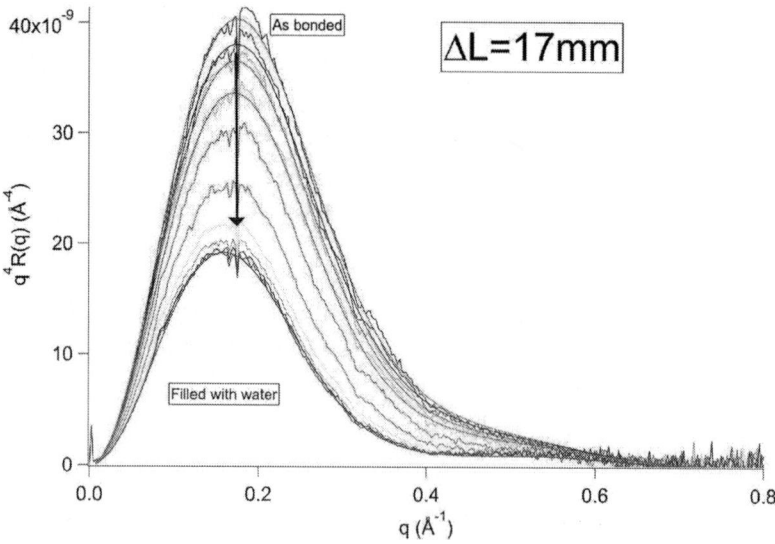

Figure 2. Evolution of the reflectivity of a Si/Si hydrophilic bonding interface with time. The reflectivity contrast (plotted here as $q^4R(q)$) decreases as the water content of the interface increases. The measurement point is located 17mm from the bevel and the time lag between two curves is typically 6000s.

Derivation of the water front advance law

The progression of the water in a narrow gap can be derived from classical fluid mechanics. It is supposed here that the fluid movement is given by a balance between the negative Laplace pressure and the viscous drag of the liquid flowing between the plates. Writing the Navier Stokes equation

$$-\nabla P + \eta \Delta v = 0 \qquad [1]$$

The flow velocity is described using a Poiseuille distribution assuming v=0 at the walls (z=±h/2).

$$v_x = \frac{3}{2}v_F \frac{\left(\left(\frac{h}{2}\right)^2 - z^2\right)}{\left(\frac{h}{2}\right)^2} \qquad [2]$$

The pressure gradient $\partial P/\partial x$ is given calculating the Laplace pressure $P = \gamma /(h/2)$ assuming a zero contact angle and assuming a constant gradient $\dfrac{\partial P}{\partial x} = -\dfrac{P}{\ell}$

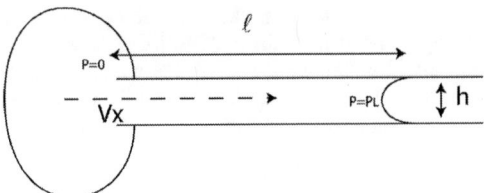

Figure 3: Sketch of the capillary flow in the gap between the two wafers. Due to the hydrophilic surfaces, the Laplace pressure is strongly negative and tend to suck the water into the gap.

The equation reads finally

$$\frac{\partial \ell^2}{\partial t} = \frac{\gamma h}{3\eta} \qquad [3]$$

This equation is the Lucas-Washburn equation (7) stating that the square of the wet length varies linearly with time, analogous to a diffusion equation. This dependence has been experimentally verified both from the X-ray measurements and the defect rim progression (Fig.4).

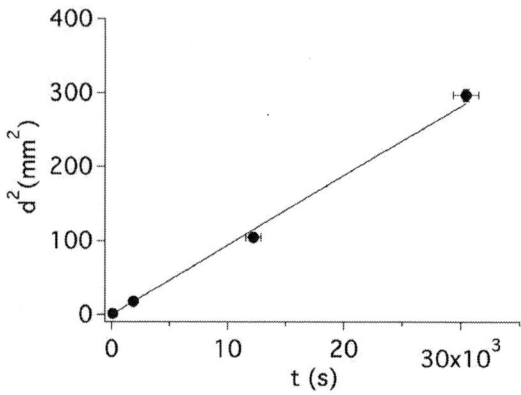

Figure 4: Experimental dependence of the waterfront progression as a function of time.

The slope coefficient for the dependence is on the order of $10^{-8} m^2/s$, to be compared to the right-hand side of equation [3]. Assuming standard value for water surface tension ($\gamma=72mN/m$) and viscosity ($\eta=10^{-3}Pa.s$) this would give a gap width of typically 0.45nm, i.e. about half the value measured by X-ray reflectivity. This fact can be understood observing that, after bonding, the gap is already partially hydrated, a fact confirmed by X-ray reflectivity. A simple explanation for our results is that this adsorbed water is not mobile, reducing the actual width over which the water can flow. Of course, the description of the gap as two separated flat planes is far from ideal and other factors like the presence of bridges through the gap at the asperity contact points have also to be taken into account, reducing the flow possibilities.

Acknowledgments

We wish to thank BM32 staff for assistance during the synchrotron experiment. Assistance from F. Madeira and C. Morales is gratefully acknowledged.

References

1. Rieutord, F., L. Capello, R. Beneyton, C. Morales, A.-M. Charvet, and H. Moriceau, *Semiconductor Wafer Bonding 9: Science, Technology, And Applications ECS Transactions*, edited by H. Baumgart, S. Bengtsson, H.

Moriceau, K.D. Hobart, T. Suga, and C. Colinge, 3:205–15. ECS Transactions, 2006.
2. Ventosa, C., C. Morales, L. Libralesso, F. Fournel, A.M. Papon, D. Lafond, H. Moriceau, J.D. Penot, and F. Rieutord, *Electron. and Solid State Lett.* **12** (2009): H373.
3. Vincent, S., I. Radu, D. Landru, F. Letertre, and F. Rieutord. *Appl. Phys. Lett.* **94** 101914 (2009).
4. N. R. Tas, J. Haneveld, H. V. Jansen, M. Elwenspoek, and A. van den Berg, Appl. Phys. Lett. **85,** 3274 (2004).
5. M. Tedjini et al., *Wafer bond conference* (2015).
6. M. Tedjini, F. Fournel, H. Moriceau, V. Larrey, D. Landru, O. Kononchuk, S. Tardif, F. Rieutord, *Appl. Phys. Lett.* , submitted (2016).
7. E.W. Washburn, Phys. Rev. **17**, 273 (1921).

168

Locally Measuring the Adhesion of InP Membranes Directly Bonded on Silicon

G. Patriarche[a], K. Pantzas[a], E. Le Bourhis[b], G. Beaudoin[a], and A. Talneau[a]

[a] Centre de Nanosciences et de Nanotechnologies, CNRS, Univ. Paris-Sud, Université Paris-Saclay, C2N – Marcoussis, Marcoussis 91460, FRANCE
[b] Institut P', CNRS – Université de Poitiers – ENSMA, UPR 3346, Futuroscope Chasseneuil 86962, FRANCE

Instrumented nano-indentation is combined ex situ with Atomic Force Microscopy and Scanning Transmission Electron Microscopy in a nano-scale analog of the Double-Cantilever Beam experiment to evaluate the adhesion of InP membranes directly bonded on Silicon. This method is shown to provide reliable and highly localized measurements of the surface bonding energy of InP on Si. The localized nature of the measurements is taken advantage of to investigate the strength of the bonded interface on areas of the Si substrate that include nano-patterned features typically found in advanced Photonic Integrated Circuits.

Introduction

Over the past few years direct bonding of III-V semiconductors on Si has emerged as a promising alternative to hetero-epitaxy for the hybrid integration of active – amplification, emission in the direct gap III-V - and passive – guiding, switching in the indirect gap Si - optical functions in next-generation photonic integrated circuits (PICs) [1-9]. Such PICs offer a variety of advantages, foremost among which is the dense integration of advanced optical functions using sub-100nm patterns in the Si guiding layer. The processes used to pattern the Si may, however, deteriorate the quality of the hybrid bonded interface. Given the localized nature of such patterns it is highly desirable to have a technique to evaluate this quality on or in the immediate vicinity of these patterned regions.

The quality of a bonded interface, i.e. the strength of adhesion between the III-V and Si is expressed in terms of the surface bonding energy – the energy per unit area required to adiabatically and reversibly separate the two materials. This surface bonding energy can be measured in a variety of experiments during which the hybrid bonded stack is deformed until the interface yields. Among these experiments, the double-cantilever beam experiment (DCB) has been repeatedly shown to yield reliable measurements in the case of Si on Si and InP on Si bonding [4,5]. Nonetheless, given the centimetric samples required to carry out the measurement, it is better suited to wafer-scale measurements of the surface bonding energy.

In the present contribution, the authors discuss a nano-scale analog to the DCB experiment. In this case, instrumented nano-indentation is used to locally deform the InP membrane. Within a certain range of applied indentation loads, the InP membrane is elastically debonded in the vicinity of the indent, forming a blister next to each facet of the indent. The geometry of the blister is recorded using atomic force microscopy and, using well established models for the DCB experiment, the surface bonding energy is measured. Scanning transmission electron microscopy is also used to understand the underlying mechanism responsible for the debonding and to correlate the geometrical features of the blister with those of the buried debonding crack that are required to measure the surface bonding energy [10-12].

The method is applied to InP membranes bonded to Si using a variety of bonding methods. The application of the method on InP bonded on sub-100nm patterned Si is also discussed. Finally, the method is applied to other hybrid stacks of interest, such as InP bonded on GaAs.

Theory

In the DCB experiment [13], a thin blade is inserted between the two bonded materials, initiating a crack that propagates along the bonded interface until it reaches a certain equilibrium length. This length L and the separation between the two materials – imposed by the thickness of the inserted blade δ – can be shown to relate to the surface bonding energy G in the following manner:

$$G = \frac{3\delta^2}{8L^4}\left(\frac{E_{InP}h_{InP}^3 E_{Si}h_{Si}^3}{E_{InP}h_{InP}^3 + E_{Si}h_{Si}^3}\right)$$

Here, h is the thickness of each of the two materials and E its Young's modulus. This equation holds if all stresses are relaxed at the crack tip – a hypothesis that is not true in most experimental cases. A variety of models have been proposed to take into account stress at the crack tip. Among these models, the semi-empirical formula of Gillis and Gillman [14]:

$$G = \frac{3\delta^2}{8L^4}\frac{F_{InP}^2/E_{InP}h_{InP}^3 + F_{Si}^2/E_{Si}h_{Si}^3}{(F_{InP}^3/E_{InP}h_{InP}^3 + F_{Si}^3/E_{Si}h_{Si}^3)^2},$$

Where

$$F_i = 1 + 3c\frac{h_i}{L} + \frac{3}{5}(1+\nu_i)\frac{h_i^2}{L^2},$$

Has been shown to yield the best results. This formula is used in the present paper.

Experiment

For the puprposes of the paper, ~400nm thick InP membranes were grown on top of an InGaAs etch-stop layer grown lattice-matched to InP substrates using metal-organic chemical vapor deposition. The membranes were bonded on Si wafers using direct oxide-free bonding and direct oxide-mediated bonding with either a thin (5 to 10nm) or thick (~60nm) SiO2 intermediate layer. References [8,9,4] provide details on the oxide-free, thin and thick oxide-mediated boding procedures, respectively. After bonding, the InP substrates and InGaAs etch-stop layers were selectively removed using wet chemical etching.

The InP membranes were then deformed in a Nano-hardness tester from CSM Switzerland using a Berkovich diamond tip. The calibration procedure suggested by Oliver and Pharr [15] was used to correct for the load-frame compliance of the apparatus and the imperfections of the shape of the indenter tip. Indentation loads between 5mN and 100mN were applied on the samples.

All AFM images were acquired using a Veeco Dimension V atomic force microscope. All images are 512 pixel by 512 pixel, for an image range of 6 μm by 6μm. Therefore, the pixel step is 15 nm. The precision along the Z direction is 1 Å, however, it is not possible to determine the blister height with that precision, as the RMS roughness of the chemically revealed surface of InP is 1nm. The blister height h_b and blister length L_b are determined from AFM profiles using the boundary conditions of the DCB problem [13]. The first condition gives $Z (X = X0) = h_b$ at X_0 such that:

$$\frac{dZ}{dX}(X = X_0) \neq 0,$$

$$\frac{d^2Z}{dX^2}(X = X_0) = 0.$$

The second condition gives $L_b = X_b - X_0$, where X_b satisfies:

$$\frac{dZ}{dX}(X = X_b) = 0,$$

$$\frac{d^2Z}{dX^2}(X = X_b) = 0.$$

Details on the resolution and precision of the blister features determined in this manner and the resulting uncertainty on the measurement of the surface bonding energy are given in Reference [12].

All lamellae used for STEM were prepared focused ion beam (FIB) etching and ion milling. Prior to FIB etching, the samples were coated with a 50nm thck carbon layer, followed by a 100nm thick silicon nitride layer. These layers serve to protect the sample surface and the shape of the blister during the preparation of the TEM lamella. The cross-sections were carefully position to cross the center of the indent. Scanning transmission electron microscopy experiments were performed in an aberration-corrected JEOL 2200FS microscope, operating at 200kV with a probe current of 150pA, and a probe size of 0.12nm at the full width at half maximum. The convergence half-angle of the probe was 30mrad and the detection inner and outer half-angles of high-angle annular dark field (HAADF) detector were 100mrad and 170mrad, respectively. To accurately determine the debonding-crack length and debonding-crack opening, 4096 by 4096 piwel STEM images were acquired at a magnification of 120000. The acquisition time for the images was 80s, during which no drift was observed in the image. In these conditions the image pixel step is 3 Angstrom.

Results and Discussion

Indentation of the InP membranes at high loads revealed a second characteristic pop-in, distinct from the pop-in corresponding to the elasto-plastic transition in InP and occurring at 0.5mN. The second pop-in occurs at loads ranging between 60mN and 80mN. The average load and the corresponding deviation are linked to the type of bonding (oxide-free, thin or thick oxide-mediated). A statistical Weibul analysis performed on this second pop-in gives qualitative information on the fragility of the bonded stack. It does not, however, allow one to measure the surface bonding energy.

Lower indentation loads – typically 10mN to 30mN – were found to be better suited when measuring the surface bonding energy. Indeed, within this range of loads an uncracked blister forms next to each facet of the indent. A STEM image of the cross-section of a blister obtained at a load of 10mN is shown in InP bonded oxide-free to Si is shown in Figure 1. The cross-section shows two distinct regions in the InP membrane: one immediately beneath the indent, dense with dislocations induced by the indenter; a second one next to this zone where the InP membrane is elastically debonded from the the Si substrate. At these loads the Si substrate is not deformed. A closer inspection of the in the two regions in InP reveal that InP in the dislocation-dense zone is tilted by as much as 8° with respect to the underlying Si – to which it is still firmly bonded – whereas InP beyond this zone is aligned to within 0.1° with the underlying Si. This observed rotation is the motor for the observed debonding. Indeed, dislocations in InP slip along (111) planes. As they do not cross over into Si, they accumulate at the bonded interface. Given the symmetry of the triagonal Berkovitch tip, more dislocations tend to accumulate on facets directly opposite an edge of the indenter tip. The edge component of this dislocation induces a torque that ultimate reaches a value high enough to elastically debond the InP membrane from the Si substrate far from the indent. As stress between the densely dislocated portion of the membrane and the remainder is relaxed through the dislocations, the portion of the membrane from the last dislocation to the debonding crack tip fulfills the hypotheses of the DCB experiment.

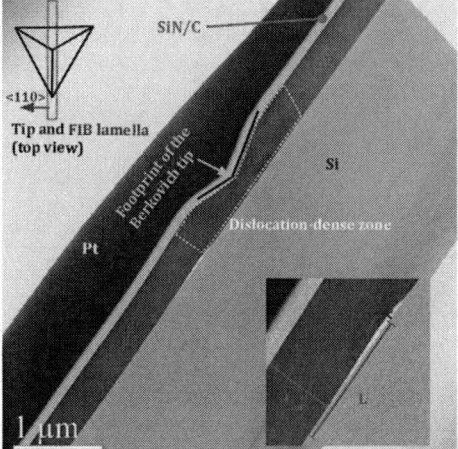

Figure 1. Low-magnification BF-STEM image of an InP/Si hybrid bonded stack. Direct oxide-free bonding was used in the present case. The image shows a cross-section of the blisters that forms during the nano-indentation experiment. The imprint of the indenter tip is clearly visible in the image, as is the dislocation-dense zone, located immediately below the imprint. The inset shows a magnification of the deboniding crack that starts at the left edge of the dislocation-dense zone. From Reference [10].

The explanation presented above can be verified by comparing the measured rotation against the expected rotation, derived from the follozing derivation. In the limit of small angles, the mean lateral spacing D between adjacent dislocations in a tilt grain boundary is given by

$$D = \frac{b}{\theta},$$

where b is the magnitude of the edge component of the dislocations' Burders vector and θ is the rotation angle of the membrane. In the case of InP this equation becomes

$$D = \frac{a_{InP}}{\sqrt{2}\theta}.$$

In the dislocation dense zone of an InP membrane indented at a load of 10mN, the average spacing between adjacent dislocations 5 Angstrom, yielding an angle of 8°. The average rotation of the InP membrane with respect to Si was measured from atomically resolved images in the same membrane using Geometric Phase Analysis. Geometric Phase Analysis yields an angle of 8.1°, in good agreement with the expected value. More details on this derivation can be found in Reference [10].

AFM images of blisters are shown in Figure 2. The blisters are shown to be symmetric around the median of the facet they are next to. The blisters in each image are not identical across all three facets of the indenter. This disymetry is linked to the geometry of the Berkovitch tip. Indeed, the tip is a pyramid with equilateral triangle at its base, i.e. the indenter has a trigonal symmetry, whereas the InP lattice has a tetragonal symmetry. Therefore, only one of the three facets of the indenter is most closely aligned with one of the 110 crystallographic directions in InP. The facet that is the most closely aligned will induce the most dislocation and result in the biggest blister. If the indenter tip is changed – if a tetragonal Vivker's tip is, for instance, used – the blisters may not be as large or, even, created. Small misalignments are still observed as the tip is manually aligned with respect to a cleaved facet of the InP membrane within 1 or 2°.

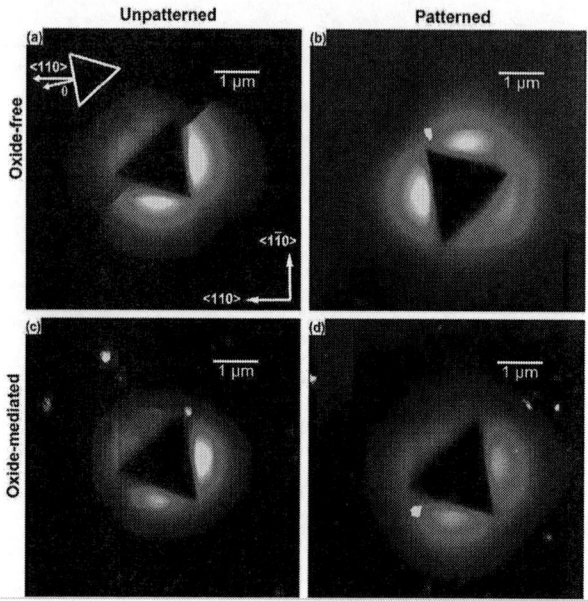

Figure 2. AFM images (6 x 6µm) of debonding blisters in InP bonded (a) oxide-free on bare Si, (b) oxide-free on patterned Si, (c) thin oxide-mediated on bare Si, and (d) thin oxide-mediated on patterned Si. The height scale for all images is 55nm. After Reference [12].

In Rerence [11], the authors showed that a direct relation exists between the geometric parameters of the blister (blister height, blister radius) and the buried debonding crack opening and debonding-crack length. More specifically, the height of the blister is exactly the height, while the deboning crack length L relates to the blister radius Lb as follows:

$$L = L_b - 3ch_{InP}$$

This discrepancy is related to the fact that stresses exist at the debonding crack tip that elastically deform the InP membrane beyond the length L. A more detailed discussion on this is given in Reference [12].

Using the AFM images and the equations described in Section Theory, the surface bonding energies for InP bonded to Si were measured and found are ~1J/m² in both direct oxide-free and oxide-mediated bonding on bare Si. A higher bonding energy is expected to be attained at higher annealing temperatures – such temepatures may, however, be undesirable in CMOS-compatible process flows for the targeted PICs. Moreover, in the case of InP, where the fracture energy is 1.5J/m², energies higher than this limit cannot be recorded using methods that debond the membrane, as a membrane that will deform to accommodate for such energies will yield once the crack opening and debonding crack length impose an elastic deformation energy higher than 1.5 J/m². Nonetheless, a surface bonding energy equal or higher than $1 J/m^2$ indicates strong covalent bonding that result in a mechanically stable interface, suitable for most PIC applications.

In the case of InP membranes bonded to patterned Si, the debonding blisters where found to extend farther away from the indent for an equivalent height. This observation, a priori, indicates a lower debonding energy. This is not, however, true as the surface bonding energy has to be related to the actual surface available for bonding. When bonding to patterned Si – the patterns are square arrays of holes – the surface available for bonding is approximately 78.7% of the Si surface. The surface bonding energy measured in this case, however, is 57% lower than the one measured on bare Si for the same bonding process. A closer inspection of cross-sections of the samples in STEM revealed that the edges of the patterns where rounded off, and that actual surface available for bonding is indeed closer to this value.

Conclusion

The method presented in this paper, combining instrumented nano-indentation in conjunction with AFM or STEM, has been shown to provide reliable, precise, and localized measurements of the surface bonding energy of InP bonded to Si in a variety of bonding schemes. The localized nature of the measurement method render particularly useful in evaluating the adhesion of InP to nano-patterned Si – configurations that prevail in advanced PICs. The underlying mechanism for debonding relies on mechanical properties that InP shares with other III-V semiconductors. The method presented here can, therefore, be applied to other III-V/Si or III-V/III-V hybrid bonded stacks.

Acknowledgments

The authors would like to gratefully acknowledge funding from the CNRS RENATECH network and the Agence Nationale de la Recherche projects COHEDIO and ANTIPODE.

References

1. G. Roelkensn, J. Broukaert, D. Van Thourhout, R. Baets, R. Notzel, M. Smit, *J. Electrochem. Soc.*, **153**, G1015 (2006).
2. M. Lamponi, S. Keyvaninia, C. Jany, F. Poingt, F. Lelarge, G. de Valcourt, G. Roelkens, D. Van Thourhout, S. Messaoudence, G. H. Duan, *IEEE Photonics Technol. Lett.*, **24**, 76 (2012)
3. P. Maszara, G. Goetz, A. Caviglia, J. B. McMitterick, *J. Appl. Phys.*, **64**, 4943 (1988)
4. F. Fournel, L. Continni, C. Morales, J. Da Fonseca, H. Moriceau, F. Rieutord, A. Barthelemy, I. Radu, *J. Appl. Phys.*, **111**, 104907 (2012)
5. D. Pasquariello, K. Hjort, *J. Electrochem. Soc.*, **147**, 2343 (2000).
6. Q. Y. Tong, Q. Gan, G. Hudson, G. Fountain, P. ENquist, *Qppl. Phys. Lett.*, **84**, 732 (2004)
7. D. Liang, D. C. Chapman, Y. Li, D. C. Oakley, T. Napoleone, P. W. Joudawlkis, C. Brubaker, C. Mann, H. Bar, O. Raday, J. E. Bowers,*Appl. Phys. A*, **103**, 213 (2011)
8. A. Talneau, C. Roblin, A. Itawi, O. Mauhuin, L. Largeau, G. Beaudoin, I. Sagnes, G. Patriarche, *Appl. Phys. Lett.*, **102**, 212101 (2013)
9. A. Itawi, K. Pantzas, I. Sagnes, G. Patriarche, A. Talneau, *J. Vac. Sci. Technol. B*, **32**, 0221201 (2014)
10. K. Pantzas, G. Patriarche, E. Le Bourhis, D. Troadec, A. Itawi, G. Beaudoin, I. Sagnes, A. Talneau, *Appl. Phys. Lett.*, **103**, 081901 (2013)
11. K. Pantzas, E. Le Bourhis, G. Patriarche, A. Itawi, G. Beaudoin, I. Sagnes, A. Talneau, *Eur. J. Appl. Phys.*, **65**, 20702 (2014)
12. K. Pantzas, E. Le Bourhis, G. Patriarche, D. Troadec, G. Beaudoin, A. Itawi, I. Sagnes, A. Talneau, *Nanotechnology*, **27**, 115707 (2016)
13. D. Maugis, *Contact, Adhesion and Rupture of Elastic Solids*, Springer, Berlin (1997).
14. P. Gillis and J. Gilman, *J. Appl. Phys.*, **35**, 647-658 (1964)
15. W. C. Oliver and G. M. Pharr, *J. Mater. Res.*, **7**, 1564-1583 (1992)

Chapter 5

Photonic Integration & Layer Transfer Technologies

Heterogeneous Photonic Integration by Direct Wafer Bonding

Michael L. Davenport[a], Lin Chang[a], Duanni Huang[a], Nicolas Volet[a], and John E. Bowers[a]

[a] Department of Electrical and Computer Engineering, University of California, Santa Barbara, CA 93106, USA

Direct wafer bonding is a powerful technique for combining semiconductor materials that cannot normally be monolithically integrated. It allows combination of heterogeneous materials with different properties onto a single substrate, enabling the production large scale and multi-functional photonic circuits.

Introduction

The silicon (Si) photonics platform has emerged as a promising integration platform for use in many applications, particularly datacenter communications, which is expected to produce demand for large numbers of low cost photonic integrated circuits (PICs). Si-based PICs are almost universally based on Si-on-insulator (SOI) wafers, which are enabled by wafer bonding technology. Si is useful as a waveguide material due to its high index contrast, low intrinsic absorption, and mature manufacturing technology. However, it does not strongly absorb or emit light in wavelength ranges of interest for data transmission. Heterogeneous integration of InP-based materials with Si using wafer bonding technology is able to overcome this limitation and has allowed the realization of a wide range of photonic devices on Si, including lasers, photodetectors, and modulators with high performance (1).

Vertical Integration

Wafer bonding may also be used to produce complex vertically-integrated circuits. A crystalline Si layer transferred to an ultra-low loss silicon nitride (SiN) waveguide allows for integration of highly confined Si photonic waveguide components such as couplers, compact wavelength multiplexers, and micro-ring modulators with high performance and low insertion loss passive components in SiN.

After bonding of the Si layer to the SiN waveguide substrate, the wafer is functionally an SOI wafer, with the SiN waveguide embedded in the buried oxide. Any device that can be realized on an SOI wafer can be integrated with the SiN waveguide, including InP-based heterogeneously integrated devices. Multiple die bonding of InP chips can add lasers, amplifiers, modulators, and photodetectors to the circuit. Lateral tapers of the waveguides allows coupling between the layers with below 0.5 dB loss.

This technology was used to combine a SiN arrayed waveguide grating (AWG) wavelength multiplexer with high-speed 50 Gb/s heterogeneous Si/InP photodetectors to form a 400 Gb/s wavelength division multiplexing (WDM) receiver (2). A cross section of the photodetector is shown in Figure 1.

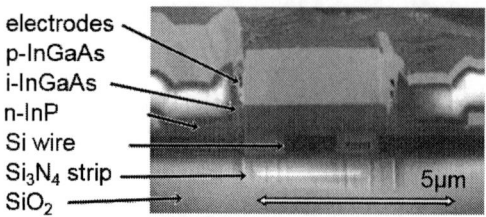

Figure 1: Cross section of a heterogeneous $Si_3N_4/Si/InP$ device

Magnetic Materials

One of the primary advantages of wafer bonding technology is the ability to integrate two materials together with large lattice mismatch. Using the heterogeneous integration approach, we are able to complement the superior wave-guiding qualities of Si and SiN with other desirable optical properties such as efficient optical non-linearities and optical nonreciprocity. The reciprocal nature of Si and other dielectric materials makes it very difficult to realize nonreciprocal devices such as optical isolators and circulators. Optical isolators are a one-way street for photons, and widely used to prevent spurious backreflections from entering a laser cavity, which can degrade or even destroy the laser. Optical circulators are a roundabout for photons in which light is routed in an ordered path, and each input is mapped to exactly one output. Both types of devices require a nonreciprocal medium, which can be achieved by bonding a magneto-optic material such as cerium substituted yttrium iron garnet (Ce:YIG) onto Si.

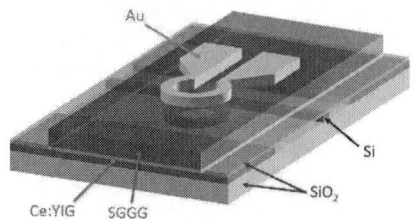

Figure 2: Schematic of the heterogeneous isolator

Our design for an optical isolator uses an all-pass Si microring with the bonded Ce:YIG die on top. The schematic of the device is shown in Figure 2. The bond is performed by contacting the die and the Si together following an oxygen plasma activation, and then strengthened with a 200°C anneal under physical pressure. After this bond, we perform substrate removal using a mechanical polishing technique. Since the Ce:YIG must be magnetized in order for the nonreciprocity to appear, we deposit a metallic microstrip on the backside of the Ce:YIG die. When a current is applied through the microstrip, a resonance wavelength split appears between the clockwise and counterclockwise mode in the ring. Thus, it is possible for the backwards light to be on resonance and eventually dissipated, while the forwards light is off resonance and passed. Here, we see a 0.16-nm resonance wavelength split as well as 32 dB of optical isolation (3), which is shown in Figure 3.

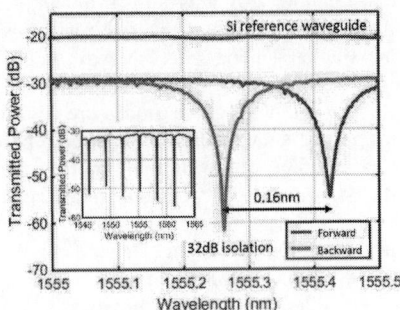

Figure 3: Spectral transmission of the forward and backward propagating light

This design can be expanded to optical circulators by using an add-drop configuration instead of the all-pass microring, as shown below in Figure 4. Here, light flowing from Port 1 to Port 2 is passed through, but light entering from Port 2 is instead dropped to Port 3. Following this analysis, the forwards propagating path in this device is 1→2→3→4→1. This is also shown by the highlighted entries in the table. By reversing the sign of the current and magnetic field, it is possible to reconfigure the circulator and change the flow to 4→3→2→1→4. This reconfigurability is a novel aspect that is made possible by photonic integration, and may lead to new applications in optical switching and interconnects (4).

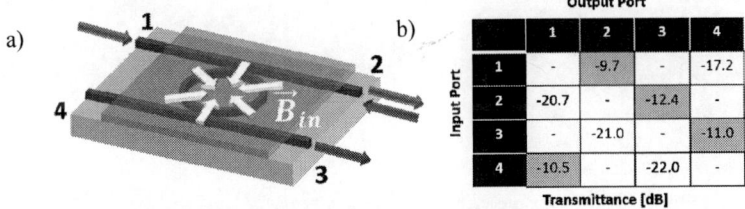

Figure 4: a) Schematic of the circulator, and b) scattering matrix values for the device

Non-linear Optical Materials

Lithium niobate (LN) is a material of paramount importance for non-linear optics, thanks to its large non-linear coefficient, broad transparent window (0.4-5.0 μm) and ferroelectric properties (5). Wavelength converters based on LN have been widely used in different area, e.g. telecommunication (6), quantum optics (7) and self-referencing of optical frequency combs (8). Recently, LN has attracted a lot of attention for use in photonic integrated circuits (PICs), in particular on Si platform, because of the lack second-order optical non-linearity in Si. At the same time, a compact scale of integrated waveguides results in high photon density that effectively enhances non-linear interaction.

The integration of LN in PICs is generally hindered by two major difficulties. One is the waveguide technology. The most commonly used LN waveguides are based on diffusion (9). They only provide a low index contrast and the optical mode is relatively large. Etching LN is another way to form waveguide but it suffers from high loss. More importantly, none of these techniques is compatible with photonic integration. Recently, the LN-on-insulator (LNOI) technology has been successfully demonstrated by ion injection and heterogeneous bonding (9). This allows the implementation of LN thin films with sub-micron thickness on various material platforms. Here based on this technology, we demonstrate a heterogeneous SiN-LN waveguide, with the cross section shown in Figure 5(a). Due to the high index contrast of the structure (~0.6) and the sub-micron thickness of the LN film, the waveguide modes are confined into an area that is more than one order of magnitude smaller than previously reported (10). This directly relates to more than an extra order of magnitude improvement for the conversion efficiency of non-linear effects. Loss of 0.3 dB/cm is demonstrated, which is a record-low value for LN waveguides.

Another important technology for LN is the domain engineering. By periodically inverse the domain to satisfy the quasi-phase matching conditions, the efficiency of non-linear effects can be improved (11). However, the previous bulk poling methods for LN are not suitable for PICs. Here we demonstrate a surface poling method for thin film, which can successfully achieve the domain engineering directly on chip. As shown in Figure 5 (b), a thin-film periodically-poled LN waveguide is fabricated by combining this and the heterogeneous waveguide (12). A second-harmonic generation conversion efficiency 4 times higher than previous bulk PPLN wavelength converters. Processing of this approach is much simpler and compatible with different platforms, as long as LN film is heterogeneous integrated.

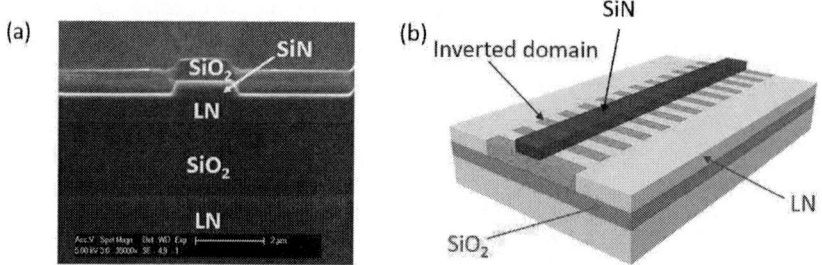

Figure 5: (a) Cross-section SEM of fabricated Si_3N_4-LNOI waveguide, (b) schematic of thin film PPLN wavelength converter.

Conclusion

Wafer bonding technology allows the realization of complex and highly integrated photonic devices. The ability to combine multiple materials such as multiple-die bonding for integration of lasers, modulators, and detectors enables PICs with a wide variety of functions. Vertical integration through bonding of high-quality thin films, such as lithium niobate, crystalline Si, and thermal SiO_2 can be used to construct integrated waveguides with useful properties and high performance. This technology has played an important role in the emergence of the field of Si photonics.

Acknowledgments

This work was supported by DARPA MTO DODOS contract (HR0011-15-C-055), DARPA MTO EPHI contract (HR0011-12-D-0006), and Air Force SBIR contract FA8650-15-M-1920 with Morton Photonics. N. Volet acknowledges support from the Swiss National Science Foundation.

1. M. J. R. Heck *et al.*, *IEEE J. Sel. Top. Quantum Electron.*, **19**, 6100117 (2013).
2. M. Piels, J. F. Bauters, M. L. Davenport, M. J. R. Heck, and J. E. Bowers, *J. Light. Technol.*, **32**, 817–823 (2014).
3. D. Huang P. Pintus, C. Zhang, Y. Shoji, T. Mizumoto, and J. E Bowers, *Opt. Fiber Commun. Conf.* (2016).
4. D. Huang, P. Pintus, C. Zhang, Y. Shoji, T. Mizumoto, and J. E Bowers, *Conf. Lasers Electro-Optics* (2016).
5. Boyd, R. W. Nonlinear Optics, 3rd Ed. (Academic Press, Burlington, 2008).
6. L.K. Oxenløwe, F. Gomez Agis, C. Ware, S. Kurimura, H.C.H. Mulvad, M. Galili, K. Kitamura, H. Nakajima, J. Ichikawa, D. Erasme, A.T. Clausen and P. Jeppesen, *Electron. Lett.* **44**, 5 (2008).
7. E. Saglamyurek, J. Jin, V.B. Verma, M.D. Shaw, F. Marsili, S.W. Nam, D. Oblak and W. Tittel, *Nature Photon.*, **9**, 83–87 (2015).
8. J. D. J. Ost, T. H. Err, C. L. Ecaplain, V. B. Rasch, M. H. P. P. Feiffer and T. J. Kippenberg *Optica,* **2** (2015).
9. L. Chen, Q. Xu, M. G. Wood, and R. M. Reano, *Optica*, **1**, 112 (2014).
10. L. Gui, Ph.D thesis, (2010).
11. G. D. Miller, PhD thesis (1998).
12. L. Chang, Y. Li, N. Volet, L. Wang, J. Peters and J. E. Bowers *Optica,* **3**, 531–535 (2016).

184

Modified Surface Activated Bonding Using Si Intermediate Layer for Bonding and Debonding of Glass Substrates

K. Takeuchi[a], M. Fujino[a], T. Suga[a]

[a] Department of Precision Engineering, University of Tokyo, Tokyo 1138656, Japan

> This study reports the bonding and debonding technology based on the Surface Activated Bonding method for the handling of the thin glass substrates. The glass substrates are bonded by The Si intermediate layer oh under 10 nm thick. From the XPS analysis, the OH groups are formed in the bonding interface and the Si layer's surface is oxidized by the reaction with the OH groups and water at high temperature. Additionally the Si intermediate layer remains on the deposited glass surface. AFM observation also shows the smooth debonded surfaces. From these results, the debonding at the bonding interface is confirmed.

Introduction

The handling of the thin glass substrates is a key issue for the fabrication of the thin display devices such as Liquid Crystal (LC) displays. In order to form the Thin Film Transistors (TFTs) of Si on the glass substrates, the high temperature process is necessary and the thin glass substrates will be damaged in the TFT process. Therefore the thin glass substrates should be bonded to the carrier glass and handled in the fabrication process. Additionally, the bonded thin glass substrates must be released from the carrier glass after the fabrication process. However, the current bonding methods for the glass are not suitable for the debonding of the glass substrates. The adhesive bonding is very popular way to bond [1]-[3], however, it cannot withstand the high temperature process. The hydrophilic bonding is also widely employed for the glass bonding, but the bond strength increases by heating process due to the decomposition of the OH groups in the bonding interface so that the debonding of the glass substrates after the high temperature [4]-[6]. For these reasons, the glass thinning process is widely employed for the display fabrication [7]-[11]. However, the HF etching for the glass thinning is high cost and high environmental burden.

In the previews study, we proposed applying SAB method to the bonding and debonding of the glass substrates for the handling of thin glass substrates [12]. In the conventional SAB method for the glass [13], [14], the bond strength of the glass substrates is so high that the debonding of the glass substrates after the high temperature process is not possible. In order to realize the bonding and debonding after high temperature process, we modified the conventional SAB process. In this modified SAB process, we form the Si intermediate layer on only the one side of the glass substrates. Additionally, we exposed the bonding surfaces to the nitrogen gas that contains water to form the OH groups on the bonding surfaces. The bond strength by this method decreases after heating. However, the effect of the OH groups on the bonding surfaces is not evaluated and the bonding mechanism is not clear. Therefore, in this study, we investigated the oxidization process of the Si intermediate layer and the debonding property.

Method

Materials

We employed the non-alkali glass wafer of 10 cm diameter and 0.5 mm thick for the bonding experiment. The surface roughness RMS of the glass wafer is under 0.5 nm.

Experimental process

The bonding experiments are conducted as shown in Fig.1. First, one of two glass wafers is introduced into the vacuum chamber and the Si layer is deposited on the glass surface by the Ar ion beam sputtering. Meanwhile, another glass wafer is held on the jig in the groove box that is filled with the nitrogen gas that contains water. After deposition of Si, the deposited glass wafer is removed from the vacuum chamber and introduced into the groove box. Next, both wafers are set onto the jig and exposed to the nitrogen gas for one hour. After that the glass wafers are introduced into the vacuum chamber not to trap the air in the bonding interface. Finally the wafers are brought into contact and pressed 5 kN force for five minutes. The bonding is conducted by the crack opening method using the mechanical force.

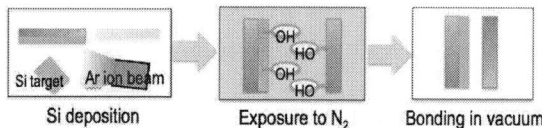

Figure 1. The bonding procedure

Evaluation

To evaluate the bonding property after the high temperature process, the bonded wafer is heated at 450°C for 90 minutes. We conducted the X-ray Photoelectron Spectroscopy (XPS) analysis on the debonded surfaces of the bonded glass wafers. Additionally, the observation of the debonded surface morphology by Atomic Force Microscopy (AFM) was conducted to evaluate the residue of the Si intermediate layer.

Results

XPS analysis

We show the Si 2p peak of the XPS analysis on the debonded glass surfaces before and after heating in Fig.2 and Fig.3. On the deposited side glass surface, as shown in Fig.2, the bulk Si peak is main components of the debonded surface before heating. After heating the Si 2p peak shifts to the higher binding energy. Basically the OH groups are decomposed by heating at 450°C and form H_2O and SiO_2 [15], [16]. These OH groups and water are derived from the exposure and the absorbed water in the glass wafer. Therefore, this peak shift means that, before heating the Si intermediate layer is not oxidized, and after heating the OH groups and the water on the glass surfaces react with the Si layer and form SiO_2 on the Si layer surface.

On the other hand, the opposite side to the Si-deposited surface shows only the SiO_2 and Si-OH bonds before and after heating as shown in Fig.3. It is indicated that no residue of the Si intermediate layer is attached on the glass surface.

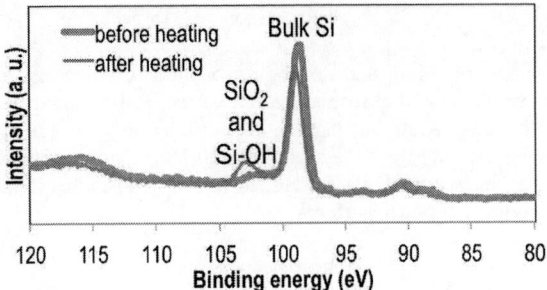

Figure 2. The XPS result of Si 2p peak on the debonded glass surface of the Si-deposited side before and after heating at 450°C for 90 minutes

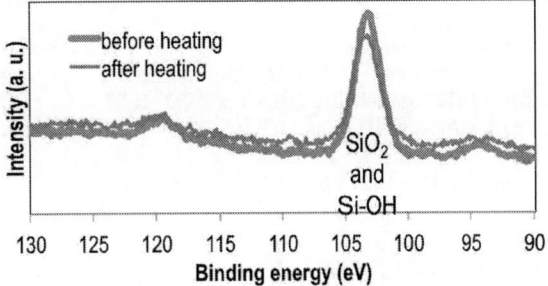

Figure 3. The XPS result of Si 2p peak on the debonded glass surface of the opposite side to the Si-deposited wafer before and after heating at 450°C for 90 minutes

AFM observation

The AFM observation results are shown in Fig.4. The surface roughness of the debonded surfaces are under Ra 0.5 nm and as smooth as the untreated glass surface. This results indicates that the Si intermediate layer is not broken and no residue of the Si layer remain on the opposite side surface to the deposited wafer.

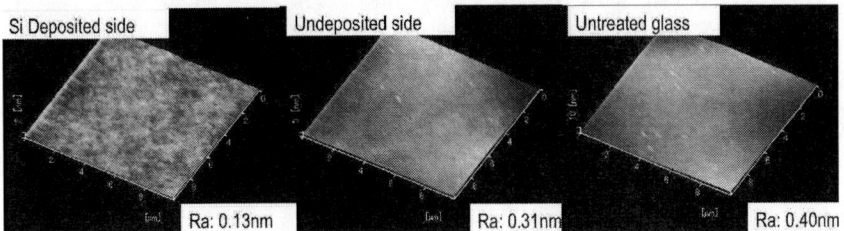

Figure 4. The AFM images of the glass surface and the surface roughness (Left) the debonded surface of the Si-deposited side after heating (Middle) the debonded surface of the opposite side to the Si intermediate layer after heating (Right) the untreated glass wafer

Summary

We established the bonding and debonding method for the glass substrates on the basis of SAB method using the Si intermediate layer. The Si intermediate layer is not oxidized before heating. After heating the OH groups and the absorbed water in the bonding interface react with the Si layer and the Si layer is oxidized. The Si layer does not remain on the opposite side glass wafer to the Si-deposited wafer after debonding, resulting in the smooth debonded surfaces.

Acknowledgments

The bonding experiments were carried out in collaboration with Lan Technical Service Co. Ltd.

References

1. W.-W. Shen, H.-H. Chang, J.-C. Wang, C.-T. Ko, L. Tsai, B. K. Wang, A. Shorey, A. Lee, J. Su, D. Bai et al., Electronic Components and Technology Conference (ECTC), 2015 IEEE 65th. IEEE, 2015, pp. 1652–1657.
2. E. Bosman, J. Missinne, B. Van Hoe, G. Van Steenberge, S. Kalathimekkad, J. Van Erps, I. Milenkov, K. Panajotov, T. Van Gijseghem, P. Dubruel et al., Selected Topics in Quantum Electronics, IEEE Journal of, vol. 17, no. 3, pp. 617–628, 2011.
3. T. Higuchi, Y. Matsuyama, K. Ebata, D. Uchida, and S. Kondo, SID Symposium Digest of Technical Papers, vol. 43, no. 1. Wiley Online Library, 2012, pp. 1372–1374.
4. Y. Le Tiec, C. Ventosa, N. Rochat, F. Fournel, H. Moriceau, L. Clavelier, F. Rieutord, J. Butterbaugh, and I. Radu, Microelectronic Engineering, vol. 88, no. 12, pp. 3432–3436, 2011.
5. F. Fournel, L. Continni, C. Morales, J. Da Fonseca, H. Moriceau, F. Rieutord, A. Barthelemy, and I. Radu, Journal of Applied Physics, vol. 111, no. 10, p. 104907, 2012.
6. C. Sabbione, L. Di Cioccio, L. Vandroux, J.-P. Nieto, and F. Rieutord, Journal of Applied Physics, vol. 112, no. 6, p. 063501, 2012.
7. J. Y. Byun and K. W. Lee, SID Symposium Digest of Technical Papers, vol. 37, no. 1. Wiley Online Library, 2006, pp. 1786–1788.
8. J.-W. Chwu, Y.-C. Liu, L.-J. Chou, J.-Y. Wu, C.-C. Lin, L.-Y. Yeh, B. A. Sventek, W. Lin, A. Hsu, and R. Chen, Japanese Journal of Applied Physics, vol. 46, no. 10R, p. 6688, 2007.
9. J.-W. Chwu, J.-L. Hsu, L.-J. Chou, Y.-C. Liu, C.-C. Lin, C.-W. Chen, M.-S. Chen, and F.-Y. Gan, Japanese Journal of Applied Physics, vol. 47, no. 1R, p. 355, 2008.
10. Y.-C. Liu, J.-Y. Wu, S.-C. Wang, J.-W. Chwu, C.-C. Lin, M.-S. Chen, and T. Huang, SID Symposium Digest of Technical Papers, vol. 39, no. 1. Wiley Online Library, 2008, pp. 249–251.
11. Y.-C. Liu, L.-Y. Yeh, J.-W. Chwu, L.-J. Chou, C.-C. Lin, M.-S. Chen, and T. Huang, Japanese Journal of Applied Physics, vol. 47, no. 3R, p. 1618, 2008.
12. K. Takeuchi, M. Fujino, T. Suga, Electronic Components and Technology Conference (ECTC), 2016 IEEE 66st. IEEE, 2016, pp. 1284–1289.

13. M. Howlader, S. Suehara, and T. Suga, Sensors and Actuators A: Physical, vol. 127, no. 1, pp. 31–36, 2006. R. Smith, *Electrochem. Solid-State Lett.*, **10**, A1 (2007).
14. R. Kondou and T. A. Suga, Electronic Components and Technology Conference (ECTC), 2011 IEEE 61st. IEEE, 2011, pp. 2165–2170.
15. T. Plach, et al. *Journal of Applied Physics*, 2013, 113.9: 094905.
16. C. Ventosa, et al. *Journal of Applied Physics*, 2008, 104.12: 123524.

190

Nanomechanical Analysis of Polydimethylglutarimide Based Lift Off
Resist used for Temporary Bonding and Film Transfers

Yousuf S. Mohammed[1,2], Takashi Matsumae[3], Andrew D. Koehler[5], Tadatomo Suga[3],
Helmut Baumgart[2,4], Karl D. Hobart[5], and A.A. Elmustafa[1,2,*]
[1] Department of Mechanical & Aerospace Engineering, Old Dominion University,
Norfolk, Virginia 23529 USA
[2] The Applied Research Center-Thomas Jefferson National Accelerator Facility, Newport
News VA 23606 USA
[3] The University of Tokyo, Department of Precision Engineering, School of Engineering,
Hongo 7-3-1 Bungyo-ku, Tokyo, Japan
[4] Department of Electrical & Computer Engineering, Old Dominion University, Norfolk,
Virginia 23529 USA
[5] U. S. Naval Research Laboratory, Washington, DC 20375 USA
[*]Corresponding author:aelmusta@odu.edu

Six samples of polydimethylglutarimide based lift off resist (LOR)
were prepared with combinations of two film thicknesses and three
prebake temperatures. The LOR thicknesses are 500 nm and 1000
nm and the prebake temperatures are 80°C, 105°C, 150°C. The
nanomechanical properties were measured using nanoindentation
to determine the modulus and hardness of the six samples. The
experiment retained useful information regarding the effects of
film thickness and prebake temperatures on the hardness and
modulus of the LOR. It was found that the hardness increases as
the film thickness decreases and hardness also increases as the
prebake temperature increases.

Introduction

Temporary bonding has attracted renewed interest for the integration of dissimilar
materials in device structures, which cannot be achieved by epitaxial growth techniques
alone when the lattice mismatch between the dissimilar materials is too large. In this
quest polydimethylglutarimide (PMGI) based lift-off (LOR) resist has been found
suitable to serve as a temporary sacrificial adhesive for clean film exfoliation, layer
transfer and temporary bonding applications [1]. Furthermore polydimethylglutarimide
(PMGI) lends itself easily for temporary bonding purposes, because subsequent complete
de-bonding can be readily achieved using an n-methyl-2-pyrrolidone (NMP)-based
solvent [2,3,4].

In this study we focus on the nanomechanical analysis of PMGI using the
Nanoindenter XP by Agilent with continuous stiffness measurement (CSM) attachment.
This investigation is aiming to elucidate the relationship between bonding parameters,
bond strength and the nanomechanical properties of Young's modulus and hardness of
polydimethylglutarimide (PMGI) as a function of prebake temperature and PMGI resist
film thickness. The influence of the spin-on process and the pre-bake treatment on the
structure and properties of the PMGI adhesive layers is discussed and correlated to the
resulting hardness, modulus and final bond strength. All Si wafers were initially cleaned
with standard clean 1 ($H_2O:NH_4OH:H_2O_2=5:1:1$). The solvent, which is used with PMGI

lift-off resist is cyclopentanone with a boiling point of 131°C. The polydimethylglutarimide (PMGI) lift off resist exhibits a glass transition point (T_g) of 180-190 °C. We have chosen to conduct a series of prebakes at 80°C, 105°C, and 150°C each for 5 min. This series of different thermal treatments modulated the resist mechanical properties, hardness and modulus.

Experimental

Sample Fabrication

LOR films of varying thickness and with different pre-bake temperatures were prepared in order to study the mechanical characterization of the thin films. Film thicknesses of 500 nm and 1000 nm were deposited on Si substrates using a standard spin-on process. For this experiment the surface contamination on the Si substrate wafers was removed using Standard Clean 1 (H_2O: NH_4OH: H_2O_2 = 5:1:1) to enhance LOR adhesion to the Si surface. After surface cleaning there is some native oxide remaining on the Si wafers. Following the surface cleaning process, the wafers were spin coated with LOR 5A (from MicroChem), which is based on PMGI, and subsequently pre-baked in order to partially evaporate the solvent in the resist. The actual spin speeds on the order of several thousand rpm were optimized to achieve 1000 nm and 500 nm resist thickness for this investigation. Thicker lift-off resist thickness is not practical for a load-depth analysis with a Berkovich tip, because the nanoindenter is depth limited. Then all samples were pre-baked at 80°C, 105°C, 150°C for 5 minutes in order to partially evaporate the cyclopentanone solvent and thereby to increase the LOR viscosity.

Mechanical properties

A continuous stiffness method (CSM) in the depth control mode was applied to measure the hardness and Young's modulus (elastic modulus) using an XP nanoindenter. Indentations were performed on the 500 and 1000 nm thick films of LOR. The CSM method enables continuous evaluation of the mechanical properties of materials as a function of contact depth as detailed in reference [5]. Several 2 μm deep indents were performed on each film to study the film properties. The allowable drift rate and the strain rate for loading were specified as 0.05 nm/s and 0.05 s^{-1}, respectively. A three-sided pyramidal Berkovich diamond indenter tip was used in testing these films. During the loading, the material undergoes elastic and plastic deformation.

Results and Discussion

Indentations of 2 μm deep were performed on the 500, and 1000 nm thick films. Six samples were prepared combining two film thicknesses of 500 and 1000 nm with three prebake temperatures of 80°C, 105°C, 150°C to investigate the influence of film thickness and prebake temperatures on the nanomechanical properties of LOR photo resist. The initial portion of the 2 μm indentation provides the mechanical properties of the LOR, while the deeper portion of the indentation helps study the gradual Si substrate influence on the films properties.

Figure 1 shows hardness versus h_c/t_f which is the ratio of indentation depth to film thickness, for all six samples. During the initial portion of the indentation the hardness values are steady for a short period and then continue to increase. The h_c/t_f value of 1.0 on the X-axis represents the boundary between the LOR and the Si substrate.

A magnified view of the data of Figure 1 between h_c/t_f of 0.0 and 0.5 is shown in Figure 2 Here we see two trends, first, the hardness trend due to film thickness and second, the trend due to pre-bake temperature. The solid lines represent the 500 nm films and the dotted lines represent the 1000 nm films Different coloring scheme is used to represent different prebake temperature. The hardness increases as the film thickness decreases and a softening effect is observed with a drop in temperature.

The hardness of the 500 nm thick films at at h_c/t_f of 0.1 are $\approx$0.4, 0.45 and 0.5 GPa for the prebake temperatures of 80°C, 105°C, 150°C respectively. Therefore, the hardness is proportional to the prebake temperatures. This is due to the evaporation of the solvent from the LOR. The higher prebake temperature of 150°C already exceeds the boiling point of the cyclopentanone solvent at 131°C. In a previous study with Thermal Gravimetric Analysis (TGA) we found the LOR precursor solution lost 93% of its original mass starting at a prebake temperature of 100°C because of solvent vaporization. So the higher the temperature, the more solvent is evaporated and hence the viscosity and the hardness of the LOR increases.

A similar trend is also observed in the samples with 1000 nm film thickness. Here the hardness is measured as $\approx$0.38, 0.38 and 0.45 GPa for the prebake temperatures of 80°C, 105°C, 150°C respectively. There is no difference in hardness between the 80°C and 105°C prebake samples for the 1000 nm thin film.

As the indentation depth increases to 2 μm, the hardness tends to approach the hardness of the Si substrate. Figure 1 shows the hardness approaches 6.5 GPa for the 500 nm thin films. Similarly, the hardness approaches about 3.0 GPa for the 1000 nm thin films. The average hardness of Si reported in the literature is 12 GPa [6]. The explanation for the discrepancy in the hardness of Si is due to the excessive pile-up. Pile-up causes the projected contact area to increase and the hardness to drop [7]. The modulus versus normalized depth of indentation is shown in Figures 3 and 4. The modulus of the 500 thin films at h_c/t_f of 0.02 of the film thickness is $\approx$ 5 GPa.

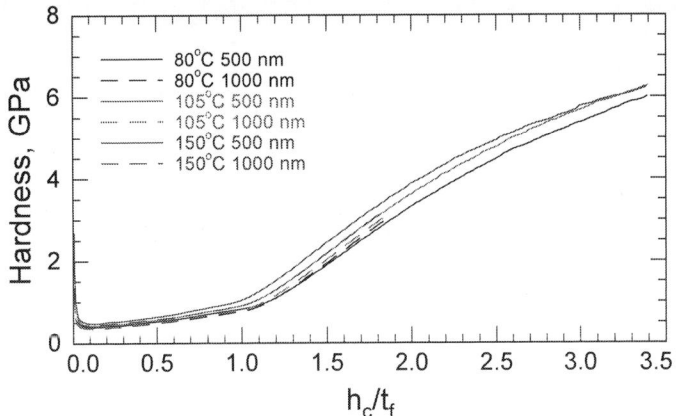

Figure 1. Hardness versus contact depth normalized to film thickness

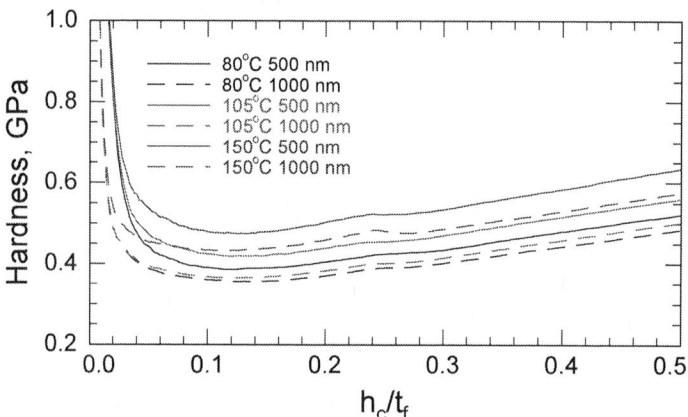

Figure 2. Hardness versus contact depth normalized to film thickness

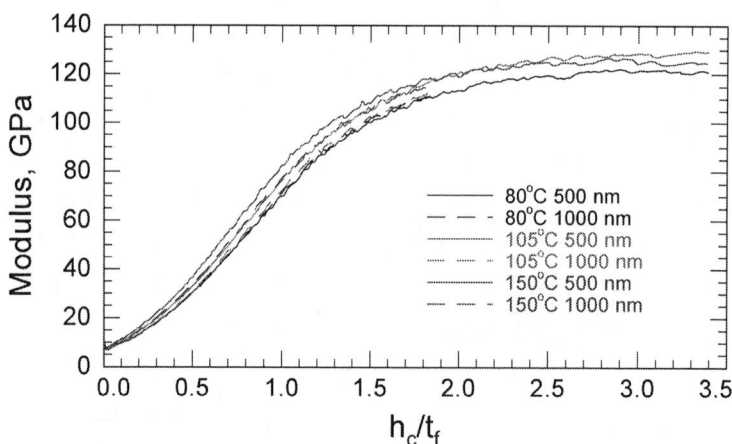

Figure 3. Modulus versus contact depth normalized to film thickness

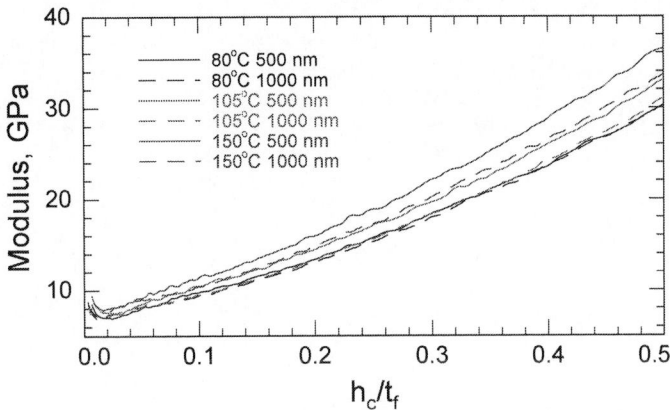

Figure 4. Modulus versus contact depth normalized to film thickness

Conclusion

The nanomechanical properties were measured using nanoindentation to determine the modulus and hardness of the LOR films. With two variations in film thickness and three variations in prebake temperatures, a total of six samples were studied. Indentations of 2 µm deep were performed on each film to study the film properties irrespective of the Si substrate influence as well as the substrate influence on the mechanical properties. The hardness increases as the film thickness decreases. The hardness also increases as the prebake temperature increases. The hardness of the 500 nm films at 0.1 depth/film thickness (hc/tf) are found to be 0.4, 0.45 and 0.5 GPa for prebake temperatures of 80°C, 105°C, 150°C respectively. For 1000 nm films, the hardness is measured as 0.38, 0.38 and 0.45 GPa for prebake temperatures of 80°C, 105°C, 150°C respectively. As the indentation depth increases to 2 µm, the hardness increases consistently due to the Si substrate. Excessive pile-up caused the projected contact area to increase which resulted in the hardness measurements to drop.

References

1. T. Matsumae, A. D. Koehler, D. Greenlee, T. J. Anderson, H. Baumgart, G. G. Jernigan, K. D. Hobart and F. J. Kub, ECS Journal of Solid State Science and Technology, 4 (7) P190-P194 (2015)
2. T. Matsumae, Andrew D. Koehler, Tadatomo Suga, and Karl D. Hobart, ECS Journal of Solid State Science and Technology, 163 (6) E159-E161 (2016)
3. T. Matsumae, Thomas Dushatinski, Tarek M. Abdel-Fattah, Tadatomo Suga, Kai Zhang, Xin Chen, Helmut Baumgart, ECS Transactions 69 (7), 99-105 (2015)
4. T. Matsumae, A. D. Koehler, J. D. Greenlee, T. J. Anderson, H. Baumgart, G. G. Jernigan, K. D. Hobart, F. J. Kub, ECS Transactions, 69 (10), 29-35 (2015)

5. W.C. Oliver, G.M. Pharr, Measurement of hardness and elastic modulus by instrumented indentation: advances in understanding and refinements to methodology, J. Mater. Res. 19 (2004) 3–20.
6. P. Mishra, S. R. Bhattacharyya and D. Ghose, Nuclear Instruments and Methods in Physics Research, Section B: Beam Interactions with Materials and Atoms 266 (2008) 1629.
7. Z. Zong, J. Lou, O.O. Adewoye, A.A. Elmustafa, F. Hammad, W.O. Soboyejo, Materials Science and Engineering A 434 (2006) 178–187.

Room Temperature Bonding with Lift-Off Resist Using the Surface Activated Bonding Method for a Layer Transfer Platform

Takashi Matsumae and Tadatomo Suga

Department of Precision Engineering, The University of Tokyo, Bunkyo-ku, Tokyo, Japan, 113-8656

A room temperature bonding of polymethylglutarimide (PMGI) is performed as a temporary bonding process. The previous study showed that temporary bonding process with PMGI achieves extremely small residues on a debonded surface, but bonding temperature reaches to 250 °C. In this study, the surface activated bonding (SAB) method is applied for bonding the PMGI to the Si wafer at room temperature for heat-sensitive materials. Using SAB, a bonded area covering over 92% of the wafer surface, with a room temperature bond strength of ~2 J/m^2, is achieved. For debonding, PMGI dissolve into N-methylpyrrolidone based solvent. Surface profile revealed that extremely small amount PMGI residues remains on the debonded surface. This process can be applied for an integration of 2D materials.

Introduction

To overcome the fundamental transistor-scaling limit, novel material integration methods such as the 3D stacking are being researched. There is a growing interest in temporary bonding as one of the key technologies for 3D stacking[1-3]. Recently, a wafer bonding method for polymethylglutarimide (PMGI) has been proposed for a layer transfer technique[4]. It has been demonstrated that the PMGI layer can be effectively bonded using thermal compression bonding. Moreover, debonding using N-Methyl-2-pyrrolidone (NMP) results in negligible amounts of the PMGI sacrificial layer as a residue on the debonded wafer surface and is therefore suitable for the transfer process[4,5].

However, the high temperature in the bonding process requires a temperature of 250°C to bond most of the loaded area. This high temperature process may damage the transferred material. Moreover, the high temperature process limits usage in applications such as integration of black phosphorous. For low temperature bonding, a surface activated bonding (SAB) process has been proposed as a room temperature bonding method[6-8]. It has been proposed that polymer materials could be bonded using this process with thinly deposited layers called nano-adhesion layers at room temperature[9]. In this study, PMGI layer is bonded to a Si substrate using the SAB method at room temperature (around 25 °C); this method is intended to be used as a layer transfer process.

Method

A PMGI layer on an Si substrate is bonded to a support Si substrate using the SAB method using a Si nano-adhesion layer (Figure 1). This method is intended to be used as a layer transfer process. The material to be transferred is required to be bonded to a support substrate during the transfer process. The proposed room temperature bonding process consists of five steps: 1. spin-coat and soft-bake, 2. bake in a vacuum, 3. ion beam sputter-deposition, 4. surface activation, and 5. bonding, as shown in Fig. 1.

At the first spin-coat and soft-bake step, a PMGI layer is formed on the target substrate. Then the support substrate and the substrate to be transferred are introduced into the load-lock chamber of the bonding machine. Both the substrates are baked at 80°C for an hour in a high vacuum ($\sim 1 \times 10^{-1}$ Pa) to degas the volatile compounds or the moisture adsorbed within the PMGI layer that prevent the adhesion of the deposited layers. Moreover, these adsorbates can cause a deterioration in the vacuum condition. After baking, both the samples are brought into the process chamber of the bonding machine containing an Ar ion beam source and an Si sputtering target. Then, the surfaces of the Si nano-adhesion layers are activated using the Ar ion beam. Finally, these surfaces are brought into contact and pressure is applied to ensure bonding. The bonding processes are conducted at room temperature, under a vacuum condition of 1×10^{-5} Pa.

Fig. 2 (a) depicts the schematic cross-sectional view of the bonded sample using the proposed method. The thickness of the PMGI layer is approximately 700 nm, as estimated from the datasheet[10]. For a debonding process, the bonded samples are soaked in N-Methyl-2-pyrrolidone (NMP). The PMGI dissolves into the NMP and a negligible residue remains on the surface of the transferred materials. The thickness of the Si nano-adhesion layer is approximately 10 nm and has been optimized for a void-free bonding of the polymer materials. To deposit the PMGI layer, a lift-off resist (LOR) 5A, from MicroChem is used in this study. The chemical formula of PMGI is depicted in Fig. 2 (b).

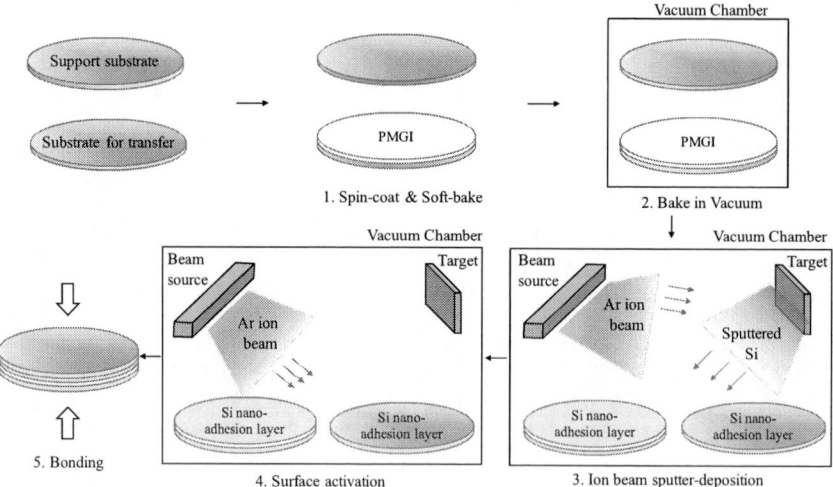

Fig. 1 Bonding procedure in the SAB method using a Si nano-adhesion layer

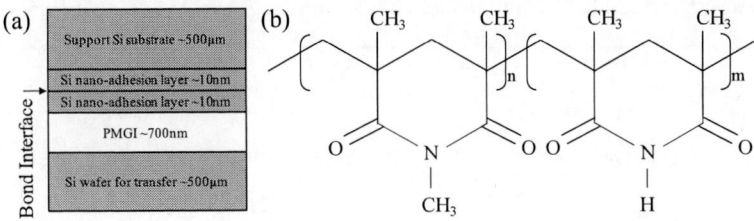

Fig. 2 (a) Schematic of the cross-sectional view of the bonded sample
(b) Chemical Structure of polymethylglutarimide (PMGI) n:m=1:1

Result and Discussion

The experimental bonding conditions are listed in Table 1. The conditions for the vacuum bake, the ion beam sputter-deposition, surface activation, and bonding have been optimized for a void-free bonding using SAB[9]. The spin-coating condition was optimized for a large-sized bonded area using the thermal compression bonding method in a previous work. In this study, the bonding of the PMGI layer was investigated for a range of soft-bake conditions. LOR 5A resist is baked after spinning at room temperature (air-dry), at 120 °C and 160 °C.

Fig. 3 displays the graphs of the bonding quality of each sample baked at different conditions; all are bonded at room temperature. Baking at 25 °C implies drying without heating after spin-coating. The size of the bonded area is measured using infrared (IR) wave transmission. The bonding strength is measured using the Maszara blade test. After comparing all the samples, it is concluded that soft-baking at 120 °C is most suitable for obtaining the strongest bond as well as the largest bonded area. Fig. 3 shows the transmission IR image of the sample soft-baked at 120 °C. 92% of the loaded area, except the edges, is bonded in this sample. The bonding strength of 2.0 J/m^2 is sufficient for usage as a transfer process.

When compared to a previous thermal compression bonding work, it requires a baking process to evaporate cyclopentanone that is the solvent in LOR 5A and has a boiling point of 131 °C. This is because a void appears if the solvent evaporates during bonding. To achieve a large bonded area of over 90%, soft-baking above 160 °C is required. On the other hand, the proposed method can achieve a considerable bonding quality when the sample is soft-baked at a temperature lower than the boiling point. Even otherwise, the baked samples achieved a bonded area of 81%. This is because the solvent cyclopentanone is evaporated in the bake-in-vacuum step.

Table 1. Experimental conditions in this study

Processes	Conditions
Spin-coating	2000 rpm, 1 min
Soft bake	25 °C (Air dry), 120 °C, 160 °C, 5 min
Bake in vacuum	80 °C, 1 h, 1×10^{-3} Pa
Ion beam sputter-deposition	1.2 kV, 400 mA, 1×10^{-1} Pa, 80 sccm
Surface activation	1.0 kV, 100 mA, 1×10^{-2} Pa, 70 sccm
Bonding	25 °C, 5 MPa, 5 min

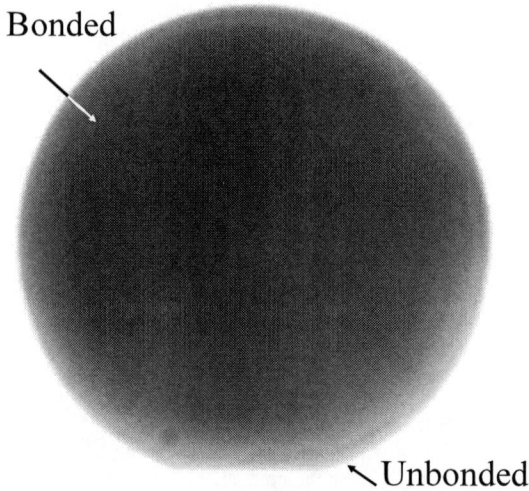

Fig. 3. Infrared transmission image of bonded sample. The pattern that looks like a cross at the center is a resist residue on the surface of bonded sample. It doesn't relate to unbonded area.

Layer transfer applications require not only strong bonding but also clean surface after debonded process. In this study, residue after debonding process is investigated using atomic force microscope (AFM). Some particles of polymer residues on the surface after debonding aren't desireble for the transfer application. For debonding, PMGI layer dissolves in N-methylpyrrolidone based solvent. Figure 4 shows the AFM profile of the Si surface after the debonding process. On the debonded surface, extremely small amount PMGI residues are detected. Surface profile after debonding process using atomic force microscope. The value of root means suqare (RMS) is 4.0 nm. This surface flatness indicate a few residue adhere to the surface after debonding process.

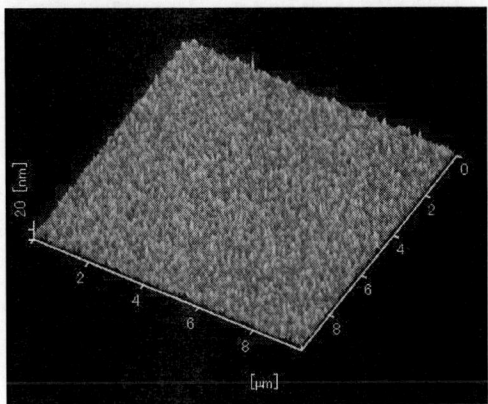

Fig. 4 Surface profile after debonding process using atomic force microscope. The value of root means suqare (RMS) is 4.0 nm. This surface flatness indicate a few residue adhere to the surface after debonding process.

Conclusions

In summary, we demonstrated that PMGI bonding at room temperature is capable of providing large and strong bonding interfaces. Additionally, amount of resist residues after the debonding process are extremely small. This process can be applied for 3D integration and material integration, without thermal damage. Thus, the proposed method can provide a platform for a layer transfer process suitable for 3D material integration of thin-layered systems like black phosphorous.

References

1. Y.Wang,C.Miao,B.C.Huang,J.Zhu,W.Liu,Y.Park,Y.H.Xie,andJ.C.S.Woo,   IEEE Trans. Electron Devices, 57(12), 3472 (2010).
2. J. D. Caldwell, T. J. Anderson, K. D. Hobart, G. G. Jernigan, J. C. Culbertson, F. J. Kub, J. L. Tedesco, J. K. Hite, M. A. Mastro, R. L. Myers-Ward, C. R. Eddy, P. M. Campbell, and D. K. Gaskill, MRS Proceedings, 1259 (2010).
3. R. Dong, Z. Guo, J. Palmer, Y. Hu, M. Ruan, J. Hankinson, J. Kunc, S. K. Bhattacharya, C. Berger, and W. A. de Heer, J. Appl. Phys., 47, 094001, (2014).
4. T. Matsumae, A. D. Koehler, J. D. Greenlee, T. J. Anderson, H. Baumgart, G. G. Jernigan, K. D. Hobart and F. J. Kub, ECS Journal of Solid State Science and Technology 4, (7), 190-194. (2015).
5. Takashi Matsumae, Andrew D. Koehler, Tadatomo Suga and Karl D. Hobart, Journal of The Electrochemical Society, 163 (6), 159-161, (2016)
6. H. Takagi, K. Kikuchi, R. Maeda, T. R. Chung, and T. Suga, Appl. Phys. Lett. 68, 2222 (1996).
7. A. Shigetou, T. Itoh, M. Matsuo, N. Hayasaka, K. Okumura, and T. Suga, IEEE Trans. Adv. Packag. 29 (218), 22, (2006).

8. T. Akatsu, N. Hosoda, T. Suga, and M. Rühle, J. Mater. Sci. 34, 4133, (1999)
9. T. Matsumae, M. Fujino, and T. Suga, Japanese Journal of Applied Physics, 54(10), 101602. (2015).
10. MicroChem, LOR / PMGI Data Sheet, (2008) [Revised 2016]

Thin layer transfer using room temperature wafer-level bonding process.

Karine Abadie[a], Frank Fournel[b,c], Christophe Morales[b,c], Hubert Moriceau[b,c], and Markus Wimplinger[a]

[a] EVGroup, DI E. Thallner Strasse 1, 4782 – Sankt Florian/Inn, Austria
[b] Univ. Grenoble Alpes, F-38000 Grenoble, France
[c] CEA, Leti, Minatec Campus, F-38054 Grenoble, France

> This article deals with the transfer of a thin Si film (100nm) on top of a Si substrate using a room temperature covalent bonding. No annealing is performed post direct bonding. Before to detail this process, the equipment used for the bonding process is characterized in terms of particle and metal contamination. The etching occurring during one of the sub process of the surface preparation is also studied. Characterizations performed after layer transfer are presented. They include C-SAM inspections, SP2-HAZE measurements and TEM.

Introduction

Direct wafer bonding has already been demonstrated as an efficient solution to create a linkage between two different surfaces. One of the main limitations is the temperature required to stabilize the bonding interface and its strength. For example, to completely stabilize a Si/Si bonding realized at room pressure, an annealing step done at 1000°C or above is required. In the past, alternative low temperature processes, using surfaces plasma activation before contacting were developed for SiO_2 to Si and SiO_2 to SiO_2 bonding, lowering the annealing temperature to 300°C and enabling a sufficient bond strength for many applications [1]. Another possible option to realize a room temperature Si/Si bonding is to use a covalent bonding process [2]. The EVG®580 ComBond® was developed with this aim by bringing the wafers in contact in an high vacuum (HV) environment, after activation in vacuum of both surfaces in the ComBond® Activation Module (CAM) [3,4]. This paper aims to present observations made after the transfer of a thin layer using this novel direct covalent Si/Si bonding process. The bonding process is realized in an EVG®580 ComBond® equipment installed at CEA-Leti. At first, the activation process is characterized in terms of particle and metal contaminations. Etching phenomenon occurring during this activation is also studied. Then an appropriate full bonding process (activation + bonding) is developed in order to realize an Si/SOI bonding with a high bond strength allowing further processing without any post bonding annealing. The process to transfer the Si thin film existing at the top of the SOI onto the Si substrate is also described. Finally, the layer transfer quality is characterized and some measurements are performed in order to study the bonding interface.

Preliminary tests

The EVG®580 ComBond® consists of several modules including loadports, a pre-aligner, a CAM module and a bonder module. In order to perform bonding processes, two wafers can be loaded at the same time into the equipment. After pre-alignment, the wafers are successively activated inside the CAM module. This sub-process is designed for removing the native oxide on the wafers. The two wafers are then transferred into the bond module for the last step: the two surfaces are brought in contact with a certain loading force. The different process parameters (activation conditions, load force and duration during bonding) can be adapted depending on the purpose of the bonding process, but also depending on the types of surfaces to be bonded. In this study, the work focuses on silicon to silicon bondings, and it is performed in 200mm.

At first, the equipment is qualified in terms of particle and metal contaminations. Regarding the particle contamination, two main measurements could be done: with or without activation. Indeed, the front face of the surface (polished face to be bonded) can be quite "disturbed" and/or contaminated by particles during the activation sub process. During all the other sub processes (pre-alignment and bonding), the front face is never touched by any part of the equipment and as the equipment operates under a HV environment, the particle contamination generated by the equipment environment should be quite low. However one known source of potential contamination is the in and out sequence in the loadport. As shown on, no particles contamination is shown traveling the wafer in all the different module without activation sequence. The EVG®580 ComBond® tool is then completely neutral regarding the particle contamination without activation sequence. However after ComBond® surface activation, a small contamination could be detected. Indeed, using a SurfScan equipment (SP2) from KLA-Tencor, measurements of particles with a threshold at 90nm are performed before and after activations of single wafers. The process parameters used for the activation are the same than for the activation process used for the layer transfer, presented later in this article.

TABLE I. Particle count before and after activation process on blanked 200mm silicon wafers into the CAM module.

Si surfaces	Part. 0,09-0,115 µm	Part. 0,115 – 0,1382 µm	Part. 0,1382 – 0,1712 µm	Part. 0,1712 – 0,2121 µm	Part. 0,2121 – 0,2628 µm	Part. 0,2628 – 0,3257 µm	Part. 0,3257 – 0,4035 µm	Part. 0,4035 – 0,5000 µm	Part. Saturated	Total part.
Before	8	1	1	0	0	0	0	0		10
After	11	12	3	4	4	2	3	1	3	43

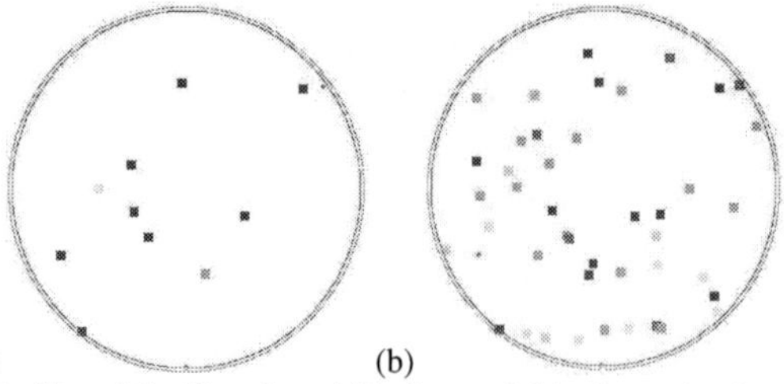

Figure 1. Profiles of the Si surface (a) before and (b) after activation process.

As it can be observed in table I and figure 1, particles contamination generated during activation process can be observed but stays limited to only 33 added particles of size equal or larger than 90nm. Further characterization are on-going in order to determine the nature of this contamination which could be also "only" a surface modification detected by the SP2 and not a real particle contamination.

In addition to these excellent results in particle contamination, the detection of metals contamination demonstrates that the whole bonding process is metal free with a specification of 1.10^{11} at/cm² for all metallic elements. To test the CAM module, Si wafers are activated using the same recipe than the one used for layer transfer later, and measurements are performed with a TXRF equipment from RIGAKU. Handling wafers are also tested in the pre-aligner, and the bond chamber. Results demonstrated that most of the elements are below the detection threshold (Co, Cr, Cu, Mg, Mn, Na, Ni, Ti, V, Zn) of the TXRF equipment (see table II), except for Al and Fe, which are detected with a magnitude of 10^{11} at/cm² after activation on the front face of the wafer. The root cause of the contamination with those two elements is probably identified and solutions are under study.

TABLE II. Metal ion contamination on the frontside of a blanked silicon wafer after activation into the CAM module, using the same process parameters that the one used for layer transfer.

Elements	Al	Co	Cr	Cu	Fe	Mg	Mn	Na	Ni	Ti	V	Zn
After activation	5,49E+11	< 4,7E9	< 9,1E9	< 4,8E9	1,94E+11	< 2,074E11	< 7E9	< 2,074E11	4,88E+10	< 2,04E10	< 1,36E10	< 7,3E9

The activation process itself is also characterized, especially in terms of etching rate. At first, for a given set of recipe parameters corresponding to the ComBond® Activation Module (CAM), an SOI wafer and an oxidized silicon wafer are activated during the same duration. Si thickness removed from the SOI and SiO_2 thickness removed from the oxidized silicon wafer are measured using an ellipsometer. Same results are observed: about 9nm of the material is etched in both cases. It is concluded that the etching rate generated by this sub-process is the same on Si and SiO_2.surfaces. SRIM simulations, based on a similar process with Si and SiO_2, demonstrate that both materials give the same oxide removal rate. Thus further tests are performed on oxidized silicon wafers, assuming that results can be applied to silicon surfaces.

An oxidized silicon wafer is activated successively several times and with different durations: 30s, 300s, 120s, and again 30s. Between two activations, ellipsometry measurements are performed in order to calculate the SiO_2 etching rate. The amorphous Silicon layer induced by the activation is taken into account. This thickness is simply added to the SiO_2 monocrystalline measured thickness considering the thickness variation of the silicon amorphization as negligible. The measurements are performed on 49 points over the wafer surface in order to realize a representative mapping of the etching. As shown in figure 2, during the first 30s, in average, 85Å of SiO_2 are removed, and so the etching rate reaches 170Å/min. During the next 300s, 415Å of SiO_2 are removed in average, the etching rate is then 83Å/min. During the next 120s, 333Å of SiO_2 are removed and the etching rate becomes 111Å/min. And to finish when again 30s of activation are performed, the SiO_2 thickness removed is about 90Å and so the etching rate is 180Å/min, which is closed to what is observed at the beginning of the sequence. Thanks to the 49 points measured, the standard deviation is also calculated. This value is well different depending on the activation time: the shorter the activation time is, the smaller the standard deviation is. And the more the activation time is, the lower the etching homogeneity is. These measurements and calculations are presented in figure 2 and 3.

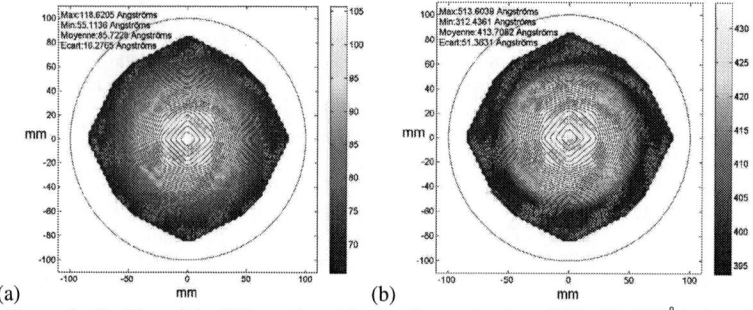

(a)   mm   (b)   mm

Figure 2. Profiles of the SiO_2 surface (a) after first activation of 30s (V=170Å/min, std. dev.=16Å (b) after 30s + 300s activation (V=83Å/min, std. dev.=51Å).

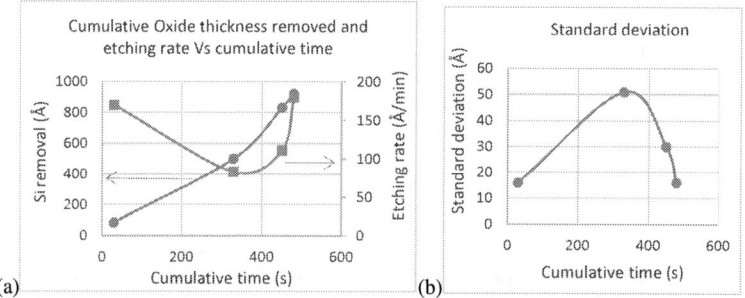

(a)   (b)

Figure 3. (a) Cumulative oxide thickness removed during test sequence (round) and corresponding etching rate (square) (b) Standard deviation calculated depending on the activation cumulative duration

Based on these observations, it seems that successive activations have no impact on each other regarding the removal rate: the etching rate observed during the last 30s of activation is quite similar to the first etching rate observed at the beginning. Noteworthy, the beginning of the activation seems to have a bigger impact on the silicon etching rate. There are already some ideas on the root cause for this phenomenon. Further tests and optimization are carried out to confirm these assumptions.

Samples preparation

After these preliminary tool qualifications, a bonding process with high bonding energy (>5 J/m²) for Si/Si 200 mm wafers is developed. Indeed, the EVG®580 ComBond® operates in HV environment, which allows creating covalent bondings after activation of both surfaces in the CAM module. This sub process is performed in order to remove the native oxide and to minimize the damages induced in the sub-surface Si lattice. The bonding energy is measured using a double cantilever beam (DCB) technique under anhydrous or humid atmosphere [5] using a blade thickness of 300μm.

By adjusting the process parameters (ComBond® activation parameters, load force and duration during bonding), it is possible to have a direct action on the bond strength. With optimized activation parameters, the bonding energy is higher than $5J/m^2$ as a breakage of the samples can be observed during the blade insertion for the DCB measurement. Concerning the load force during bonding, no impact on the bonding energy is observed. The bonding inside the ComBond tool is then very fast and spontaneous. In these conditions, there is no need for further annealing after the bonding process to strengthen the interface, and the bonded pair can be directly further processed with a grinding step and a chemical etching for example.

Noteworthy, concerning the bonding energy measurements, similar results are observed in both humid and anhydrous measurement atmospheres (even with measurable bonding energy). The water stress corrosion has no impact on the bond pairs realized with this process. This result emphasizes the absence of silanol bonds at the bonding interface.

Using the previous process, bondings of SOIs (Si 100nm - SiO_2 200nm) with Si wafers are performed. The goal is to transfer the Si top layer of the SOI onto a new substrate and to characterize both the bonding and the transfer qualities. After bonding process into the EVG®580 ComBond®, the SOI is grinded down to 20μm. The remaining Si is removed by a wet etching using TMAH solution at 12.5%. HF chemistry is used at last to remove the 200nm buried oxide layer which was used as a convenient etch stop layer for the TMAH etching. At the end of the process, only a thin film of 100nm of Si is transferred on top of a 725μm thick silicon wafer.

Samples characterization

In this study, bonding quality is assessed using C-mode Scanning Acoustic Microscopy (C-SAM) which is performed just after bonding. As shown on figure 4, no defects are observed.

Figure 4. C-SAM images realized after bonding of a Si wafer with an SOI wafer (Si 100nm - SiO_2 200nm) using the EVG®580 ComBond®.

In order to control the bonding and the layer transfer processes, more than particle contamination measurements, a haze measurement using the SP2 deserve to be performed on the SOI surface before bonding and after layer transfer. Indeed, the haze measurement allows to characterize easily non transferred areas coming from bad bonding and/or particle contamination. Using a SOI wafer with a silicon film of 100nm or less, non-transferred areas are not always detected with classical SP2 particle measurement. As shown on figure 5b and c, no holes or additional defects are detected after such EVG®580 ComBond® layer transfer. Thus, very good quality of ultra-thin (100nm) transferred layer on 200 mm full wafers is obtained. For comparison, a bad transfer with several non-transferred areas is also shown on figure 5a.

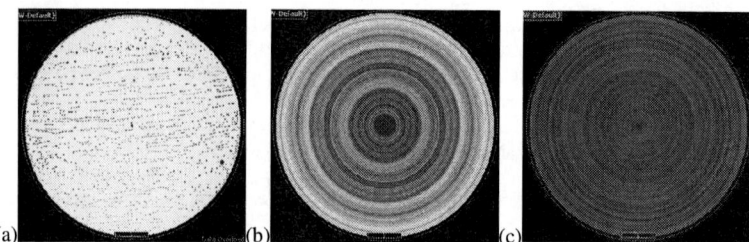

Figure 5. SP2 measurements in haze mode (a) after layer transfer with bad quality, many holes can be observed in the layer transferred, (b) SOI surface before bonding process , (c) after transfer of the thin Si top layer (100nm) of the SOI on the Si substrate with the EVG®580 ComBond®.

Nevertheless some very small irregularity into the Si surface can be observed on the right lower side (see figure 6c) after layer transfer. This location of the thin layer is then observed with an Atomic Force Microscope (AFM).

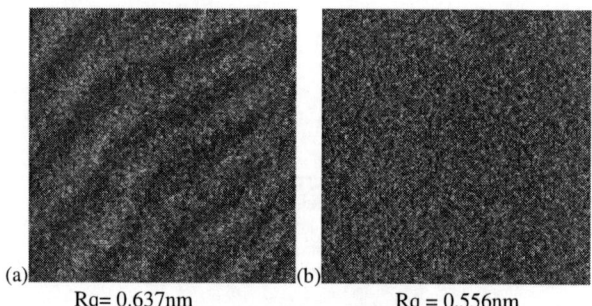

Rq= 0.637nm   Rq = 0.556nm

Figure 6. AFM measurement (a) on the area presenting small irregularities on the SP2 – haze image, (b) in the middle of the layer transfer

As it can be observed on figure 6, the roughness after layer transfer is slightly higher than on a bulk SOI silicon thin film (<0.2nm) and the roughness is even a little higher on the area presenting small irregularities on the haze mapping. This area is then not real defect inside the layer, but is due to small difference of roughness.

In order to compare this layer transfer process to a more classical one, a second layer transfer of another SOI thin (100nm) silicon layer onto an oxidized silicon wafer is performed. But this second transfer is done using a standard hydrophilic direct bonding process. Thus, a bonding process with N_2 plasma activation is performed in an EVG850LT equipment, and a post bonding annealing at 1100°C for 2h is used in order to stabilize the bond pair. After annealing, a C-SAM observation is made to control the bonding quality: only very small defects can be observed on the left lower side (see figure 7). These defect are the classical defect coming from the bonding wave dynamics [6]. They are obviously not present in the EVG®580 ComBond® bondings.

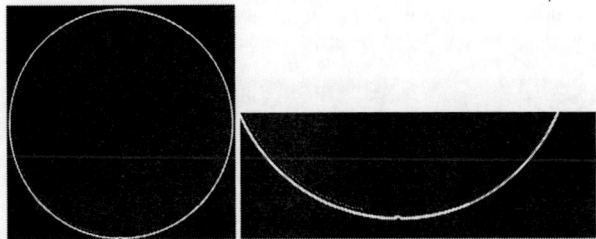

Figure 7. C-SAM images realized after bonding of an Si wafer with 145nm of thermal oxide grown at the top with an SOI wafer (Si 100nm - SiO_2 200nm) using an EVG850LT + annealing at 1100°C

The same grinding and chemical etching processes are realized in order to remove the SOI backside silicon substrate and the buried oxide layer. Only the first SOI silicon thin layer of 100nm remains on the oxidized silicon substrate. The layer transfer quality is also characterized with the SP2 haze mode. Comparable images to those obtained after the layer transfer performed with the EVG®580 ComBond® are observed, except the small irregularities at the surface. Obviously the edge defects leads to non-transferred areas which can be seen on figure 8b. But, these defects automatically lower the gain of the Haze image. The image is then darker and small irregularities like the one of figure5c might not be seen. Both process might then be consider as equivalent showing the good performances of the EVG®580 ComBond®.

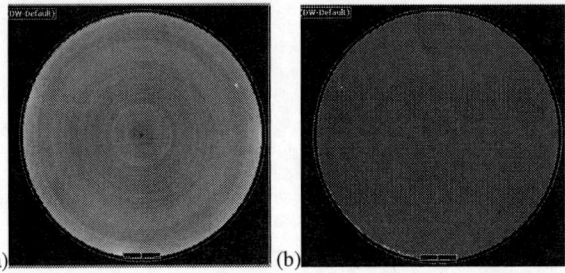

Figure 8. SP2 measurements in HAZE mode (a) SOI surface before bonding process with a SiO_2 substrate, (b) after transfer of the thin Si top layer (100nm) of the SOI on the substrate.

Additional measurements are performed in order to deeply characterize the bonding interface. The wafer with the layer transfer made using the EVG®580

ComBond® is diced in several beams. Some of the beams are annealed during 2h in nitrogen at different temperatures (400°C, 500°C and 600°) in order to recrystallize the amorphous layer created at the bonding interface during the activation. After annealing, the small thermal oxide generated during the annealing in a non-perfect nitrogen atmosphere is removed by chemical treatment. Ellipsometry is then used to evaluate at first the thickness of the amorphous layer remaining at the bonding interface after annealing. To go further into the characterization, from this first observation, samples are selected and also characterized by Transmission Electronic Microscopy (TEM). It allows to measure the thickness of the amorphous layer before annealing, as well as to study the temperature at which the Si interface is fully recrystallized.

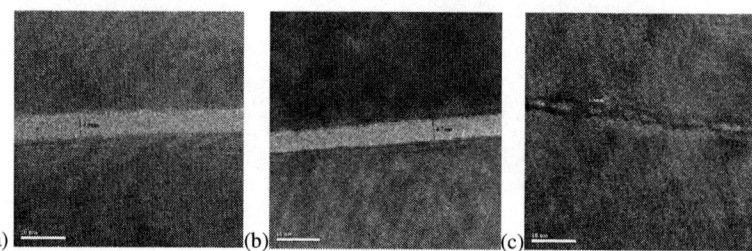

Figure 9. TEM images (a) with no annealing, (b) after annealing at 500°C, (c) after annealing at 600°C

Figure 9 presents the images realized during TEM observation at the interface of three samples: the first sample is without any annealing after layer transfer. The second sample is annealed at 500°C during 2h and the third sample at 600°C during the same time. After bonding process, the amorphous layer is about 4.9 nm thick. The annealing realized at 500°C has almost no impact on the recrystallization of the amorphous Si, as the thickness measured on this sample is 4.7 nm. However, after annealing at 600°C, the Si is partially recrystallized with some fully recrystallized areas and some remaining amorphous zone of 1.8nm maximum thickness.

Conclusion

After preliminary characterizations of the EVG®580 ComBond®, a covalent bonding process is used to transfer a thin Si layer of 100nm on top of a Si substrates. No post bonding annealing is required to stabilize the bonding process before processing of the bonded pair with grinding and chemical etching. Moreover, the surface quality of the layer transfer is characterized and no major voids or defect are detected into the thin layer. Partial recrystallization of the amorphous Si generated at the bonding interface during bonding process is observed after annealing at 600°C with fully recrystallized areas.
The successful transfer of a thin Si layer onto a new substrate shown in this work demonstrates that such a bonding process would deserve to be applied to many other thin layer transfer techniques.

References

1. T. Plach, K. Hingerl, S. Tollabimazraehno, G. Hesser, V. Dragoi and M. Wimplinger, J. Appl. Phys., 113, 094905 (2013)
2. H. Takagi, K. Kikuchi, R. Maeda, T. R. Chung and T. Suga, "Surface Activated Bonding of Silicon Wafers at Room Temperature", Appl. Phys. Lett. Vol. 68, No.16 (1996), pp. 2222-2224.
3. C. Flötgen, N. Razek, V. Dragoi, M. Wimplinger, "Novel Surface Preparation Methods for Covalent and Conductive Bonded Interfaces Fabrication", ECS Trans. 64(5), The Electrochemical Society, pp. 103-110, 2014
4. C. Flötgen, N. Razek, V. Dragoi, "Functional Interfaces Fabrication Through Room Temperature Wafer-Level Bonding", Proc. of WaferBond Conference, 2015, pp. 55-56
5. F.Fournel, L. Continni, C. Morales, J. Da Fonseca, H. Moriceau, F. Rieutord, A. Barthelemy, and I. Radu, "Measurement of bonding energy in an anhydrous nitrogen atmosphere and its application to silicon direct bonding technology", Journal of Appl. Phys. 111 (2012), pp. 104907
6. A. Castex,a,z M. Broekaart,a,* F. Rieutord,b K. Landry,a and C. Lagahe-Blancharda, "Mechanism of Edge Bonding Void Formation in Hydrophilic Direct Wafer Bonding", ECS Solid State Letters, 2 (6) P47-P50 (2013) P47 2162-8742/2013/2(6)/P47/4/$31.00 © The Electrochemical Society
7. EVG®580 is a Trade Mark of EVGroup
8. ComBond® is a Trade Mark of EVGroup

212

Chapter 6

Poster Session

214

Optical Isolator with Si Guiding Layer Fabricated by Photosensitive Adhesive Bonding

H. Yokoi[a,b], S. Choowitsakunlert[a], K. Kobayashi[a], and K. Takagiwa[a],

[a] Graduate School of Engineering and Science, Shibaura Institute of Technology,
Tokyo 135-8548, Japan
[b] SIT Research Center for Green Innovation, Tokyo 135-8548, Japan

An optical isolator employing a nonreciprocal guided-radiation mode conversion is described. The optical isolator composes of a magneto-optic waveguide with a Si guiding layer, which can be realized by photosensitive adhesive bonding. The optical isolator was designed at a wavelength of 1.55 μm. Relationship of waveguide parameters was clarified for isolator operation. Dependence of the thickness of an adhesive layer was investigated for the design of the optical isolator.

Introduction

In optical communication systems, optical nonreciprocal devices are indispensable for the protection of active photonic devices from unwanted reflected light. In the near-infrared region, magnetic garnet crystals are necessary components of an optical nonreciprocal device because of their transparency and large magneto-optic effect. For isolators that use TE-TM mode conversion, nonreciprocal and reciprocal mode converters must be installed (1,2). Therefore, complicated magnetization control is needed. Moreover, phase matching between the two modes is imperative, which results in the necessity for precise control of waveguide parameters.

An optical isolator employing a nonreciprocal phase shift is attractive because there is no need for phase matching or complicated control of the direction of magnetization (3,4). A nonreciprocal phase shift occurs in transverse magnetic (TM) modes traveling in magneto-optic waveguides when the magnetization is aligned transversely to the light propagation direction in the film plane. When a magneto-optic layer is used as a cladding layer in the magneto-optic waveguide, the nonreciprocal phase shift attains its maximum value with a high-refractive-index guiding layer (5). A magneto-optic waveguide with a Si guiding layer is a practical and promising candidate for a nonreciprocal phase shifter in the optical isolator (6,7). In order to realize a magneto-optic waveguide with the Si guiding layer, novel integrating technology must be developed.

By utilizing the nonreciprocal phase shift, an optical isolator employing a nonreciprocal guided-radiation mode conversion is realized. The optical isolator employing the nonreciprocal guided-radiation mode conversion is attractive, because it has simple structure and easy control of direction of magnetization (8,9). In this paper, we report on an optical isolator with the Si guiding layer employing the nonreciprocal guided-radiation mode conversion. The magneto-optic waveguide with the Si guiding layer is fabricated by photosensitive adhesive bonding between Si and magnetic garnet crystals. The nonreciprocal phase shift in the magneto-optic waveguide was calculated at a wavelength of 1.55 μm. The optical isolator was designed at a wavelength of 1.55 μm. Dependence of the thickness of an adhesive layer was discussed for the design of the optical isolator. Photosensitive adhesive bonding between Si and a garnet crystal was investigated.

Device Structure

Figure 1 shows an optical isolator with a Si guiding layer employing a nonreciprocal guided-radiation mode conversion (10). A magnetic garnet cladding layer is bonded with the Si guiding layer by photosensitive adhesives. A cerium-substituted yttrium iron garnet (Ce:YIG) is used as the magnetic-garnet cladding layer (11). Faraday rotation coefficient of Ce:YIG is approximately -7.85×10^3 rad/m at a wavelength of 1.55 μm (8,12). The optical isolator is comprised of a straight rib waveguide with a Ce:YIG/Si/SiO$_2$ structure. An external magnetic field is applied to the magneto-optic waveguide in film plane. TM modes travelling in the magneto-optic waveguide have distinct propagation constants for the forward- and the backward-travelling waves owing to a nonreciprocal phase shift. By adjusting waveguide parameters, the following relationship is satisfied:

$$\beta_b^y < \beta_c^x < \beta_f^y \qquad [1]$$

where β_f^y and β_b^y denote the propagation constants of the forward- and the backward-traveling TM modes and β_c^x denotes the cutoff of TE modes. Only the backward-traveling TM modes can couple to the TE radiation modes so that the device acts as a TM-mode optical isolator.

The optical isolator employing the nonreciprocal guided-radiation mode conversion was designed at a wavelength of 1.55 μm. At first, the nonreciprocal phase shift was calculated for a magneto-optic waveguide with a Ce:YIG / gap (adhesive layer) / Si / SiO$_2$ structure. The refractive indices of Ce:YIG, Si, and SiO$_2$ are 2.22, 3.50, and 1.44, respectively, at 1.55 μm. Figure 2 shows the calculated nonreciprocal phase shift with various thicknesses of the adhesive layer depending on the thickness of the Si guiding layer. When the thickness of the adhesive layer is 0 nm, the magneto-optic waveguide has a structure of Ce:YIG / Si / SiO$_2$. In this case, when the Si guiding layer thickness is 200 nm, the nonreciprocal phase shift has its maximum. When the gap is 25 nm, 50 nm, and 100 nm, the thicknesses of the Si guiding layer for the maximum nonreciprocal phase shift are 210 nm, 220 nm, and 230 nm, respectively. When the gap width increases, the amount of the nonreciprocal phase shift diminishes. In order to satisfy the relationship of the propagation constants denoted by expression [1], it is desirable that propagation of the lightwaves in the magneto-optic waveguide incurs larger nonreciprocal phase shift. When the magneto-optic waveguide with the Si guiding layer is fabricated by photosensitive adhesive bonding, it is required that the adhesive layer between the Si guiding layer and the Ce:YIG cladding layer is as small as possible.

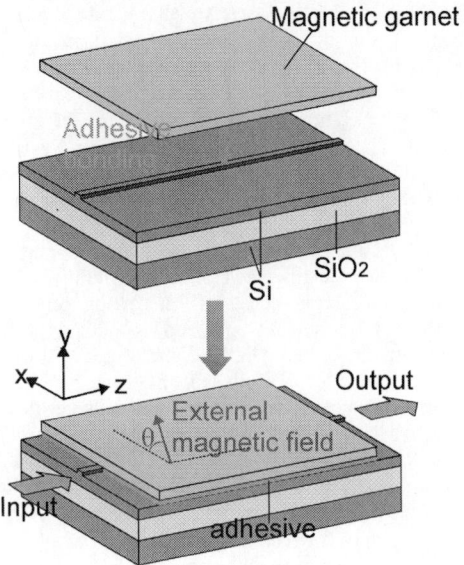

Figure 1. Schematic diagram of optical isolator employing nonreciprocal guided-radiation mode conversion.

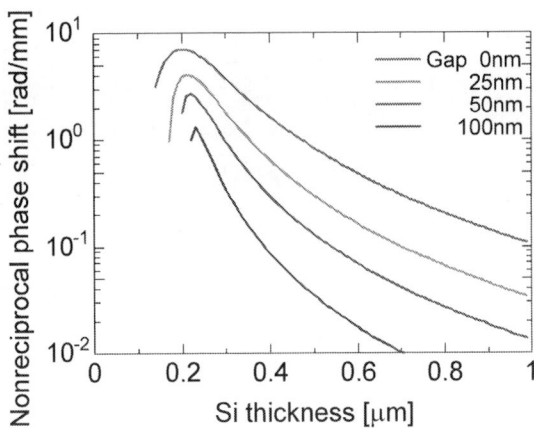

Figure 2. Calculated nonreciprocal phase shift in magneto-optic waveguides with a Si guiding layer.

The optical isolator employing the nonreciprocal guided-radiation mode conversion was designed. Figure 3 shows a cross-sectional diagram of the magneto-optic waveguide fabricated by photosensitive adhesive bonding. When the thicknesses of the Si guiding layer and the gap layer are fixed, the nonreciprocal phase shift is determined, as shown in Fig. 2. The propagation constant β_c^x, the cutoff of TE modes, is obtained when the rib height is set. The propagation constants β_f^y and β_b^y can be calculated when the rib width is given. From the expression [1], the rib width for the isolator operation is determined. Higher isolation ratio is expected with longer waveguides because this device utilizes the guided-radiation mode conversion.

Figure 4 shows the relationship of the waveguide parameters for the isolator operation. Once the rib height is set, the range of the rib width for the isolator operation, which means that the propagation constants satisfy the relationship denoted by expression [1], is given. The isolator operation is limited in the magneto-optic waveguide whose rib width ranges between filled circles and open circles. When the rib width is less than the filled circles, both the forward- and backward-traveling waves couple to the TE radiation modes. When the rib width is larger than the open circles, neither the forward- nor backward-traveling waves couple to the TE radiation modes. For example, when the gap width is 0 nm, that is, the magneto-optic waveguide has a structure of Ce:YIG / Si / SiO$_2$, the rib width for the isolator operation ranges between 2.26 and 2.38 μm at the rib height of 86 nm. When the gap width is 25 nm, the rib width for the isolator operation ranges between 2.20 and 2.26 μm at the rib height of 99 nm. When the gap width is 50 nm, the rib width for the isolator operation ranges between 2.09 and 2.12 μm at the rib height of 104 nm. It is confirmed that the larger gap width brings about the smaller tolerance of the rib width, which can be attributed to the fact that the amount of the nonreciprocal phase shift decreases when the gap width increases. The smaller gap width is required for large tolerance of the waveguide parameters.

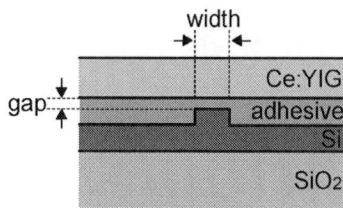

Figure 3. Cross-sectional diagram of the magneto-optic waveguide fabricated by photosensitive adhesive bonding.

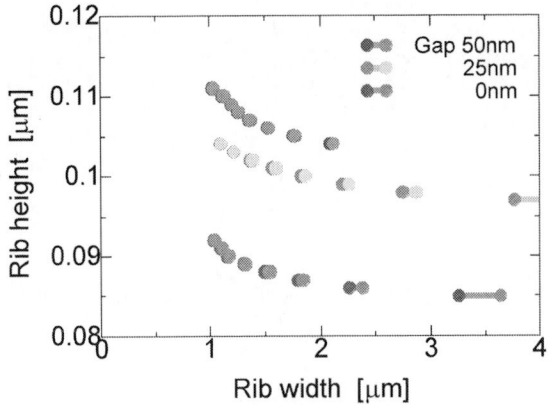

Figure 4. Relationship of waveguide parameters for isolator operation.

The magneto-optic waveguide in the optical isolator employing the nonreciprocal guided-radiation mode conversion is realized by photosensitive adhesive bonding. As a preliminary experiment, adhesive bonding between Si and $Gd_3Ga_5O_{12}$ (GGG) was attempted by use of photosensitive adhesive (TOKYO OHKA KOGYO CO., LTD.). Figure 5 shows a cross-sectional SEM image of the bonded sample. The adhesive bonding between Si and GGG was successfully achieved with the heat treatment at 160°C. However, the adhesive layer thickness is larger than 100 nm so that further experiments must be conducted to accomplish the bonding between Si and the garnet crystals with smaller gap width.

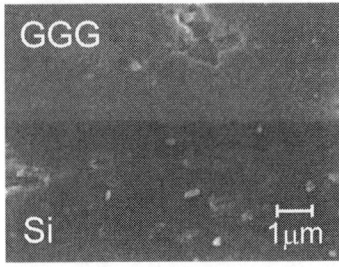

Figure 5. Cross-sectional SEM image of Si / GGG fabricated by photosensitive adhesive bonding.

Conclusion

The optical isolator employing the nonreciprocal guided-radiation mode conversion was designed. The optical isolator has the magneto-optic waveguide with the Si guiding layer, which is realized by photosensitive adhesive bonding. Relationship between rib height and rib width was clarified for the magneto-optic waveguide with various thicknesses of the adhesive layer. Photosensitive adhesive bonding between Si and GGG was demonstrated.

Acknowledgments

This work was partially supported by the SIT Research Center for Green Innovation.

References

1. J. P. Castéra and G. Hepner, *IEEE Trans. Magn.*, **13**, 1583 (1977).
2. K. Ando, T. Okoshi, and N. Koshizuka, *Appl. Phys. Lett.*, **53**, 4 (1988).
3. Y. Okamura, H. Inuzuka, T. Kikuchi, and S. Yamamoto, *J. Lightw. Technol.*, **4**, 711 (1986).
4. T. Mizumoto, S. Mashimo, T. Ida, and Y. Naito, *IEEE Trans. Magn.* **29**, 3417 (1993).
5. H. Yokoi, *Opt. Mater.*, **31**, 189 (2008).
6. H. Yokoi, T. Mizumoto, and Y. Shoji, *Appl. Opt.*, **42**, 6605 (2003).
7. R. L. Espinola, T. Izuhara, M.-C. Tsai, and R. M. Osgood, Jr., *Opt. Lett.*, **29**, 941 (2004).
8. T. Shintaku and T. Uno, *J. Appl. Phys.*, **76**, 8155 (1994).
9. H. Yokoi, K. Yamaguchi and Y. Uchiumi, *Jpn. J. Appl. Pyhs.*, **50**, 078001 (2011).
10. H. Yokoi, K. Sasaki and T. Aiba, *Jpn. J. Appl. Phys.*, **48**, 062202 (2009).
11. T. Shintaku, T. Uno and M. Kobayashi, *J. Appl .Phys.*, **74**, 4877 (1993).
12. M. Gomi, S. Satoh, and M. Abe, *Jpn. J. Appl. Phys.*, **27**, 1536 (1988).

Determination of Band Structure at GaAs/4H-SiC Heterojunctions

J. B. Liang[a], S. Shimizu[a], M. Arai[b], and N. Shigekawa[a]

[a] Graduate School of Engineering, Osaka City University, Sumiyoshi, Osaka 5588585, Japan
[b] New Japan Radio co., Ltd., Fujimino, Saitama 3568510, Japan

The effects of thermal annealing process on the interface in p+-GaAs/n-4H-SiC heterojunctions fabricated by using surface-activated bonding (SAB) were investigated. It was found by measuring their current-voltage (I-V) characteristics that the reverse-bias current and the ideality factor were extracted to be 7.57×10^{-7} A/cm^2 and 1.33, respectively, for the junctions annealed at 400 °C. The flat-band voltage obtained from capacitance-voltage (C-V) measurements was found to be 1.29 eV, which is almost consistent with the turn-on voltage extracted from I-V characteristics. These results suggest that the SAB-based GaAs/4H-SiC heterojunctions are applicable for fabricating high-frequency power devices.

Introduction

Heterojunctions are widely used for fabricating high performance electrical and optical device. The heterojunction fabricated by conventional epitaxial growth generally introduces a large number of crystalline defect densities at the interface due to the large differences in crystal lattice and thermal expansion coefficient (1, 2). To solve these difficulties a variety of wafer bonding methods have been employed to fabricate semiconductor hetero-structure. Among them the most effective candidate is surface activated bonding (SAB) (3-9). In this method, surfaces of substrates are activated by the irradiation of Ar fast atom beams prior to bonding and consequently bring into contact in a vacuum condition. It enables us to fabricate heterojunctions composed of dissimilar materials with large mismatched lattice constants. We previously fabricated p-Si/n-SiC and p-Si/n-GaAs heterojunctions and investigated their electrical properties (7, 10). It was found that the energy band diagram of p-Si/n-GaAs and p-Si/n-SiC junctions revealed type-II and type-I features, respectively. Furthermore, their conduction band discontinuities were calculated to be 0.59 and 0.3 eV, respectively. If SAB-based GaAs/SiC heterojunction are realized, the conduction band discontinuity of less than 0.3 eV should be expected, which should be suitable for fabricating devices with excellent high-frequency, high-temperature, and high-power performances that have not yet been achievable by each component material alone.

In this work, we fabricated p$^+$-GaAs/n-SiC heterojunction by using SAB method and have investigated the annealing temperature dependence of the band structure of GaAs/SiC heterojunction. The influence of the thermal annealing process on the electrical properties of the junctions were systematically investigated by measuring their current-voltage (I-V) and capacitance-voltage (C-V) characteristics. The applicability of

GaAs/SiC heterojunctions for functional devices was explored based on these measurements.

Experiments

p^+-GaAs (100) epitaxial substrates (400 nm, ~ 1 × 10^{19} cm^{-3} epitaxial layer / substrate ~ 1 × 10^{18} cm^{-3}) and n-4H-SiC epitaxial substrates (2.8 μm, 1.1 × 10^{17} cm^{-3} epitaxial layer / 0.5 μm, > 2 × 10^{18} cm^{-3} buffer layer / substrate ~ 1 × 10^{19} cm^{-3}) were used for the bonding experiment. Before bonding Al/Ni/Au multilayers were evaporated on the backside of n-SiC substrates. The ohmic contacts of n-SiC substrates were formed by a rapid thermal annealing at 1000 °C for 60 s in N_2 gas ambient. p^+-GaAs epitaxial substrates and n-SiC epitaxial substrates were bonded to each other by using SAB (3, 6). After the bonding, AuZn/Ti/Au multilayers were evaporated on the bottom surfaces of p-GaAs substrates. The p^+-GaAs/n-SiC heterojunctions were annealed separately at 100, 200, 300, and 400 °C for 60 s in N_2 gas ambient. All the samples were diced into 4 mm^2 pieces. The characteristics of these samples were also investigated while the junction was unannealed. Their I-V and C-V characteristics were measured using an ADCMT 6242 Source Measurement Unit and an Agilent E4980A Precision Impedance Analyzer, respectively.

Results

The I-V characteristics measured at room temperature are shown in Fig. 1. The GaAs/SiC heterojunction diodes showed good rectification properties. The ideality factor (n) for the forward bias voltages between 0.3 and 0.5 V was extracted to be 1.31, 1.38, 1.33, 1.32, and 1.33 for the unannealed junction and the junctions annealed at 100, 200, 300, and 400 °C, respectively.

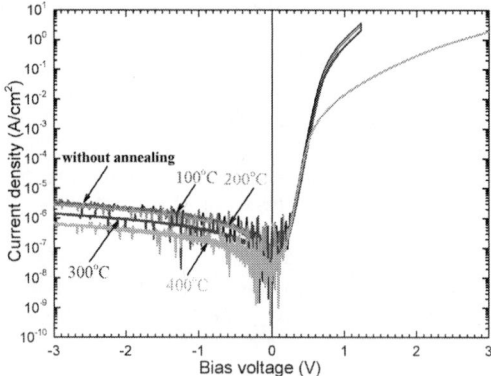

Figure. 1. I-V characteristics of p^+-GaAs/n-SiC heterojunctions without being annealed and annealed at 100, 200, 300, and 400 °C measured at room temperature.

It was found that the ideality factor (n) remained almost unaffected by the annealing temperature. In addition, the series resistance of the junctions was extracted from the slope of the measured *I-V* characteristics for the forward bias voltages between 0.8 and 1.2 V. We found that the resistance increased after annealing above 300 °C, which is attributed to the oxidation of the evaporated metal for the higher annealing process. In addition, the turn-on voltage of the junctions defined at the current of 100 mA/cm^2 dramatically increased from 0.73 to 1.55 V as the annealing temperature increased up to 400 °C. Moreover, the magnitude of the current increased as the junctions were more deeply reverse biased. Furthermore, as the annealing temperature increased, the magnitude of the reverse-bias current at - 3 V significantly decreased from 3.38×10^{-6} to 7.57×10^{-7} A/cm^2. The values of parameters for the respective junctions are summarized in Table I.

TABLE I. The reverse-bias current, turn-on voltage, ideality factor, resistance, and conduction-band offset of p$^+$-GaAs/n-SiC junctions.

Annealing temperature	Reverse-bias current (A/cm^2)	Turn-on voltage (V)	Ideality factor	Resistance ($\Omega \cdot$cm^2)	ΔE_c (eV)
Without annealing	3.38×10^{-6}	0.73	1.31	0.08	0.17
100 °C	3.44×10^{-6}	0.76	1.38	0.08	0.17
200 °C	2.82×10^{-6}	0.75	1.33	0.09	0.17
300 °C	1.51×10^{-6}	0.78	1.32	0.14	0.16
400 °C	7.57×10^{-7}	1.55	1.33	0.41	0

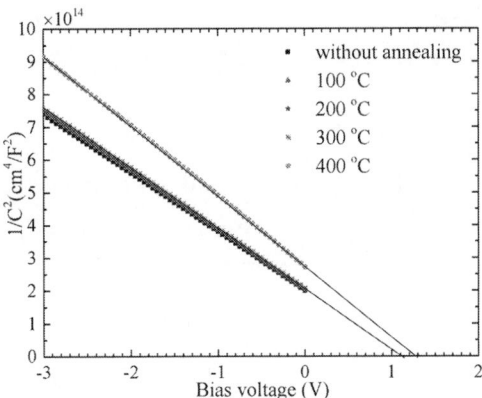

Figure 2. *C-V* characteristics of p$^+$-GaAs/n-SiC heterojunctions without being annealed and annealed at 100, 200, 300, and 400 °C measured at room temperature.

The $1/C^2$-V characteristics measured at room temperature and a frequency of 100 kHz are shown in Fig. 2. The characteristics indicated a straight line and the flat-band voltages (V_d) were found to be 1.12, 1.12, 1.12, 1.13 and 1.29 V for the unannealed junction and junctions annealed at 100, 200, 300, and 400 °C, respectively, by linearly extrapolating $1/C^2$ to zero. It is noteworthy that the flat-band voltages remained constant as the annealing temperature increased up to 200 °C and then increased as the annealing temperature increased from 300 to 400 °C. Using the slopes of $1/C^2$-V characteristics, the donor concentrations of the n-SiC epitaxial layer were estimated to be 8.14×10^{16}, 8.02×10^{16}, 7.98×10^{16}, 7.93×10^{16}, and 6.94×10^{16} cm^{-3}, for the unannealed junction and junctions annealed at 100, 200, 300, and 400 °C, respectively, which are close to the norminal value of 1.1×10^{17} cm^{-3}, as determined by Ni/n-SiC Schottky diode.

The *I-V* characteristics of the heterojunction annealed at 400 °C measured at various temperatures are shown in Fig. 3(a). The respective curves revealed a more marked asymmetric nature at lower temperature. Furthermore, as the temperature was raised, the magnitude of the current for the reverse bias voltages increased while their slope remained almost invariant to temperature. The dependence of the ideality factor and turn-on voltage as a function of temperatures in the range of 88 - 473 K are shown in Fig. 3(b). It can be seen that both parameters exhibited strong temperature dependence and the turn-on voltage decreased and the ideality factor increased with increasing the ambient temperature.

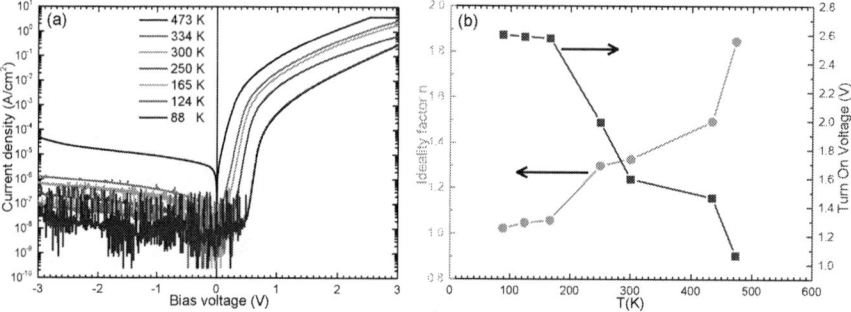

Figure 3. (a) *I-V* characteristics of p$^+$-GaAs/n-SiC heterojunctions annealed at 400 °C measured at various temperatures and (b) Temperature dependence of flat-band voltage and ideality factor (n) as function of temperatures in the range of 88 - 473 K.

Discussion

We observed that the turn-on voltage (~ 0.75 V) was smaller compared with the flat-band voltage (~ 1.2 V) and the magnitude of the current increased for larger reverse bias voltage for respective curves. Similar result was observed in the p-GaAs/n-GaN (11) and p-SiC/n-GaN (12) junctions fabricated by wafer fusion and molecular beam epitaxy, respectively, which was attributed to the scheme of interface states assisted tunneling. The interface states should be formed at the boding interface of the p$^+$-GaAs/n-SiC heterojunction and distributed in the amorphous layer of the bonded interface, which is due to damages fabricated during the Ar plasma irradiation in the SAB process according to our previous report. Furthermore, the 2.5 nm thick amorphous layer formed at the bonded interface was reported in the GaAs/SiC wafer bonded by SAB (13). If the electron transport property is dominated by carrier recombination or tunneling through the interface sates, the ideality factor should be close to 2. However, the extracted ideality factors (n) are as low as ~ 1.3 irrespective of the annealing condition.

Note that the linear extrapolation of $1/C^2$ is based on the assumption that no electric charges are placed at the GaAs/SiC interfaces. In this scheme, V_d is given by the difference in the work functions between p$^+$-GaAs and n-SiC and is expressed as

$$qV_d = E_{gp} - \Delta E_c - \delta_{n-SiC} - \delta_{p^+-GaAs} \qquad (1)$$

where q is the elementary charge, is the energy gap of p$^+$-GaAs, is the conduction-band offset, and and refer to the position of the Fermi energies relative to the valence-band maximum in p$^+$-GaAs and that relative to the conduction-band minimum in n-SiC, respectively. Using eq. (1), the values of were determined for the unannealed junction and junctions annealed at 100, 200, 300, and 400 °C and are summarized in Table I. It is noteworthy that is ≈ 0 eV, which means that is ≈ 1.8 eV for heterojunctions annealed at 400 °C. The energy band diagrams of the heterojunction annealed at 400 °C based on the estimation is shown in Fig. 4(a). Their diagram suggests that minority electrons in p$^+$-GaAs layers could be transported into n-SiC layers without the inference of heterointerfaces, which means that the GaAs/SiC heterojunctions are suitable for fabrication of high frequency power devices (such as heterojunction bipolar transistors).

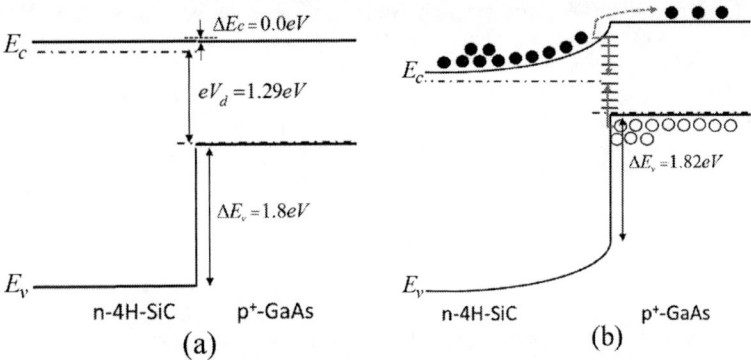

Figure 4. (a) Flat-band diagrams of p$^+$-GaAs/n-SiC heterojunctions and (b) Band diagrams of p$^+$-GaAs/n-SiC heterojunctions under the forward bias.

We have to note that the ideality factor continuously increased and the turn-on voltage decreased as the ambient temperature was raised, as shown in Fig. 3(b). This phenomenon can be explained on the basis of Fig. 4(b), where the band diagram of the GaAs/SiC heterojunction under the forward bias is shown. The electrons accumulated near the interface of the GaAs/SiC heterojunction drop into the interface states and recombine with the trapped holes. The rate of the capture of holes from the valence band increases with increasing temperature (14). The bias voltage and the temperature dependence of the reverse-bias current suggest that the electrons in the valence band edge of the p$^+$-GaAs were thermally excited to the interface states, and then tunnel into the conduction band of the n-SiC for the reverse bias. In addition, the activation energy was estimated to be 0.25 eV from the temperature dependence of the reverse-bias current.

The features in the p^+-GaAs/n-SiC heterojunction after annealing at 400 °C that (1) the reverse-bias current at − 3V decreased to 7.57×10^{-7} A/cm^2, (2) the obtained turn-on voltage is consistent with the flat-band voltage extracted from C-V measurement, and (3) the determined ideality factor (1.33) is close to 1 indicated that the impacts of the interface states on the conductive properties of the GaAs/SiC heterojunction is comparatively small and the diffusion current mechanism mainly dominates the transport properties of carriers across the interface. These results suggest that the SAB-based GaAs/SiC heterojunctions are expected to play a significant role for fabricating high-power and high-frequency devices.

Conclusion

The lattice-mismatched p^+-GaAs/n-SiC heterojunctions were realized by using the surface-activated bonding without heating. The influence of thermal annealing process on the interface of the p^+-GaAs/n-SiC heterojunctions was demonstrated. The reverse-bias current at − 3 V decreased with increasing the ambient temperature, the value was finally reduced to 7.57×10^{-7} A/cm^2 after annealing at 400 °C. The flat-band voltage extrapolated from the *C-V* characteristics, which is close to the turn-on voltage obtained from the *I-V* characteristics. The conduction-band offset extracted from the *C-V* characteristics was found to be 0 eV. This result suggested that the band profile of the GaAs/SiC heterojunction was likely to be preferable for fabricating collector junction in heterojunction bipolar transistors. Thus, the SAB technology in combination with the termal annealing is likely to be useful for fabricating devices for high-power applications.

Acknowledgements

This work was partly supported by "Creative Research for Clean Energy Generation Using Solar Energy" project in Core Research for Evolutional Science and Technology (CREST) programs of the Japan Science and Technology Agency (JST).

References

1. S. F. Fang, K. Adomi, S. Iyer, H. Morkoc, H. Zabel, C. Choi, and N. Otsuka, *J. Appl. Phys.*, **68**, R31 (1990).
2. V. K. Yang, M. Groenert, C. W. Leitz, A. J. Pitera, M. T. Currie, and E. A. Fitzgerald, *J. Appl. Phys.*, **93**, 3859 (2003).
3. H. Takagi, K. Kikuchi, R. Maeda, T. R. Chung, and T. Suga, *Appl. Phys. Lett.*, **68**, 2222 (1996).
4. S. Essig, O. Moutanabbir, A. Wekkeli, H. Nahme, E. Oliva, A. W. Bett, and F. Dimroth, *J. Appl. Phys.*, **113**, 203512 (2013).
5. M. M. R. Howlader, T. Suga, F. Zhang, T. H. Lee, and M. J. Kim, *Electrochem. Solid-State Lett.*, **13**(3), H61 (2010).
6. M. M. R. Howlader, T. Watanabe, and T. Suga, *J. Appl. Phys.* **91**, 3062 (2002).
7. J. Liang, S. Nishida, M. Arai, and N. Shigekawa, *Appl. Phys. Lett.*, **104**, 161604 (2014).

8. N. Shigekawa, M. Morimoto, S. Nishida, and J. Liang, *Jpn. J. Appl. Phys.*, **53**, 04ER05 (2014).
9. N. Shigekawa, J. Liang, N. Watanabe, and A. Yamamoto, *Phys. Status Solidi C*, **11**, 644 (2014).
10. J. Liang, T. Miyazaki, M. Morimoto, S. Nishida, N. Watanabe, and N. Shigekawa, *Appl. Phys. Express*, **6**, 021801 (2013).
11. C. Lian, H. Xing, Y. Chang, and N. Fichtenbaum, *Appl. Phys. Lett.*, **93**, 112103 (2008).
12. H. Miyake, T. Kimoto, and J. Suda, *Jpn. J. Appl. Phys.*, **53** 034101 (2014).
13. E. Higurashi, K. Okumura, K. Nakasuji, and T. Suga, Jpn. *J. Appl. Phys.*, **54**, 030207 (2015).
14. S. M. Sze, *Semiconductor Devices Physics and Technology*, p. 49, Wiley& Sons, New York, (1985).

228

Surface Preparation and Eutectic Wafer Bonding

M. W. Heller[a], M. Zoberbier[b], T. Fujita[c], M. Eichler[d]

[a] Kionix, Inc., Ithaca, New York 14850, USA
[b] SUSS MicroTec Lithography GmbH, 85748 Garching, Germany
[c] Rohm Co. Ltd., Ukyo-ku, Kyoto 615-8585, Japan
[d] Fraunhofer-Institut für Schicht- und Oberflächentechnik IST, 38108 Braunschweig, Germany

Bond surface preparation is an integral part of all eutectic wafer bond processes currently used in high volume production. It is not unusual to have a requirement conflict between the bond surfaces and the device itself, especially in case of MEMS inertial sensors. A typical example of the issue at hand is the need of an anti-stiction coating material for the device and the interference of the material with the eutectic bond process. This paper reports on the novel usage of localized plasma in conjunction with an improved permanent wafer bonder to reconcile the conflicting needs in case of eutectic Aluminum-Germanium wafer bonding.

Introduction

As of today, glass frit bonding is still widely used in high volume inertial sensor production for consumer and automotive applications. It is predominantly used for wafer-level packaging of accelerometers whereas nearly all of the current generation gyroscopes on the market are using alternative wafer-level packaging solutions like eutectic solder bonding (Invensense, Bosch, Maxim, NXP, Analog Devices) or metal thermocompression bonding (ST Microelectronics). One of the main reasons for the staying power of glass frit bonding is its excellent compatibility with commercially available anti-stiction coatings based on organic self-assembled monolayers (SAM). These coating types are widely used in the inertial MEMS sensor industry to improve the robustness of accelerometers against latching type failures. Properly designed gyroscopes on the other hand do not require anti-stiction coating as latching type failure modes can be safely avoided by the inherently much larger mechanical restoring forces.

The main thrust to abandon glass frit stems from its limitation in achievable minimum seal width, the cost of the frit material and the ongoing uncertainty regarding the RoHS and REACH compliance as the overwhelming majority of glass frit paste used in high volume sensor production contains Lead. Lead free pastes with low melting points have been developed in the last few years by companies like Ferro, AGC and Schott, but they have not been successful in replacing their Lead containing counterparts in high volume production so far.

Eutectic wafer bonding processes require clean surfaces to accomplish a successful wafer bond. Excessive amounts of native oxides and other organic contaminants on the surfaces intended to form the eutectic seal can be detrimental to bond formation and strength. Depending on the seal metallization layer stack configuration and the device, the removal of the native oxide layer and other contaminants from the surface can be a challenging process integration issue.

Aluminum-Germanium Eutectic Wafer Bonding

Aluminum-Germanium (AlGe) eutectic wafer bonding is the most widely used wafer level packaging process for MEMS gyroscopes in high volume production today. The history behind its usage for wafer bonding dates back over 20 years. A patent describing the usage of an Aluminum-Germanium layer stack to bond two wafers together, no mention of MEMS though, was first filed in 1991 by G. Schuster, et. al. (1). The first MEMS specific publication was by P. M. Zavracky in 1995 and the first mention of using an AlGe eutectic wafer bond process to electrically connect a MEMS inertial sensor wafer to an ASIC wafer can be found in a German patent application from 1995 by G. Flach et. al. (2-3). One of the main advantages of AlGe compared to other eutectic wafer bond processes like Gold-Silicon (AuSi) is its compatibility with a standard in-house CMOS ASIC wafer fab or outside foundries like TSMC, Globalfoundries, X-Fab. No overly burdensome countermeasures and procedures to limit process line cross contamination need to be implemented. The eutectic point for AlGe has a melt temperature in the vicinity of 690K with an Al/Ge weight percent ratio of 48.4/51.6 according to literature (4). This is about 60K higher than the temperature of the AuSi eutectic point but not too high for having an ASIC compatible wafer bond process (5-7).

conventional surface preparation

The goal of seal surface preparation before bond is to minimize the amount of surface oxides and other contaminants negatively impacting the wafer bond result as well as the process stability. Depending on the exact seal metal layer stack configuration, one has to prepare either Aluminum, Germanium or both types of surfaces. Germanium forms mainly Germanium-dioxide (GeO_2) with a suboxide layer (GeO_x, $x<2$) (8). The GeO_2 is water soluble, whereas the suboxide is not (8). A dip in HF with a sufficiently long DI water rinse afterwards enables also the removal of the suboxide layer (8). The wafers should be stored under Nitrogen in the bond wait queue to prolong the time until the native oxide reforms. It has been reported that using a HBr based etchant and Nitrogen storage reformation of the native oxide layer on Germanium can be inhibited up to 24 hours (8). Aluminum forms a hard native oxide layer up to several nanometers thick (9-10). One documented solution is using a Hydrogen containing gas mixture and try to reduce the oxide in-situ during the wafer bond process. The efficiency of the process in the available temperature range is undocumented though (5). Another possibility to deal with the native oxide layer on Aluminum is by using brute bond force to break thru the passivation layer. Published bond recipes for AlGe therefore usually mention bond forces in the 40-60 kN range, the exact amount of seal area per wafer and therefore the actual pressure on the seal area is rarely documented though (7-8). A further process option to remove the native Alumina layer is using a reduced pressure Argon plasma before bonding to remove the oxide using an ion milling/Argon sputtering process. The wafers in the bond queue should be kept under Nitrogen atmosphere to slow down the reformation of the native oxide layer after using such a process. The maximum queue time allowed depends on the Aluminum surface condition going into the surface preparation process and the process condition itself. It is a well-known fact that an Aluminum surface subjected to a gas phase etch (GPE) process using anhydrous Hydrogen fluoride (AHF) is terminated with Fluorine, effectively forming a mixture of Aluminum fluoride and some amount of Aluminum oxyfluoride. The XPS surface analysis results of Aluminum post AHF GPE and its state after two different reduced pressure plasma based surface preparation processes is summarized in Table I. Process #1 clearly improves the condition over the initial state showing an increased amount of Aluminum and reduced amount of Oxygen. The Fluorine amount is roughly the same. Process #2 leads to a huge amount of Carbon on the surface, only low amounts of Aluminum and Oxygen are detected and the Fluorine is reduced by 10%. Process #1 was chosen as base process to prepare the Aluminum seal metal before bonding. The XPS depth profile in Figure 2 after Process #1 clearly shows an improved surface condition compared to Figure 1 showing the initial surface state. Cap shear testing was done to evaluate the influence of Process #1 on the bond strength. The results in Table II as well as the wafer appearance after grinding and cap dicing indicate that Process #1 improves the seal surface condition for bonding. There is a queue time limitation though. Wafers should be bonded to as soon as possible after preparation, best within 24 hours. The main drawback of Process #1 is the non-selective / un-patterned nature of the surface modification. It cannot selectively modify only the seal surface but modifies the whole wafer. As a consequence, the process cannot be used after an anti-stiction coating has been deposited on a wafer as it would be removed.

TABLE I. XPS analysis of different Aluminum surface conditions.

condition	Al (at%)	O (at%)	F (at%)	C(at%)
post AHF GPE state	16.9	18.6	52.6	1.9
Process #1	26.7	10.1	53.9	5.0
Process #2	3.5	3.1	43.2	50.2

TABLE II. average cap shear strength for a specific seal design as result of process condition.

condition	seal type A - average cap shear strength / kgf
Post AHF GPE state	2.6
1h post Process #1	6.5
48h post Process #1	3.2

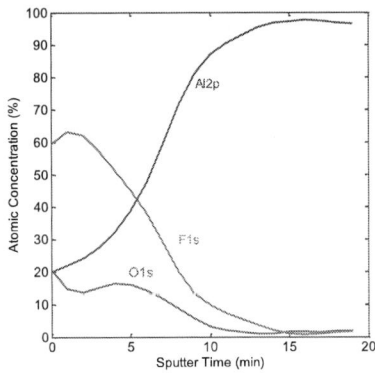

Figure 1. XPS depth profile post AHF GPE state.

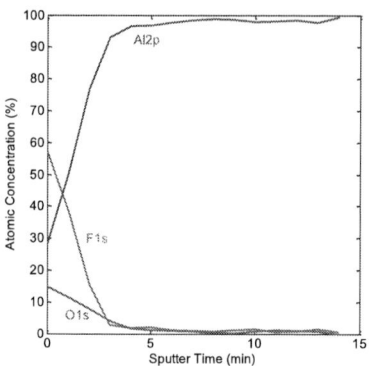

Figure 2. XPS depth profile post process #1.

high bond force and anti-stiction coating

Based on the assumption that high force should be able to break through the anti-stiction layer and enable normal AlGe eutectic seal formation, a DOE based on process of record recipe was set-up to test the hypothesis on eutectic bond short loop wafers. Wafer bonding was accomplished using a Suss MicroTec XB8 high force bonder with improved bond force uniformity compared to the previous generation and a maximum bond force of 100kN. Another change from the previous tool generation is the dual zone heating configuration enabling better control of the temperature distribution within the bond chuck. The DOE parameters were anti-stiction coating and bond force. The deposited anti-stiction coating is used by Kionix in the accelerometer production process. The parameters as well as the results are shown in Table III. Even with a bond force of 100kN no sufficiently strong bond could be formed. Wafer pairs with bond condition B03 and B04 were easily split apart by razor blade insertion test indicating unacceptable low bond strength.

TABLE III. high force Bond DOE.

Bond	anti-stiction coating	bond force	bond result
B01	no	POR	OK
B02	no	100kN	OK
B03	yes	POR	FAIL – wafer delamination
B04	yes	100kN	FAIL – wafer delamination

Wafer pairs B03 and B04 showed abnormal behavior in the post bond IR inspection. Circular voids in varying size could be found all over the wafer in the squished eutectic melt as shown in Figure 3. Inspection after splitting the B03 and B04 wafer pairs apart clearly showed the interference of the anti-stiction coating with the eutectic bond process as can be seen in Figures 4,5 and 6 from wafer B04. The coating does not dissolve cleanly into the eutectic melt but keeps acting as a separation barrier introducing a weak interface thru the seal and inhibiting the squished melt to interact with a Silicon dioxide layer beneath the coating.

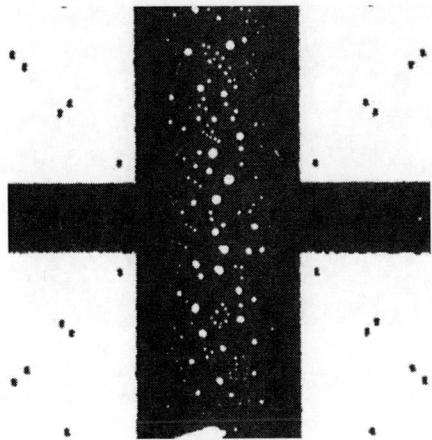

Figure 3. post bond IR image of wafer pair B04 showing circular voids.

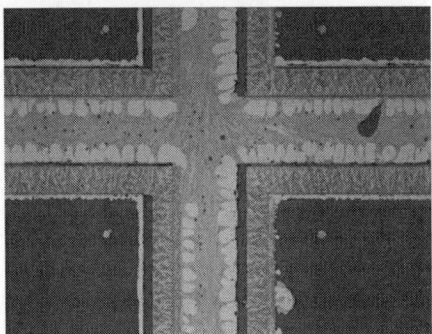

Figure 4. optical image uncoated side of wafer pair B04 after razor blade test.

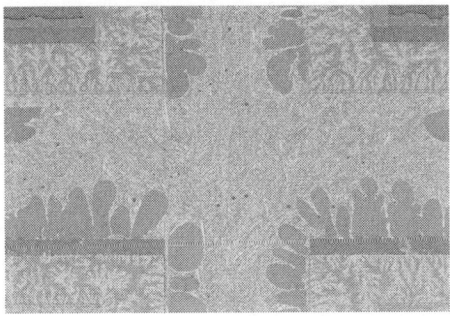

Figure 5. SEM image uncoated side of wafer pair B04 after razor blade test.

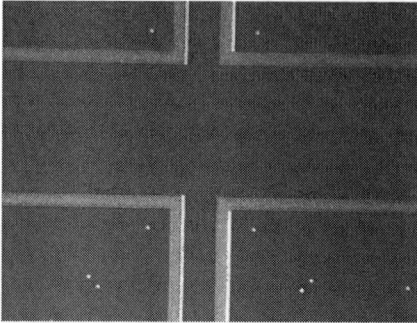

Figure 6. optical image of the coated wafer pair B04 side showing no signs of any real adhesion between the eutectic melt and the coated oxide surface.

Localized Plasma Process for Eutectic Wafer Bonding

Selectively removing the anti-stiction coating from the seal areas immediately before bonding or depositing it everywhere besides and preparing the surface at the same time would be a very convenient solution. Several concepts have been patented over the last few years to selectively remove the anti-stiction coating before bonding, none of them specifically target improving the bond surface characteristics besides the removal of the anti-stiction coating from the seal area (11-12).

Dielectric Barrier Discharge - Atmospheric Pressure Plasma

A dielectric barrier discharge (DBD) is a cold plasma able to run at atmospheric pressure levels (AP plasma). An electrode coated with a dielectric is brought into close proximity to a substrate and a AC voltage is applied between the electrode and substrate igniting a plasma (13-14). Compared to reduced pressure plasmas used in conventional wafer processing equipment like RIE etchers, an AP plasma cannot provide high energetic ion bombardment. On the upside DBD are capable of creating localized micro plasma environments as the mean free path length is only about 100nm at 1 bar. The first application using DBD in a wafer bond process was the surface preparation for low temperature Silicon-Silicon fusion bond processes (15-16). An option for the current generation of Suss mask or bond aligners called SELECT adds to the tool the possibility to generate a structured DBD (17-18). A structured glass mask with a transparent electrically conductive coating on the backside side acts as an electrode with a dielectric. The wafer on the aligner chuck acts as the counter electrode. The DBD can used with different gas types or mixtures of gasses to modify the surface in the plasma areas accordingly. There are two main ways the DBD can be used. One is called plasma printing or plasma stamping; the other one local plasma treatment. In case of plasma printing the active volume is inside a recess in the dielectric. Local plasma treatment has the plasma active either between the flat dielectric and raised wafer areas or between a flat dielectric or inside recessed features on the wafer. Both situations are illustrated in Figure 8.

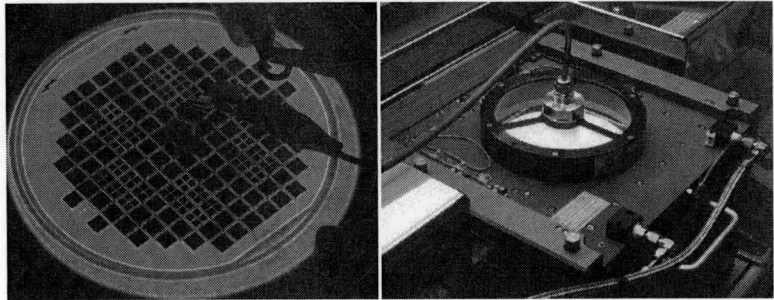

Figure 7. left: DBD using a structured dielectric material, right: the SELECT option installed in a Suss mask aligner. The thick cable connected to the glowing cap supplies the AC voltage for plasma ignition to the glass electrode below. The two flexible metal lines deliver the working gas for the plasma

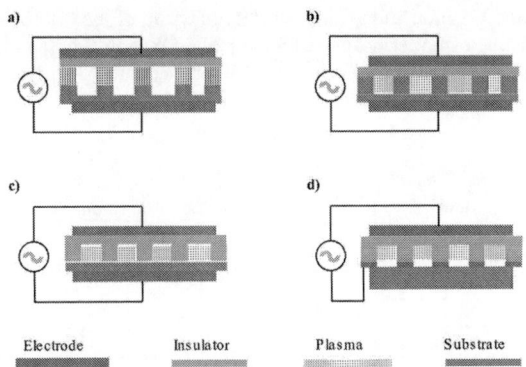

Figure 8. a), b) depicting local plasma treatment   c), d) plasma printing.

Modification of AlGe seal surfaces by plasma printing

The Aluminum surface modification capability of different gas mixtures and combinations thereof using different process times/condition were investigated. The gas mixtures consist of a noble carrier gas and other reactive gases, usually up to two different species. Mixtures A1, A2, A3 and A4 consist of the same species in different mixing ratios. Accordingly, mixture B1 has different species and their composition ratio is different from B2. The other main process variable is process time. The goal of the investigation was the assessment of a suitable set of process conditions to remove the anti-stiction coating currently used in production at Kionix and improve the surface condition prior to bonding. The Aluminum surface was investigated with a combination of XPS as well as EDX analysis to get a more complete picture. The resulting change in surface condition measured by EDX and/or XPS for the various experiments are summarized in tables IV to VII following below.

TABLE IV. initial EDX and XPS results for gas mixtures A1 and B1with varying process time.

gas	time	EDX – ab initio change in %					XPS – ab initio change in %				
		C	O	F	Al	Si	C	O	F	Al	Si
A1	10s	-2.74	-38.79	14.49	2.05	-62.39	-52.58	20.77	-13.99	300.30	-79.82
A1	20s	-3.80	-40.02	20.70	1.91	-64.65	-59.85	12.83	-7.03	306.79	-73.88
A1	100s	3.71	-44.50	43.08	0.79	-64.76	-60.68	6.37	2.31	286.41	-82.06
B1	10s	10.24	-1.51	-38.35	-1.34	-0.77	-62.52	143.15	-79.11	121.12	94.72
B1	20s	-1.12	5.39	-29.73	0.24	0.63	-66.22	130.27	-70.41	127.62	93.93
B1	100s	-2.64	15.39	15.42	-0.40	-8.41	-64.45	115.09	-61.67	150.07	66.89

The XPS results for gas A1 in table IV show a strong increase in the Aluminum signal, whereas EDX does not really indicate a change. As XPS is extremely sensitive to the surface whereas EDX is more material volume sensitive, the huge change in the XPS Aluminum signal is caused by removal of a big part of the coating and contamination (Silicon, Carbon) from the seal area allowing XPS access to the Aluminum. EDX results for A1 do not really show an improvement for the Aluminum and Carbon content but a large reduction for Silicon and Oxide. The results for B1 clearly show a strong reduction of the Fluorine content in the EDX as well as the XPS results. XPS shows a strong increase in Silicon content and Oxygen and again a strong reduction of the Carbon content whereas EDX only shows a much smaller change in Silicon and Oxygen. These results indicate that the B1 mixture is able to remove the Carbon and Fluorine content from the surface whereas A1 removes the Oxygen and Silicon contamination.

TABLE V. EDX results for 2nd round of experiments with gas mixtures A1, B1 and combination.

gas	time	EDX – ab initio change in %			
		O	F	Al	Si
B1	10s	6.22	-57.85	0.56	-6.25
B1	20s	7.57	-49.59	0.33	3.75
A1	20s	-34.86	15.70	1.68	-58.75
B1 + A1	10s + 20s	-36.76	19.83	1.74	-63.75
B1 + A1	20s + 20s	-32.97	24.79	1.50	-60.00

In the second round of experiments the surface modification properties of the gas mixtures A1 and B1 were confirmed and two sequential combinations of the gas mixtures tested. The EDX results in table V are in line with the results from the results from the first experiments and the combination results further confirm the assumptions.

TABLE VI. EDX results for 3rd set of experiments varying concentration the gas mixture constituents.

gas	time	EDX – ab initio change in %			
		O	F	Al	Si
A1	10s	-14.34	30.44	0.49	-43.56
A2	10s	-35.45	21.78	1.45	-50.00
A3	10s	-38.53	26.07	1.59	-61.36

The concentration of the gas mixture constituents A1 was varied in the third DOE to judge their individual effectiveness. Based on the EDX results shown in table VI, mixture A3 seems to be the most effective choice to reduce the Oxygen and Silicon contamination.

TABLE VII. XPS results for 4th set of experiment varying concentration the gas mixture constituents.

gas	time	XPS – ab initio change in %				
		C	O	F	Al	Si
B2	10s	-70.06	127.56	-77.25	104.86	81.66

In the fourth experiment the balance of the gas mixture B1 was modified and the influence on the top surface condition investigated by XPS. The results shown in table VII do not indicate an improvement compared to the XPS results from the initial set of experiments.

TABLE VIII. EDX results for 5th round of experiments with combination of B1 and A4, varying time.

gas	time	EDX – ab initio change in %			
		O	F	Al	Si
B1 + A4	1s + 1s	2.90	-19.11	0.29	-25.76
B1 + A4	2s + 2s	-29.66	-11.11	1.50	-34.85
B1 + A4	4s + 3s	-29.38	7.56	1.27	-34.85
B1 + A4	8s + 8s	-24.41	12.89	1.03	-36.36
B1 + A4	10s + 10s	-40.09	-1.40	3.18	-81.43

The results for a third variant A4 of the gas mixture A1, tested in combination with B1 under varying process time conditions are shown in table VIII. The conclusion based on the various experiments is, that a combination process using gas mixture B1 for 10 seconds followed by gas mixture A3 or A4 for 20 seconds seems to be reasonably suited for the intended purpose of removing the anti-stiction coating and improving the seal surface bondability. An extensive bond DOE to verify the experimental results is currently in progress. Initial results seem to indicate the validity of the surface modification result interpretation.

selective deposition of anti-stiction coating by plasma printing process

The plasma printing process can also be used to selectively deposit a hydrophobic coating with anti-stiction properties on a MEMS device wafer enabling in principle the usage of the proven surface preparation Process #1 before bonding. The contact angle of four different coatings was measured with a Dataphysics OCA20 using dynamic tracking and a water flow rate of 0.06 μl/s . The contact angle measurements results are shown in figure 9. Type A has a maximum advancing contact angle of 147° and no hysteresis. Type B has a contact angle of approximately 109° and a hysteresis of about 0.3° to 1.1 °. Type C has maximum contact angle of 102° and the hysteresis rises from 5° to 35°. Type D shows discontinuities in one of the measurements because of surface roughness. In general, the contact angle seems stable at about 109° and the hysteresis is also stable at about 18°. Based on the contact angle results type A and B were selected as the alternatives to be tested against the anti-stiction coating used in production. The standard Kionix product qualification flow is currently used to assess their suitability as a replacement. Conclusive test results are not yet available.

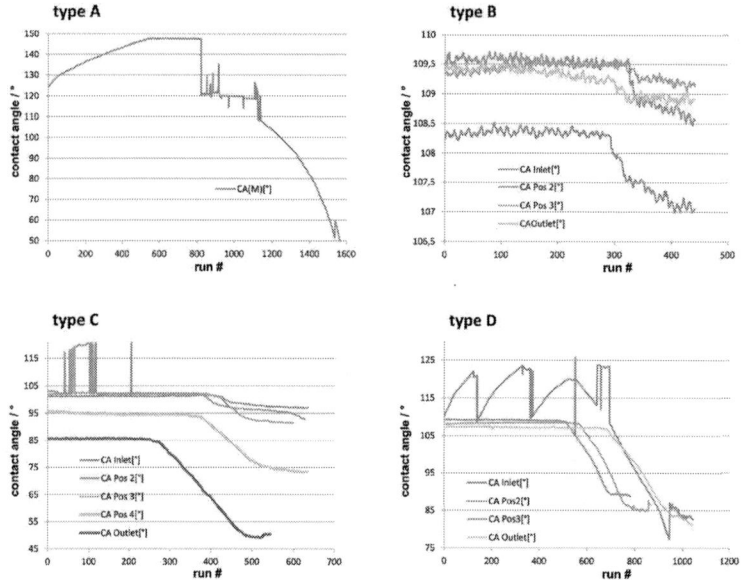

Figure 9. dynamic contact angle measurement results for four different coatings.

Conclusion

Two process integration flows are currently under evaluation as possible manufacturing solutions. The results so far indicate the real possibility of having at least one solution capable of combining anti-stiction coating and AlGe eutectic wafer bonding without the need to compromise between the coating anti-stiction performance on one hand and the bond process stability and capability requirements for a high volume production on the other. As the SELECT option is integrated into the bond aligner, the seal area surface preparation can be done in-situ during the wafer alignment step (17). The additional process time needed for the step is less than two minutes based on the results so far. By proper tool scheduling the increase in total runtime for bonding a 25 wafer lot can be kept below 15 minutes causing only miniscule throughput reduction during the wafer bond stage.

Acknowledgments

The authors would like to express their gratitude for the extensive support provided by their colleagues at Kionix, Rohm, Suss MicroTec and Fraunhofer IST.

References

1. G. Schuster and K. Panitsch, *Process for the laminar joining of silicon semiconductor slices*, US5693574 (1997).
2. P. M. Zavracky and B. Vu, *Patterned eutectic bonding with Al/Ge thin films for MEMS*, in *Proc. Micromachining and Microfabrication Process Technology*, p. 46-52, SPIE, Bellingham,WA (1995).
3. G. Flach, U. Nothelfer, G. Schuster and H. Weber, *Micromechanical acceleration sensor*,US5905203 (1999).
4. H. Okamoto, Journal of Phase Equilibria and Diffusion, **19**, 86 (1998).
5. S. S. Nasiri and A. F. Flannery, *Method of fabrication of a AL/GE bonding in a wafer packaging environment and a product produced therefrom*, US7442570 (2008).
6. S. Sood, S. Farrens, R. Pinker, J. Xie and W. Cataby, *ECS Transactions*, **33**, 93 (2010).
7. S. Sood, *Advanced Metal-Eutectic Bonding for High Volume MEMS WLP*, in *IEEE MEMS Bay Area Meeting* (2014)
8. B. Onsia, T. Conard, S. De Gendt, M. Heyns, I. Hoflijk, P. Mertens, M. Meuris, G. Raskin, S. Sioncke and I. Teerlinck, *Solid State Phenom*, **103**, 27 (2005).
9. L. P. H. Jeurgens, W. G. Sloof, F. D. Tichelaar, C. G. Borsboom and E. J. Mittemeijer, *Applied Surface Science*, **144–145**, 11 (1999).
10. J. Evertsson, F. Bertram, F. Zhang, L. Rullik, L. R. Merte, M. Shipilin, M. Soldemo, S. Ahmadi, N. Vinogradov, F. Carlà, J. Weissenrieder, M. Göthelid, J. Pan, A. Mikkelsen, J. O. Nilsson and E. Lundgren, *Applied Surface Science*, **349**, 826 (2015).
11. M. Hancer, *Selective UV-Ozone dry etching of anti-stiction coatings for MEMS device fabrication*, US8237296 (2012).
12. P. Y. Liu, L. C. Chu, H. H. Lin, S. Y. Tsai, Y. C. Hsieh, J. H. Peng, L. L. Chao, C. S. Tsai and C. W. Cheng, *Self-removal anti-stiction coating for bonding process*, US8905293 (2014).
13. M. Eichler, C. P. Klages, M. Lindmayer and R. Thyen, *Method and device for the plasma-activated surface treatment and use of the inventive method*,EP1264330B1 (2003).
14. C.-P. Klages, A. Hinze, K. Lachmann, C. Berger, J. Borris, M. Eichler, M. von Hausen, A. Zänker and M. Thomas, *Plasma Processes and Polymers*, **4**, 208 (2007).
15. M. Gabriel, B. Johnson, R. Suss, M. Reiche and M. Eichler, *Microsystem Technologies*, **12**, 397 (2006).
16. M. Eichler, P. Hennecke, K. Nagel, M. Gabriel and C.-P. Klages, *ECS Transactions*, **50**, 265 (2013).
17. M. Gabriel, S. Milde, M. Eichler and C.-P. Klages, *MikroSystemTechnik* (2009).
18. M. Eichler and M. Gabriel, *Verfahren und Vorrichtung zur selektiven Plasmabehandlung von Substraten zur Vorbehandlung vor einem Beschichtungs- oder Bondprozess*, DE102005042754B4 (2008)

240

Decreased Surface Porosity and Roughness of InP for Epitaxially Grown Thin-Film Devices: A Path to Integration of High Performance Electronics

M. Gervasoni, A. Machness, M. S. Goorsky

Department of Materials Science and Engineering, University of California, Los Angeles, California, 90095, USA

Devices based on III-V semiconductors have generated widespread interest due to their superior performance over conventional semiconductors. However, III-Vs are inhibited from mass commercialization because of high costs. Thus, there is a motivation to develop affordable, high-quality thin film III-V devices, which has been realized through exfoliation techniques. In this work, a porous InP structure with optimal morphology was fabricated for exfoliation. A relationship between surface porosity and current density was determined for obtaining a desirable porous structure. Surface porosities as low as 2% were achieved, and it was found that reducing the porosity lowered the surface roughness. Thus, porous layers with surface roughness approaching polished InP were demonstrated. The porosities and surface roughness obtained are lower than previously reported and more ideal for growing a low-defect epitaxial film. By growing higher quality epitaxial films on a reusable seed wafer, high performance and cost effective III-V devices are made possible.

Introduction

Integrating thin film III-V semiconductors into high performance electronics remains a promising route for increasing device efficiency while reducing material costs. For example, indium phosphide's (InP) direct band gap and high electron mobility are optimal properties for high-electron-mobility transistors (HEMTs) with efficiencies that surpass silicon (1). Moreover, InP thin films utilize less material, which drive down fabrication costs and increase their potential in flexible electronics. Thin film InP is commonly formed by mechanically grinding down the substrate (2). However, this process is wasteful and time consuming and such challenges have been circumvented through exfoliation techniques.

The epitaxial layer TRANsfer (ELTRAN) manufacturing process readily separates an epitaxially grown thin film device from a bulk substrate through a mechanically weak porous layer buried underneath the film (3, 4). The substrate is reused several times for subsequent layer transfers, resulting in a more economical thin film device. The porous morphology requires a top layer with small pores, low porosity, and a low surface roughness approaching that of the substrate to ensure a high quality epitaxial layer. Furthermore, a buried layer should have pore sizes and porosities large enough to readily detach the layer from the substrate. The ELTRAN process has been readily demonstrated with silicon, and research efforts have focused on employing this technique to III-V semiconductors, which are more promising materials in high performance devices but are inhibited by mass commercialization because of their cost (5, 6).

Previous research efforts have focused on III=V films transferred to a porous substrate or epitaxially grown on porous silicon. However, their lattice mismatch and difference in thermal expansion coefficients result in defects within the film and reduce device performance (7, 8). More recent work grew InP epitaxial layers on porous InP, but focused on layers with porosities greater than 60 percent (9-11). Therefore, pore morphologies for more ideal epitaxial growth of high quality layers have not been observed. This work extends on previous current density studies, and elucidates a more optimal layer for epitaxial growth at a lower current density range, respectively. A relationship between etching current and porosity is determined, as well as a relationship between porosity and surface roughness.

Experimental

Porous InP samples were prepared by electrochemically etching n-type InP wafers (1-10×10^{18} cm^{-3}, S-doped, <001> oriented, Wafer Technology Ltd.) in aqueous HCl solution using a Teflon etcher with a gold plated mesh counterelectrode, shown in Fig. 1. All samples were etched in a 5% HCl aqueous solution (ACS reagent 37%, Sigma Aldrich used as received) for 60 seconds. The etching current density was varied from 0.14 to 50 mA/cm^2 using a Keithley 220 current source. Surface morphologies were analyzed using plan-view secondary electron microscope (FEI Nova NanoSEM 230) and surface roughness was measured by atomic force microscopy (AFM).

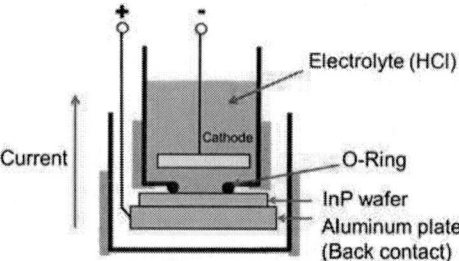

Figure 1. Schematic of electrochemical etching of InP. Image reproduced from Kou (11).

Results and Discussion

In agreement with previous reports, the porosity decreased with decreasing etching current density as shown in Fig. 2 (10, 11).

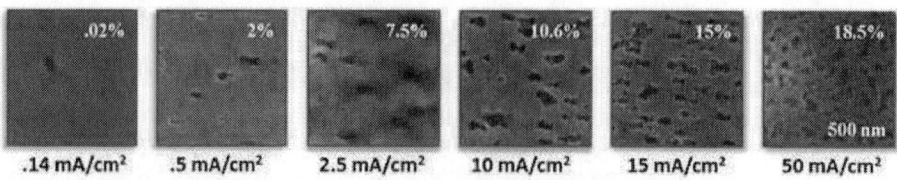

Figure 2. SEM Images of InP surfaces showing porosity at different current densities. ?

A relationship between porosity, P, and current density was determined from Fig. 3 as logarithmic and is given by the following equation

$$P = 3.2134 * \ln(J) + 5.1082 \tag{1}$$

Where P is the porosity (percent area covered by pores) and J is the current density in mA/cm^2.

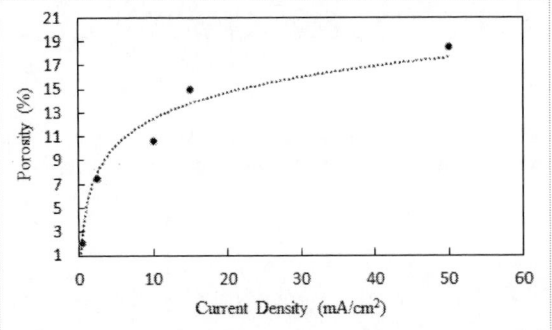

Figure 3. Surface porosity as a function of current density for InP. All samples were etched in 5% HCl for 60 seconds.

In contrast, a porosity versus current density relationship previously developed for porous silicon demonstrated a linear relationship (12). This trend discrepancy is perhaps due to the transition III-V semiconductors undergo from crystallographically oriented pores to current-line oriented pores with increasing etching current (13).

In comparison to previous studies, the surface porosities were greatly reduced, as shown in Fig. 4 which, in turn, have created a more suitable substrate for epitaxy growth. Moreover, Fig. 4 shows that porosity was reduced by two orders of magnitude by simply reducing the current density to ranges which were previously not explored.

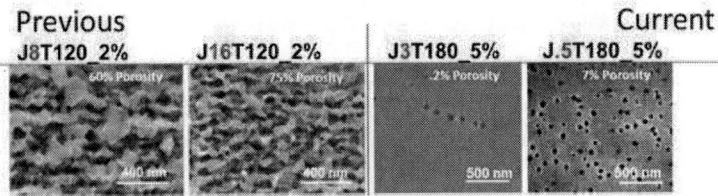

Figure 4. Previous surface porosities (left) compared to new surface porosities. Image reproduced from Kou (10). Can you add "0" to the .2%?

Previously it was observed that when forming multilayers, subsequent etching steps did not affect previous ones. However, we discovered that porosities have a low-end limit before the buried layer formation is no longer independent of the top layer formation. Fig. 5 shows how a layer with 2% porosity is not affected by subsequent etching steps while a layer with 0.14% porosity is in fact affected when a buried layer is formed. This threshold value was discovered to be between top-layer porosities of 0.02% and 2% which correspond to current densities of 0.14 mA/cm^2 and 0.5 mA/cm^2 respectively.

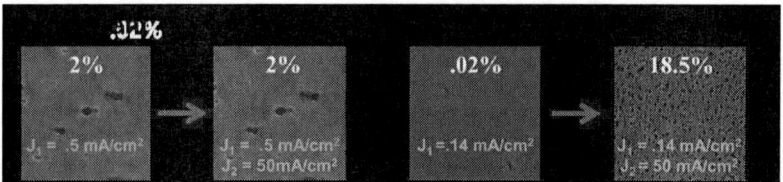

Figure 5. Low current limitation where buried layer formation is no longer independent of top-layer formation.

The pore morphology directly affects surface roughness which in turn affects the quality of the epitaxially grown film. Therefore, understanding how surface roughness changes with pore morphology is vital for achieving high quality films. Fig. 6 shows AFM images for porous InP etched with current densities of 0.14 mA/cm^2 and 50 mA/cm^2.

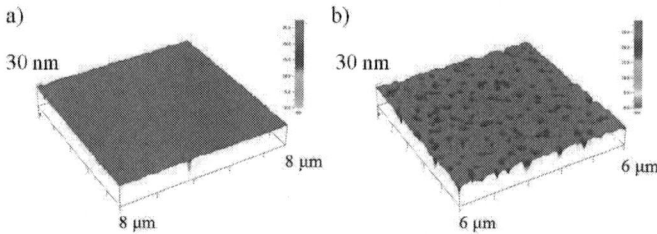

Figure 6. AFM images of top porous InP layers etched with a) 0.14 mA/cm^2 and b) 50 mA/cm^2.

By decreasing the current density, the root mean square average of height deviation (RMS) is reduced from 2.5 nm to 0.68 nm. Where previously an epitaxial layer was successfully grown on a porous layer with a RMS of 7.7 nm (9), a reduced surface roughness is expected to improve the quality of the epitaxially grown thin film .

Conclusion

This study demonstrates an even more optimal porous InP morphology for growing high quality epitaxial films that play a key role in high performance devices. A relationship between porosity and etching current density was obtained, and discrepancies between porous silicon were observed at lower etching current densities. A low-end limit current density before subsequent etching steps become dependent of each other was determined. Furthermore, surface roughness measurements confirmed that these porous InP films were more suitable for high quality thin film formation. Working in this lower etching current density range yields a promising direction for reducing defects in epitaxial films and making cost-effective, high performance devices a reality.

References

1. G. Raghavan, M. Sokolich, W.E. Stanchina, IEEE Spectrum, 37 (10), p. 47 (2010).
2. M. Nishiguchi, N. Goto, H. Nishizawa, J. Electrochem. Soc., 138 (6), p. 1826 (1991).
3. E. Yablonovitch, T. Gmitter, J. P. Harbison and R. Bhat, *Appl. Phys. Lett.*, **51**(26), pg. 2222 (1987).
4. T Yonehara, K. Sakaguchi and N. Sato, *Appl Phys. Lett.*, **64**, pg. 2108 (1994).
5. T. Takizawa, S. Arai, M. Nakahara Jpn. J. Appl. Phys., 33, p. L643 (1994).
6. M.S. Goorsky, M.B. Joshi, S.L. Hayashi, M. Jackson, Int. Conference on Compound Semiconductor Manufacturing Tech., (2008).
7. M.B. Joshi and M. S. Goorsky, *J. Appl. Phys.*, **107**, 024906 (2010).
8. M. Lajnef, A. Bardaoui, I. Sagne, R. Chtouroua, H. EzzaouiaAm. J. Applied Sci., 5 (5), p. 605 (2008).
9. D. Chen, X. Kou, S. Sareminaeini, M. S. Goorsky, ECS Transactions, 64 (5), p.49 (2014).
10. X. Kou, M.S. Goorsky, ECS Transactions, 50 (7), p. 325 (2012).
11. X. Kou, *Cleave Engineered Layer Transfer for III- V Devices via Electrochemical Etched Porous Indium Phosphide,* University of California Los Angeles (2014).
12. M. Joshi, *Fabrication of Engineered Composite Semiconductor Substrates for Flexible Solar Cell Applications*, University of California Los Angeles (2009).
13. M. Christophersen, S. Langa, J. Carstensen, I. M. Tiginyanu, H. Foll, Phys. Stat. Sol. (a), 197 (1), p. 197–203 (2003).

Transfer of ultra-thin semi-conductor films onto flexible substrates

P. Montméat [a, c], I. De Nigris Brandolisi [a, c], S. Tardif [b, c], T. Enot [a, c], G. Enyedi [a, c], R. Kachtouli [a, c], P. Besson [d], F. Rieutord [b, c], F. Fournel [a, c]

[a-] CEA, LETI, MINATEC Campus, 17 rue des Martyrs, F-38054 Grenoble, France
[b-] CEA, INAC, MINATEC Campus, 17 rue des Martyrs, F-38054 Grenoble, France
[c-] Univ. Grenoble Alpes, F-38000 Grenoble, France
[d-] STMicroelectronics, 850 rue J. Monnet, 38926 Crolles cedex, France

> This study describes a process for transferring a 200 nm Si thin film from a SOI onto a flexible substrate. The objective is to find a way for applying a tensile strain onto a very thin semi-conductor film for tensile strain engineering. The processs is achieved with 200 mm wafers and based on a polymer temporary bonding process. Grinding and etching are used for removing the SOI backside substrate and buried oxide. The process leads to a 200 mm Si thin film transferred onto a dicing tape. No crack or ripple is observed on the film and further tensile strain experiments can be conducted on the structure.

Introduction

Applying tensile strain on an indirect band gap semiconductor crystal is a very promising way to tune it into a direct band gap [1, 2]. This basic feature can be an outstanding progress for the use of classical semiconductor as silicon or germanium in interesting optoelectronic applications [3]. In the case of germanium, recent simulations demonstrates that a direct gap can be obtained with a tensile strain from 1.05 % [4] to 4.2 % [2]. From the experimental point of view, several research teams show that a significant tensile strain can be applied onto a single semi-conductor crystal by different techniques. A stress can be transferred from a silicon nitride thin film to germanium nanowires, and the direct bandgap of germanium nanowires can shift from 0.80 to 0.73 eV with a strain of 1,48 % [5]. Applying 1,9 % tensile strain is also possible onto thick structures such as a 350 nm thin germanium film. In that case, the crystal deformation is measured from Ge stressed membranes [6]. One current way to apply a tensile strain is to transfer an ultra-thin of semiconductor onto an organic flexible substrate like the transfer of a thin oxide film onto a 4 cm² film polymer [7]. The expendable polymer is then used to apply the tensile strain onto the oxide thin film.

The goal of this study is to propose a simple way to obtain a 200 mm monocrystalline ultra-thin silicon film (<200 nm) onto an expendable polymer. It consists in transferring an Unibond™ SOI silicon thin film onto a classical peak and place tape maintained by a dicing frame. This process is mainly based on a classical temporary bonding process. A tensile strain can then be applied to the polymer and then to the silicon film.

Experimental

As shown on figure 1, the process is carried out using 200 mm SOI wafer with a silicon film of 200 nm and a buried oxide layer of 400 nm. A classical temporary bonding process using a Brewer thermoplastic adhesive is then performed with a wet silicon etching step after the classical back grinding. The whole process can be split into 6 steps:

- Step A: the SOI is bonded in an EVG520 bonder with a temporary Si carrier coated with a remove layer based on a fluorinated polymer. The glue is the brewer BsiT09001A with a thickness of 40 μm.

- Step B: the stack is then grinding down to 50 μm with a DISCO tool.

- Step C and D: the last 50 μm of the SOI silicon substrate is removed by wet chemical etching as well as the buried oxide layer with a SEZ tool.

- Step E: the thinned stack is mounted onto a dicing flexible tape with an automatic laminator (EVG 850DB).

- Step F: the temporary carrier is mechanically debonded from the 200 nm thin silicon film using an automatic debonding tool (EVG 850 DB). If necessary, the glue can be removed from the Si film with a chemical cleaning.

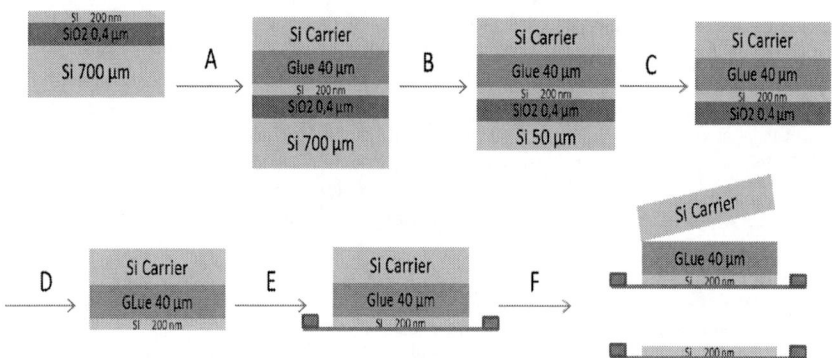

Figure 1: process for transferring a 200 nm silicon film from a SOI to a flexible substrate

Results

The acoustic pictures of the stack at various steps of the process are presented in figure 2. No bubble or void is observed after the bonding process (2a). After grinding down to 200 and 50 μm, no more delamination or default appears (2a, 2b). The chosen stack (SOI/Adhesive/remove layer/Si) is thus compatible and safe with the back grinding process.

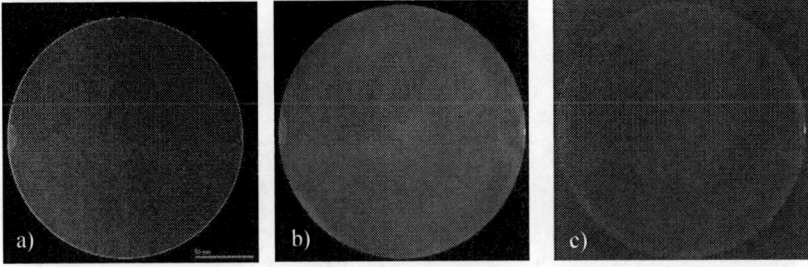

Figure 2: acoustic picture of the stack after bonding (a), grinding at 200 μm (b) aditional grinding at 50 μm (c).

The profiles of the 50 μm thin film from five processed wafers are presented in figure 3. The morphologies are well reproducible and the average thickness of the film is 46,9 μm. The Total Thickness Variation is 6,5 μm.

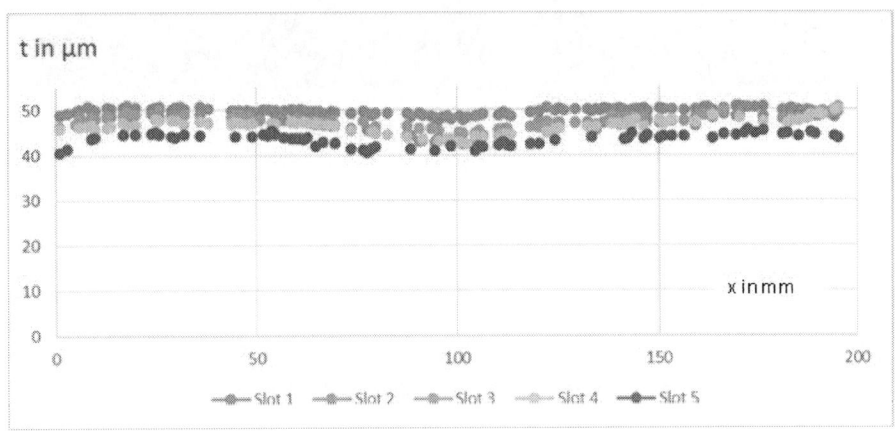

Figure 3: profiles of the 50 μm Si films after grinding.

The etching processes used for removing the Si (50 µm) and the SiO_2 (400 nm) are successful, as seen in figure 4a, no default is observed after the chemical process. No SiO_2 residual layer is observed and the Si thickness is evaluated by ellipsometry at 240 nm which is exactly the nominal thickness of the started SOI thin film.

After mounting the stack onto a frame using our standard dicing tape, the debonding of the carrier is easy and no default appears on the silicon film. The silicon thin film is presented in figure 3b and c. The film is yellow-orange as predicted (8) and exhibits no scale, ripple or crack. The adhesive from the temporary bonding process remains at this stage on the Si film as a protective layer but a standard solvent cleaning could easily remove it.

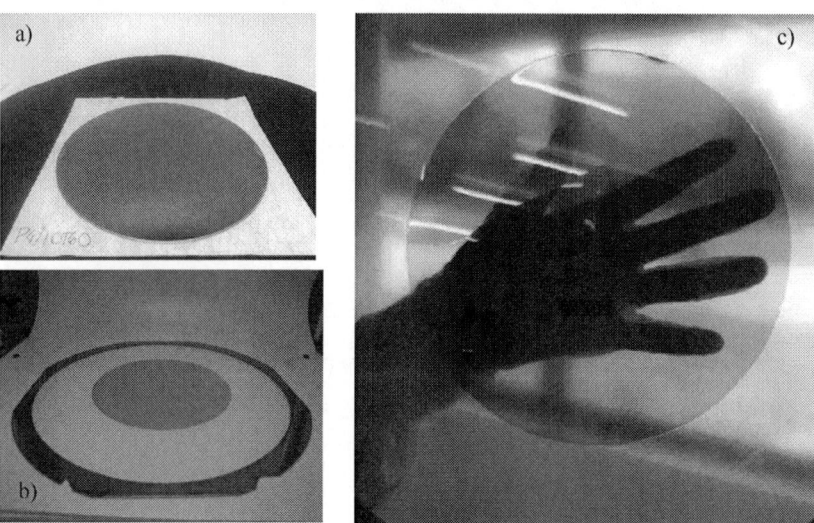

Figure 4: pictures of the stack after etching (a) and after debonding (b, c)

Preliminary strain measurements were performed using Laue microdiffraction at beamline BM32 at the ESRF, as shown in Figure 5a. Disks of 40 mm diameter were cut from the 200 mm films and installed on a bulge test apparatus. The pressure in the chamber was increased from atmospheric pressure (Fig. 5b) up to 1.85 bar resulting in an increase of the tape surface area up to 37%, as estimated from the bulge height (Fig. 5c). The corresponding Laue patterns were recorded on a CCD camera (Fig. 5d) and the deviatoric strain tensor was extracted from the Laue pattern indexation. The total strain tensor was recovered by assuming no normal stress on the free (001) surface[9]. The maximum strain that was measured during this experiment was limited to about 0.2 % biaxial tensile strain. This low strain compared to the tape expansion is due to cracks introduced during the cutting process that cleave the film into smaller "scales", as can be seen in Figure 5c. To avoid this problem, further tests will be performed on the full 200 mm film. One can note however that the absence of Bragg reflection streaking (Fig. 5d) indicates the absence of plastic relaxation within the probed film scale and reveals a good crystalline quality.

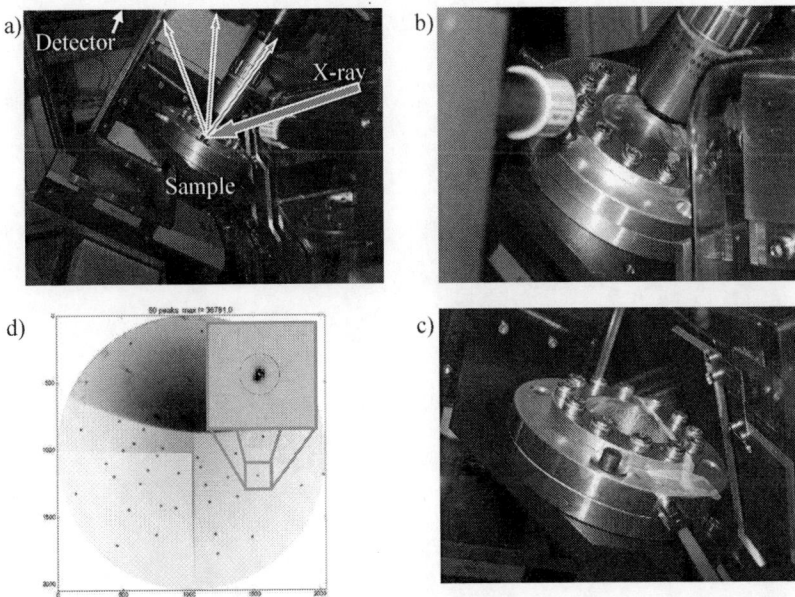

Figure 5: (a) Experimental setup of the Laue micro-diffraction setup, featuring the sample (40 mm disk) on a bulge test apparatus. The optical microscope is used for alignment purposes. (b) Initial condition of the film and (c) final condition after increasing the gas pressure in the chamber; (d) corresponding Laue micro-diffraction pattern, showing a good crystalline quality from the shape of the Bragg reflections (insert).

Conclusion

We successfully obtain a 200 nm silicon film bonded onto the tape. To our knowledge, this is the first time that a transfer of a 200 nm Si thin film is achieved onto a flexible substrate using a full 200 mm wafer. XRR and XRD characterizations of this transferred thin film are ongoing as well as potential amount of tensile stress achievable with this unique structure.

1 Ju Li et al, in *MRS Bulletin*, 39, pp 108-114, (2014)
2 Feng Zhang et al., *in Phys. Rev. Lett.*, 102, (2009)
3 Diode Lasers and Photonic Integrated Circuits, Second Edition, 2012 John Wiley & Sons, chap
4 H Tahini et al., *in Phys. Condens. Matter* , 24, (2012)

5 Kevin Guilloy et al., *in Nano Lett.*, 15 (4), pp 2429–2433, (2015)
6 A. Gassenq et al., *in Applied Physics Letters* ,107, 191904, (2015)
7 Salvatore Giovanni A.et al., *in ACS NANO*, Volume 7, Issue: 10, P8809-8815, (2013)
8 J. Henrie et al. In OPTICS EXPRESS, Vol. 12, No 7, p. 1464, (2004)
9 S. Tardif et al., accepted in J. Applied Crystallography (2016), available at arXiv:1603.06370

Chapter 7

MEMS Integration

254

**Glass Frit Wafer Bonding
Sealed Cavity Pressure in Relation to Bonding Process Parameters**

Roy Knechtel, Sophia Dempwolf, Holger Klingner

X-FAB MEMS Foundry GmbH, D-99097 Erfurt, Haarbergstrasse 67, Germany

Glass frit wafer bonding remains a very attractive process for industrial applications. The main benefit is that the glass frit is an active bonding layer, which planarizes surface roughness and topography up to the direct sealing of metal lines at the bonding interface. This allows very simple process integration. The bonding yield and bonding strength are high, while the bonding interface is reliable regarding mechanical degradation and hermeticity. The processing costs for screen printing, thermal conditioning (firing) of the printed glass paste to a solid glass are moderate, but no extra costs arise from any specially-required preparations of the electrical connecting metallization (no passivation or planarization is needed). In this publication detailed investigations and optimizations of bond process parameters regarding stable and low inner pressure of cavities sealed by glass fit bonding, considering the mechanical bonding strength and bonding behavior, are described.

Introduction

State of the art MEMS devices, such as acceleration sensors and gyroscopes, consist of very fragile, sensitive elements which should be protected by a cap at the wafer level, to allow standard chip dicing and assembly technologies. Typically, wafer bonding is used for this capping. Since, in the bonding process, the type and pressure of the gas can be controlled, the sealed gas in the MEMS device can be tuned and used for device functionality. This is, for example, very important for resonant MEMS devices such as gyroscopes. These essentially require a high Q-factor, which can only be achieved at low cavity pressures, which must be consistent from wafer to wafer in the production process and be stable over the device lifetime. The actual sealed pressure depends on many factors, such as the chamber pressure during bonding, out-gassing from the bonding material (e.g. glass frit), as well as from the sealing behaviour of the bonding process used. Finally, the bond interface has to be hermetic over a long period of time in order to guarantee reliability. This hermeticity factor is not only relevant for resonant MEMS devices sealed at low pressure, but for any wafer-bonded microsystem, because gas leaks can be a risk for trapping moisture, which ultimately could cause critical failure modes, such as corrosion and sticking. At the 12th ECS Semiconductor Wafer Bonding Symposium in 2012, an approach was introduced for measuring the MEMS cavity inner pressure using the thermal conductivity of the sealed gas, which is strongly dependent on the absolute pressure [1]. In that publication it was shown that this kind of inner cavity measurement works very well and has high potential for practical use. However, since the measurement structure itself was quite new at that time, measurement results were quite sparse at that time but in 2014 basic results about the inner cavity behaviour of glass frit bonded devices were published at 13th ECS Semiconductor Wafer Bonding Symposium [6], in the time afterward the work to improve the glass frit bonding process reading lower sealed pressure, smaller bond frames and process stability was continued and main results are presented in this paper.

Concepts for MEMS Hermeticicty Testing

At present, there are no relevant standards available to test the hermeticity of wafer-bonded MEMS chips which are characterized by an extremely small inner volume. Therefore, for hermeticity-critical applications, the essential functions of the MEMS device, mainly the Q-factor of resonant MEMS, are used to determine the hermeticity level. However, there are various challenges for measuring the Q-factor: frequency sweep and amplitude or phase shift algorithms are required, which need special test equipment and a long measurement time. Additionally, the mechanical structure must be suitable for the pressure measurement task. If the functional structure itself is used, the pressure measurement resolution will very probably be limited. Better, suitable resonant test structures require much more space on the wafer and are difficult to integrate into mask sets. Another very promising approach to measure the MEMS cavity inner pressure is to use the thermal conductivity of the sealed gas, which is strongly dependent on the absolute pressure. These kinds of measurement principle are reported in the literature [2], but this was adapted by X-FAB's R&D team to a MEMS foundry surface micromachining process [3] for inertial sensors with high demands on controlled, stable inner cavity pressures. With this process gyroscopes, acceleration sensors, vibration measurement devices as well as IMU's can be realized. Figure 1 gives an overview of the final chip structure in which moveable mechanical elements, such as seismic masses, springs and read-out capacitors are realized, electrically contacted and hermetically sealed by a silicon cap.

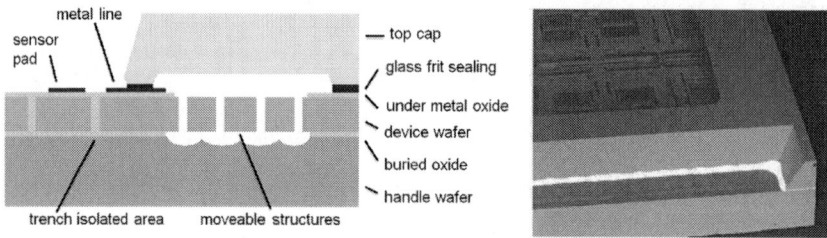

Figure 1. Cross-sectional principle of the surface micromachining process and its realization with opened cap

The technology is based on an SOI wafer, this special substrate providing many technological benefits, despite it being cost-intensive. The main advantage is that the moveable structures are made from a single-crystalline device layer of 15µm, and hence have excellent, well-defined mechanical properties and high reliability. For realisation of the complex mechanical inertial sensor structures (seismic mass, comb drives, read-out capacitors), trenches and holes are etched anisotropically in the device layer of the SOI wafer, down to the buried oxide, retaining the shape of the gyroscope mechanics. To release the mechanical structures, the sidewalls of the etched silicon structures are passivated with CVD oxide, and the buried oxide is opened at the bottom of the trenches. By using isotropic etching into the handle wafer silicon, the structures are then under-etched and their ends released. In this process, the width of the structure defines whether a structure becomes moveable or remains fixed. Subsequently, any remaining oxide is stripped from the mechanical structures by HF vapour etching, so that only pure single-crystalline silicon remains. To improve the electrical behaviour of the gyroscopes, filled insulation trenches are processed prior to fabricating the mechanical structures. These trenches separate the gyro structure as well as the bond pads from the surrounding chip, and allow defined electrical contacting of the different gyroscope elements by metal wiring. However, the main benefit of the trenches is the reduction of parasitic capacitances, which greatly increases the gyroscope performance. Furthermore, an insulation layer between the device layer silicon and the metal layer in the area surrounding the mechanical structures provides the possibility of complex electrical wiring around the mechanical structure. Finally, the mechanical structures are sealed with a capping wafer using glass frit wafer bonding [3]. The capping wafer is pre-structured with through-holes to access the bond pads after bonding, as well as cavities over the sensor structures to reduce damping. This bonding provides effective protection of the very sensitive mechanical structures at the wafer level, making standard wafer probing, dicing and assembly processes possible. Due to the planarising effect of the soft glass frit during bonding, metal lines for driving and sensing signals can be embedded in the glass, and by this hermetically sealed, so that low pressures for high Q-factors can be realised in the sensor cavity. The inner cavity pressure measurement element was adapted to this process flow as well as one modified version using cavity SOI wafers to simplify the etching of the mechanical structures and allow the realization of 3D sensors by additional out of plane movement [5].

Design and Manufacturing of Measurement Structure

The biggest challenge of glass frit bonding is the control of the pressure of the sealed MEMS devices. Resonant MEMS devices such as gyroscopes require a controlled cavity pressure which must not be higher than a certain value in consideration of the sensor design, since at higher pressures, the sensor loses functionality and precision. The functional principle of the inner pressure monitoring and measurement structure is based on the heat transition from a heated element to detection elements which is strongly dependent on the sealed pressure of the cavity in which these structures are located. In the low-pressure or vacuum range in which the sealing of MEMS devices is typically done, heat transfer by conduction of heat dominates, since, due to the low pressure, convection is irrelevant. Transfer by radiation is fairly low, since the heating of the source element is moderate. The test structure consists of a heater element (silicon resistor) and a half Wheatston-Bridge – 2 silicon resistors close to the heating element (these are dominated in their values by heat transfer from the heater, which strongly depends on the sealed pressure) and 2 reference resistors which are far away from the heater and are independent of heat transfer. Figure 2 shows the principle of this configuration, which allows a very sensitive measurement of the heat transfer, and ultimately the gas pressure depending on resistor changes.

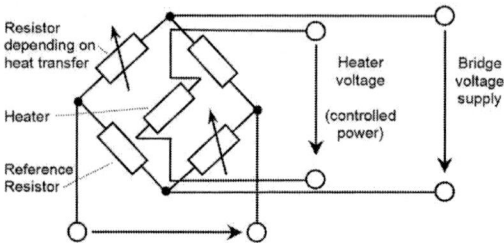

Bridge output voltage = pressure depended measurement signal

Figure 2. Schematic of the pressure measurement construction

This measurement structure was realized in the above-mentioned MEMS foundry technology. The heater and bridge resistors were etched into the device layer of an SOI wafer, and undercut by opening the buried oxide and isotropic silicon etching. The sensing and reference resistors were thermally separated by location. Double measurement resistors are used, with different spacing to the heater, to allow the compensation of tolerances. To improve the measurement characteristics, a reference element, which is not sealed by the wafer bonding and so it is at atmospheric pressure when the measurements are done, was also included.

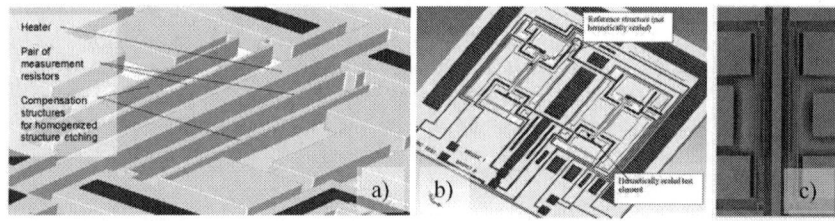

Figure 3. Construction of the test pressure measurement as-realized (reference resistors not shown) – a) 3D visualisation process simulation software – details of heater and measurement resistors, b) overall layout of the measurement structure including an unsealed reference element c) optical image of realized structure (heater and measurement resistors) – as reference: the sensing resistors have width of ~2µm

Characterization of the Pressure Measurement Structure

The pressure sensitivity of the measurement structure was characterized by using a pressure sensor calibration chamber. In this chamber the pressure can be precisely adjusted by a state of the art pressure controller in the range of 1 mbar to 5 bar. The open test structure (without wafer-bonded cap) is located in this pressure chamber. Special lead-through ducts allow electrical operation, i.e., the heating of the heating resistor using a defined power and to read out the Wheatstone bridge signal. The output voltage versus pressure is shown in Figure 4 for one test structure using 2 sets of measurement resistors with different spacings from the heater. While for standard measurements (such as production monitoring) one measurement pair would be sufficient, the second pair can be used to determine slight manufacturing tolerances in spacing, to allow high-precision measurements.

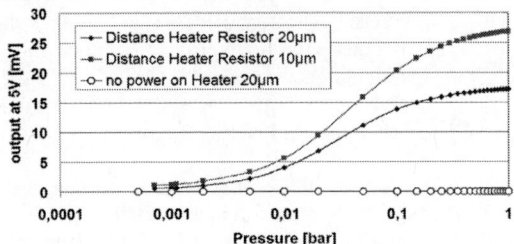

Figure 4. Pressure vs. output voltage characteristic

In Figure 5, it can be seen that in the pressure range from 1 to 100 mbar there is a very good sensitivity of the bridge output signal to the inner pressure of the cavity. This range fits well with the MEMS foundry technology which addresses this range – very low pressure for low damping and high Q-factors for resonating gyroscopes, and higher pressure for acceleration sensors which should not show resonating behaviour. In the characterization of several pressure measurement structure chips from different wafers and different wafer positions, it was shown that slight variations in the structuring process and form of the SOI wafers (specification tolerance of device layer thickness) can influence the heat transfer, and by this, the pressure measurement value. The geometry compensation using the second pair of detecting resistors can partially compensate for these effects, but cannot provide any benefit for the device layer tolerance, because both pairs will have almost the same thickness. For this reason, another measure to compensate structural tolerances was investigated. Here, the atmospheric pressure, or for high-precision measurement, a precise reference pressure of 1 bar, is used as a reference. This reference is applied to an open measurement structure (easy to realize by interrupting the bond frame in the test structure layout) as a reference. The output signal of the sealed inner cavity pressure measurement structure is finally normalized to the output signal of the open structure operated at a known pressure. This relative output signal is well related to the inner pressure to be measured, and a functional fit (Figure 5) allows an almost error free calculation of the inner pressure of the test structure, which is ultimately monitoring the bonding process.

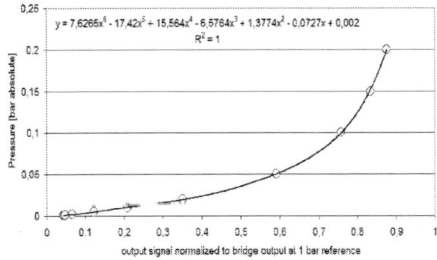

Figure 5. Pressure vs. normalized bridge signal to realize compensation of structural tolerances of the measurement structure

The structure itself is easy to integrate into production mask sets, and the described measurement approach, including the compensation of structural tolerances in combination with an easy, precise Wheaton-Bridge-Signal measurement, allows the process monitoring of the cavity inner pressure in production processing, and through this the quality assurance of the bonding process.

Glass Frit Bonding Process Optimisation

As already mentioned the biggest challenge of glass frit bonding is the control of the pressure of the sealed MEMS devices. But resonant MEMS devices such as gyroscopes require a controlled cavity pressure, which must not be higher than a certain value in consideration of the sensor design, since at higher pressures, the sensor looses functionality and precision – stability of the sealed pressure for production runs is the most important point here. The working point of the ASIC driving and evaluating for example a gyroscope sensor can be adapted to a quality factor and a pressure range, but once set in overall sensor concept, it has to be fixed and this requires stability from the MEMS sensor element side and this mean in first order a stable inner cavity pressure. At the gyroscope senor this publication is based on, the upper limit of the pressure sensor was set to 5 mbar related to the ASIC configuration.

In early production sate we have seen an higher than expected variation of the inner cavity pressure from lot to lot , wafer to wafer and over the wafer itself and it was found and confirmed in the practice that the integrated circuit (ASIC) actually cannot compensate these high variations, optimizations were needed. These optimizations, described here, were based on the following hypotheses:

- Outgassing of the glass frit material causes increasing of pressure inside of the sealed cavity.

- Outgassing happens during the bonding process itself, when the bond frame is already hermetically sealed and the temperature still high (bonding temperature and very early stage of cooling down).

- This outgassing cannot be completely prevented by optimized binder burn out before the actual bonding process in the paste firing and conditioning process [6].

- By this outgassing from the glass frit inner cavity pressure is increased, compared the chamber pressure, set in the wafer bonding recipe and cannot be tuned during the bonding anymore, since no getter material is used in the used technology.

- Variations of the glass frit height over the lot and the wafer itself is one main contribution to varying sealed cavity pressure; not all frames are sealed at the same time, this cause a different actual time on temperature situation of the sealed frames, resulting in a different amount of outgassing trapped in the cavities.

Based on these hypotheses an experimental split was set up and executed as shown in Table I, which also contains the main results, the inner cavity pressure measured with the special measurement structure described above. The two main factures to be investigated were:

1) Process temperature during bonding
2) Waiting time, which is the time at highest bonding temperature

Both conditions are combined in 4 process versions (V1...4) of the following a split plan:

TABLE I. Experimental Split plan and inner cavity pressure results

Version	#Wafer	Bonding Temperature		Bonding Time	Inner Cavity Pressure [mbar]
		Top Heater	Bottom Heater		
V0	2	standard	standard	standard	3,5
	6	standard	standard	standard	3,4
V1	4	standard	standard	7 min shorter	1,6
	7	standard	standard	7 min shorter	1,3
V2	5	5K lower	5K lower	standard	3,6
	8	5K lower	5K lower	standard	3,1
V3	9	standard	5K lower	4 min shorter	2,5
	11	standard	5K lower	4 min shorter	2
V4	10	standard	standard	2 min shorter	4,1
	12	standard	standard	2 min shorter	2,5

To get a complete picture of the bonding results the following investigations were done on all wafers:

- Inner cavity pressure measurement using 8 process control monitor (PCM) – structures per wafer – the same which are used to monitor the production process (Figure 6)
- Mechanical Dicing and evaluation dicing yield
- Stud Pull Test to measure bonding strength and fracture load of diced chips followed by fracture images evaluation

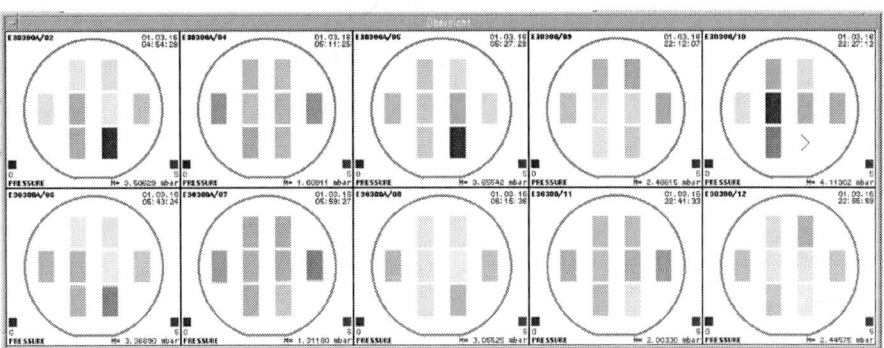

Figure 6. Inner cavity pressure distribution maps of the investigation (PCM)

All measurements were distributed over chips from different on wafer locations, in addition two wafers per variant were tested to allow first assessment about wafer to wafers process variations.

At first the results of the inner pressure cavity pressure should be discussed here using figure 7. The diagrams on the left side summarize the relation between the resulting inner cavity pressure and the bonding temperature for fixed bonding time. The bonding temperature has been reduced by 5°C for top- and bottom heater which does not help to reduce and stabilize the inner cavity pressure furthermore. Lower temperatures have not been tested because this will have major consequences for the device reliability (bonding strength). The second diagram on the right side showed that it is very beneficial to reduce the time during the bonding process at the maximum bonding temperature. The standard bonding time results in an inner cavity pressure of ~3.5 mbar. By reducing the bonding time by 60% the inner cavity pressure value could be reduced to 50% of the original value (<2 mbar).

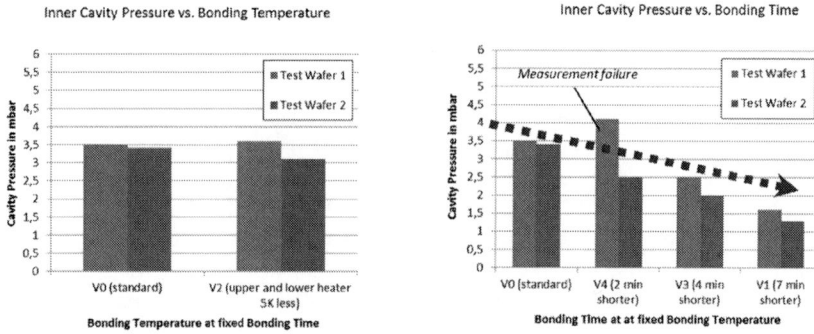

Figure 7. Evaluation of the inner cavity pressure measurement results regarding the process split conditions

Out of this inner cavity pressure evaluation the conclusion is that the outgassing is dominated by the time on the bonding tempeture, while the bonding temperture itself does not have a real influence:

-Reduction of temperature: no sinificiciant influence on inner cavity pressure
-Reduction of bonding time: significant influence on inner cavity pressure

Beside the cavity internal pressure the other important parameter is the bonding strength, reducing bonding time or temperature can reduce the bonding strength. So for the most interersting two of the variations including V1 the prefered version stud pull tests were done at diced chip samples. The fracture loads were measured and the fracture images were investigated. As figure 8 shows, the slight paramter changes are still resulting in a strong bonding. The fracture images confirmed, that the glass frit is flowing well as expected, no bubbles are found (plan areas at pictures wafers 5 are fractures at the buried oxide interface of the MEMS-SOI Wafer). With this it can concluded that the wafer bonding process adaptation for reduced inner cavity pressure provides good bonding quality and reliability as well.

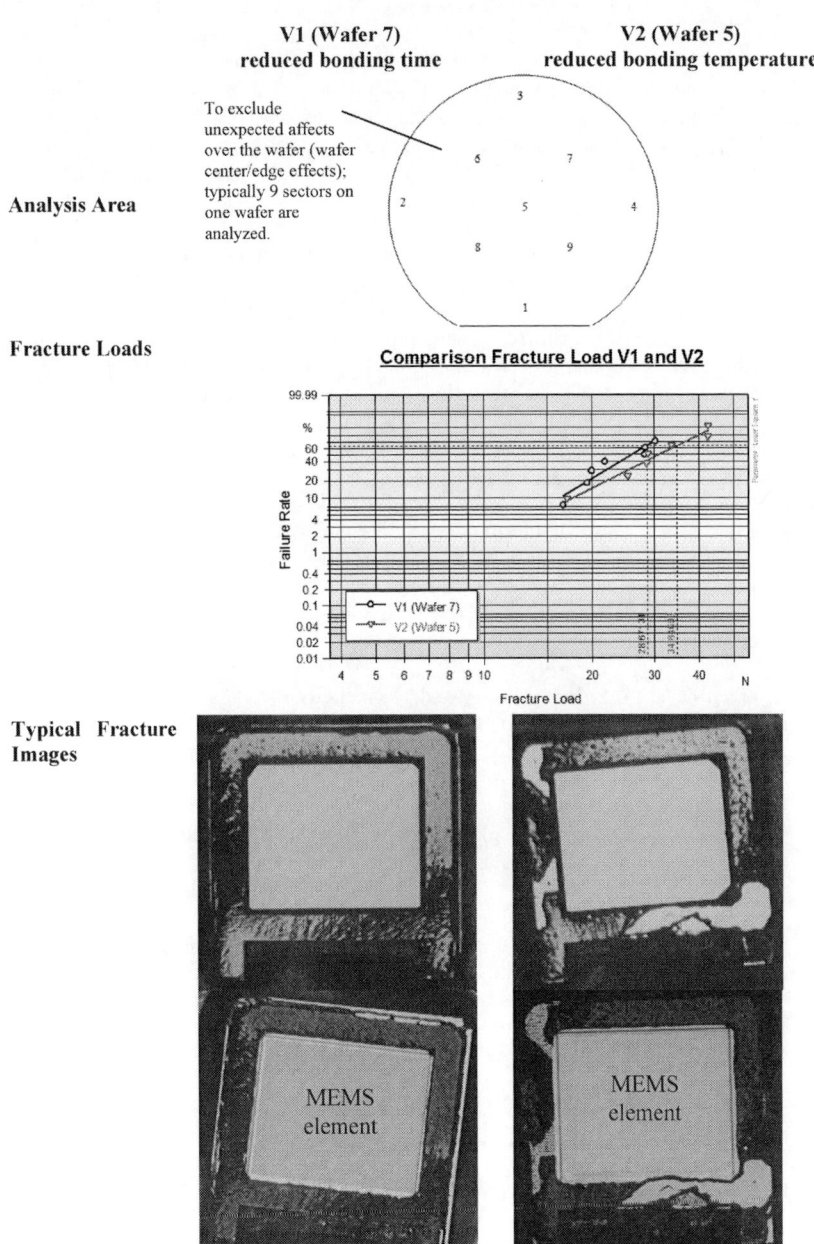

Figure 8. Bonding strength evaluations of process variations with reduced time on fixed bonding temperature vs. reduced bonding temperature at constant time

Conclusions

Using an easily to integrate heat transfer-based inner cavity measurement structure it is well possible to monitor the inner pressure of glass frit sealed cavities, e.g. on MEMS Gyroscope wafers, and to do complex bonding process optimisations. In the given example the glass frit bonding process was optimized regarding low and stable inner cavity pressure. The time on bonding temperature was found as the main influencing factor. When the bond frames are already sealed, but the glass is still on bonding temperature and soft, outgassing occurs and increases the inner cavity pressure – this is time depending, as shorter the time on temperature, as lower the outgassing amount and as lower the inner cavity pressure increasing. On the other hand a slightly decreased bonding temperature is not significantly reducing the outgassing and the inner cavity pressure. Stud pull tests confirmed, that even with minor changes in the bonding parameters the glass frit bonding is still mechanically strong enough and the glass flows well for a perfect sealing with good reliability.

References

1. R. Knechtel et. al., Monitoring Inner Pressure of MEMS Devices Sealed By Wafer Bonding, 12th International Symposium on Semiconductor Wafer Bonding: Science, Technology, and Applications 2012 Fall Electrochemical Society Meeting Honolulu Hawaii, - ECS Transactions, 50 (7) 379-386 (2012), © The Electrochemical Society

2. K. Glien et. al., Gas Permeability and Hermeticity of Glass Frit Bonded Micro Packages; Workshop on Wafer Bonding for MEMS Technologies Halle/Germany 2004

3. R. Knechtel; Single Crystalline Silicon Based Surface Micromachining for High Precision Inertial Sensors   - Technology and Design for Reliability; SPIE MST for the Millennium; Dresden Germany 2009

4. R. Knechtel: Glass Frit Bonding - An Universal Technology For Wafer Level Encapsulation and Packaging, Published by Springer-Verlag GmbH, Microsystem Technologies ISSN: 0946-7076 (Paper) 1432-1858 (Online) Issue: Volume 12, Numbers 1-2 Date: December 2005 Title: Symposium on Design, Test, Integration and Packaging of MEMS/MOEMS, Montreux, Switzerland, 12-14 May 2004; DOI: 10.1007/s00542-005-0022-x)

5. S. Dempwolf et. al., MEMS 3-Axis Inertial Sensor Process, DTIP - Design, Test, Integration & Packaging of MEMS/MOEMS; 16-18 April 2013, Barcelona, Spain

6. R. Knechtel et. al., Detailed Investigations of Inner Cavity Pressure of MEMS Devices Sealed by Wafer Bonding, 13th International Symposium on Semiconductor Wafer Bonding: Science, Technology, and Applications 2014 Fall Electrochemical Society Meeting Cancun Mexico, - ECS Transactions, 2014 64(5): 285-296, © The Electrochemical Society

Wafer-Level Hermetic Seal Bonding at Low-Temperature with Sub-Micron Gold Particle Using Stencil Printing

H. Ishida[a], and T. Ogashiwa[b]

[a] SUSS MicroTec KK, Yokohama, Kanagawa, 226-0006 Japan
[b] Tanaka Kikinzoku Kogyo K.K., Hiratsuka, Kanagawa, 254-0076 Japan

Wafer-level low-temperature hermetic seal bonding has been successfully demonstrated using sub-micron-size gold particles (0.3 μm mean diameter) with adopting a narrow-width rim structure to reduce the required total bonding force down to the range available on commercial wafer bonders. Conventional stencil printing combined with a newly developed suspended metal mask was employed as a cost-effective fabrication method to form sealing lines over the rim structure. Au-Au thermo-compression bonding using 10 μm-wide rim structure was performed at 200°C for 30 minutes with the bonding pressure of 200 MPa. An excellent hermeticity was achieved and the encapsulated pressure inside the sealed cavity was estimated to be around 100 Pa.

Introduction

Among low-temperature metal bonding, such as Au-Au, AuSn, AuIn, for wafer level MEMS hermetic packaging, Au-Au thermo-compression bonding is a promising candidate having advantages of no concern of surface oxidation and simple process control (1 - 4). To compensate the surface roughness of processed wafers, which can degrade the hermetic properties, surface compliant Au-Au bonding using sub-micron-size Au particles (called sub-μm Au particles hereafter) has been developed (5). In order to apply that process to practical use, pattern transfer method of sub-μm Au particles has also been developed later, which can allow sub-μm Au particles sealing line formation on completed device wafers at the very final step (6). In this study, we have developed another practical patterning method employing conventional stencil printing combined with a suspended metal mask as an alternative method to the pattern transfer technique. Using this technology, sealing line patterns can also be formed at the final step with increasing throughput and decreasing process cost. Since the pattern width by stencil printing is relatively wider and therefore more total bonding force is required, a rim structure was adopted to define a thin bond line width for reducing the bonding force within the range of commercial wafer bonders (e.g. the maximum force of 100 kN for 200 mm wafers).

Materials and Equipment

Sub-μm Au Particles Material

The 99.95wt% purity, spherical sub-micron-size Au particles (called sub-μm Au particles hereafter) were obtained through a wet chemical processing method by mixing chloroauric acid solution with a reducing agent. The obtained particles consisted of individual gold particles and their diameters were distributed in the range of 0.2 μm - 0.5 μm with the mean diameter of 0.3 μm. Reactivity of the Au particles, or bonding temperature in actual processes, is significantly influenced by the particle size. These sub-μm Au particles can react and connect each other even at a temperature as low as 200°C. In fact, Figure 1 shows the sintering behavior of sub-μm Au particles at 150°C for 5 minutes in air with no compression force, indicating that particles connected each other by necking reaction even at such a low temperature.

Figure 1. SEM image of sub-micron Au particles after heating at 150°C for 5 minutes in air with no compression force.

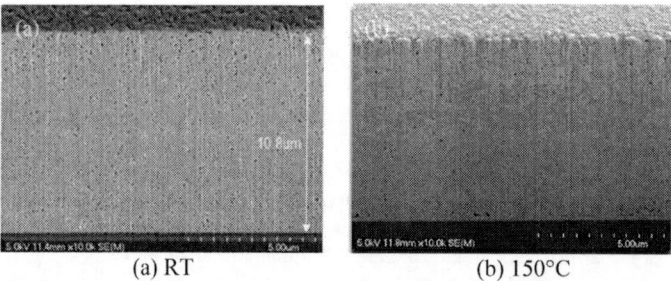

(a) RT   (b) 150°C

Figure 2. Cross sectional FIB-SEM images of sintered Au particles after applying the pressure load of 100 MPa (a) at room temperature and (b) at 150°C.

The compressive deformation characteristics of sub-μm Au particles were evaluated by using micro compression testing machine (MCT510, Shimadzu Corp.). The 200 μm diameter tool tip compressed the sintered Au particles with increasing the load at a rate of 4 grams per second. Initial height of the patterns is around 20 μm. Figure 2 shows the cross sectional FIB-SEM images of sintered Au particles after applying the pressure load of 100 MPa at different temperatures, (a) at room temperature and (b) at 150°C. At room temperature, the sintered Au particles with the initial height of 20 μm can be deformed to 10.8 μm after the compression. This densification is mainly realized due to spatial rearrangement of Au particles. On the other hand at 150°C, the compressive deformation induced a dynamic structure change to be a bulk gold accompanied with recrystallization. The photos show that the recrystallization can reduce the total number of voids and results in the isolation of voids at the same time. This phenomenon indicates a possibility of tight hermetic sealing by a low-temperature thermo-compression bonding.

Stencil Mask Printer

A high-precision screen printer (LS-25TVA, Newlong Seimitsu Kogyo Co., Ltd.; Figure 3 (a)) together with a suspended Ni metal mask (Taiyo Yuden Chemical Technology Co., Ltd.; Figure 3 (b)) was employed to form sealing line patterns on φ100mm Si wafers. The minimum feature size, i.e. line width, of the stencil metal mask is 50 μm. The stencil metal mask can be aligned to the Si wafer using alignment keys on both sides with the alignment accuracy of < +/- 20 μm. According to these specifications, 50 μm seal line patterns on 100 μm-wide rim structure can be printed, while 10 μm seal line patterns on 10 μm-wide rim can also be formed, as shown in Figure 4 (a) and (b), respectively. After the alignment between the metal mask and the Si wafer, sub-μm Au particle printing can be completed within 10 seconds per wafer, which can meet a high throughput in mass-production.

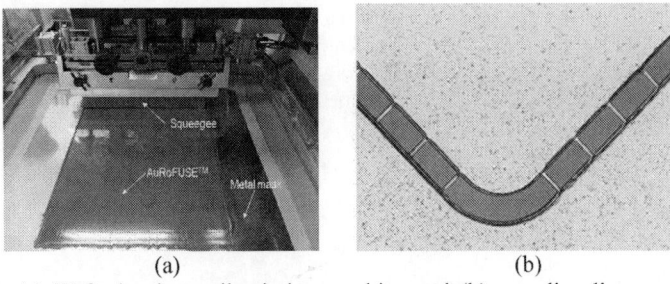

(a)   (b)

Figure 3. (a) Wafer-level stencil printing machine and (b) a sealing line opening of a suspended Ni metal-mask.

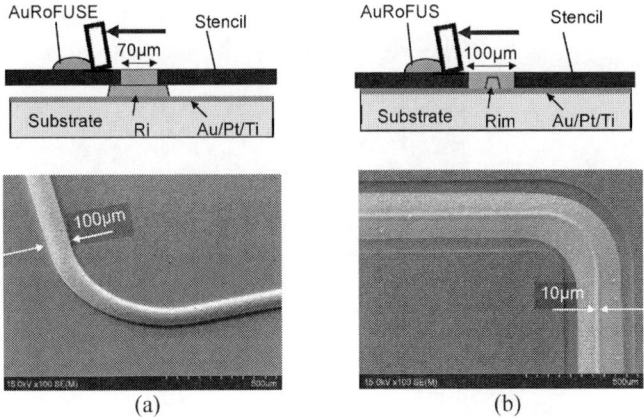

(a)   (b)

Figure 4. Schematic illustrations and SEM pictures of two types of pattern structures: (a) sub-micron Au particles are printed on top of the 100 μm-wide rim using 70 μm-wide opening stencil mask, and (b) printed with covering the 10 μm-wide rim using 100 μm-wide opening stencil mask.

Experimental

Figure 5 shows the wafer-level fabrication process of test samples. 10 μm-tall/10 μm-wide rim structures were fabricated on a φ100 mm Si wafer by deep-RIE, and Au/Pt/Cr (0.2/0.03/0.03 μm-thick) metallization layer were deposited by sputtering. Sub-μm Au particles paste (AuRoFUSE™, Tanaka Kikinzoku Kogyo K.K.) was deposited to cover rims by conventional stencil printing method using a high-precision screen printer together with a suspended Ni-metal mask described above. After aligning the rim wafer and the metal mask by using alignment keys, printing was performed and completed across the wafer within 10 seconds. Wafer-level printing on lager wafers, e.g. 200 mm wafers, has also already been demonstrated within a similar process time. Figure 6 (a) illustrates a cross-sectional view of the sub-μm Au particles coated rim structure and Figure 6 (b) shows an SEM top view of the paste coated rim pattern. Following the annealing of the rim wafer at 200°C for 2 hours, precise alignment was performed between the rim wafer and the diaphragm wafer on the bond aligner (BA8, SUSS MicroTec AG). Then, the aligned wafer pair was bonded on the commercial wafer bonder (SB8e, SUSS MicroTec AG) at a temperature of 200°C for 30 minutes with an applied pressure of 200 MPa to fabricate the structure as shown in Figure 5 (c). The bonded wafer pair was diced into single chips for evaluation of the bond strength and hermetic properties.

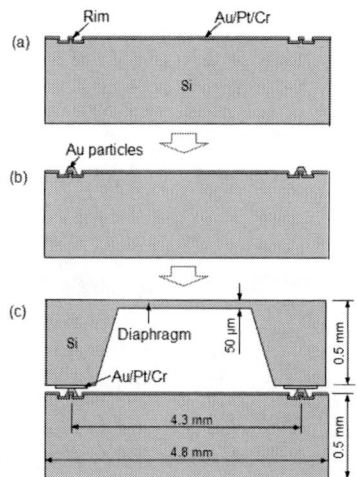

Figure 5. Schematic illustration of wafer-level fabrication process using sub-micron Au particles: (a) a rim structure formed by dry etching process and metallization by sputtering, (b) coating of sub-micron Au particles (AuRoFUSE™) by stencil printing with heating at 200°C for 2 hours, and (c) thermo-compression bonding at 200°C for 30 min in a vacuumed chamber.

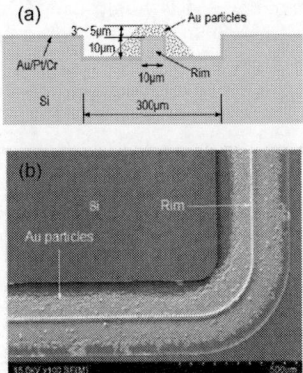

Figure 6. A rim structure covered with sub-micron Au particles. (a) Schematic cross-section illustration, and (b) SEM top view.

Result and Discussion

Bonding Performance

Figure 7 indicates cross-sectional FIB-SEM images of the rim joint after tearing off the diaphragm chip. The joint was teared at the boundary of silicon wafer of diaphragm side and its metallization layers of Au/Pt/Cr. The 0.2 μm-thick metallization layers of diaphragm side attached firmly to the densified structure of sintered sub-μm Au particles. The sintered sub-μm Au particles layer with the initial thickness of 3 - 5 μm was densely compressed on the rim structure down to the thickness of 0.6 μm during the thermo-compression bonding. The crystal grain growth was found with the recrystallization of sintered Au particles at the pressure of 200 MPa for 200°C, resulting in the reduction of voids and grain boundaries. In fact, it is difficult on Figure 7 (b) to identify the bond interface of the recrystallized structure and both sides of metallization layers, which indicates that the sub-μm Au particles recrystallized and densified into bulk structure to realize hermetic interface.

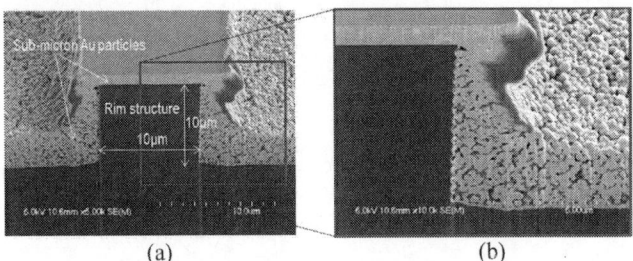

Figure 7. Cross-sectional FIB-SEM images of rim joints after tearing off the diaphragm chip. A complete picture of the rim joint (a), and a magnified view of the recrystallized structure of sintered Au particles and metallized layers (b).

The die shear strength of a singulated test chip with 4.3mm-square sealing line was measured by a bondtester (series-4000, Dage) as illustrated in Figure 8 (a). The relationship between the shear strength and the rim width is shown in Figure 8 (b). The average value of the shear strength for 10 μm-wide rim was 44 N.

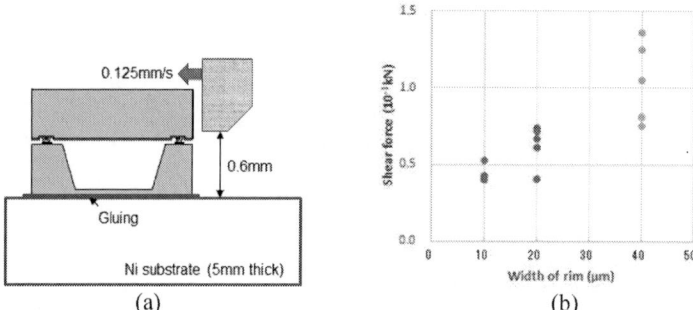

(a)   (b)

Figure 8. (a) Schematic illustration of die shear test to measure the shear strength of a singlulated test chip and (b) a relationship between the shear strength and the width of the rim.

Hermetic Sealing Properties

Figure 9 (a) is a picture of a 4.8mm x 4.8mm test chip exhibiting a concave deflection of the 50 μm-thick diaphragm under the atmospheric pressure, which indicates a good vacuum encapsulation inside the sealed cavity, and (b) shows the deflection profile of the 2.4mm x 2.4mm diaphragm measured by an optical surface profiler (Nexview, Zygo), indicating the concave deflection of 5μm. The encapsulated pressure was estimated by precisely measuring the diaphragm deflection in vacuum chamber using an optical surface profiler (MSA-500, Polytec). The bonded sample was put into a vacuum chamber, of which pressure was manually controlled, and the diaphragm deflection was measured at varied pressure. The displacement of the diaphragm against the chamber pressure is shown in Figure 10. As shown in the result, the pressure inside the sealed cavity is estimated to be about 100 Pa when the diaphragm becomes flat. The hermetic sealing performance of the sub-micron Au particles bonding was also evaluated by helium fine leak test (MUH series, Fukuda Co., Ltd.). The maximum He leak rate of the test chip was estimated in the range of 10^{-14} Pa·m^3/s, which is sufficient for most of the MEMS applications.

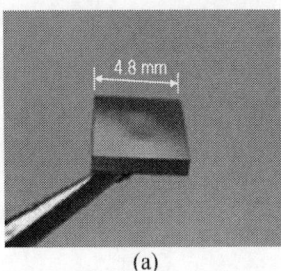
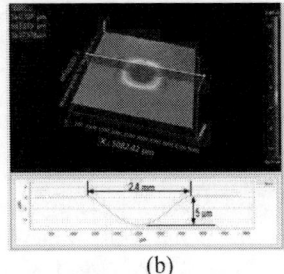

(a)   (b)

Figure 9. (a) A singulated test chip exhibiting concave deflection due to vacuum sealing and (b) the deflection profile of the diaphragm measured by an optical surface profiler (Nexview, Zygo).

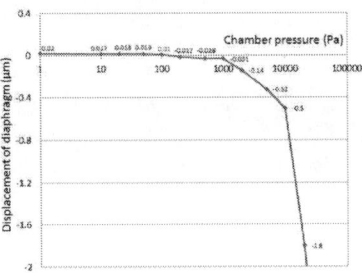

Figure 10. Displacement of diaphragm measured by an optical surface profiler (MSA-500, Polytec) at varied chamber pressures.

Conclusion

Low-temperature Au-Au hermetic seal bonding has been successfully demonstrated employing a rim structure covered with sub-micron Au particles. 3 - 5 μm-thick sub-micron Au particles layer was formed on the 10 μm-tall/10 μm-wide rims by a conventional stencil printing method combined with suspended metal mask. This stencil printing technique has a capability to form sealing line with the width of 50μm. Vacuum hermetic sealing by the sintered sub-μm Au particles on the rim structure was achieved by the thermo-compression bonding with 200 MPa at 200°C. An excellent hermetic sealing property was confirmed and the encapsulation pressure was estimated to be about 100 Pa. This patterning and bonding technique can be used for actual manufacturing of wafer-level hermetic sealing of MEMS packaging.

Acknowledgments

The authors thank A. Yanagisawa, Newlong Seimitsu Kogyo K.K., and Y. Kurata, Taiyo Yuden Chemical Technology Co., Ltd., for their kind help in performing stencil printing and developing metal masks. The authors also appreciate Dr. K. Totsu, Tohoku University, for his great support in conducting evaluation of hermetic sealing properties.

References

1. C. S. Tan, J. Fan, D. F. Lim, G. Y. Chong and K. H. Li, *J. Microelectromech Systems*, **21**, 075006 (2011).
2. N. Malik, H. R. Tofteberg, E. Poppe, T. G. Finstad, K. Schjolberg-Henriksen, *ECS Trans.*, **64**(5), 167 (2014).
3. M. Small, R. Ruby, S. Ortiz, R. Parker, F. Zhang, J. Shi, B. Otis, *2011 Joint Conference of the IEEE International Frequency Control and the European Frequency and Time Forum (FCS) Proceedings*, P. 1 (2011).
4. N. Belov, T. -K. Chou, J. Heck, K. Kornelsen, D. Spicer, S. Akhlaghi, M. Wang, T. Zhu, *2009 IEEE International Interconnect Technology Conference*, P. 128 (2009).
5. H. Ishida, T. Ogashiwa, T. Yazaki, T. Ikoma, T. Nishimori, H. Kusamori, J. Mizuno, *Transactions of The Japan Institute of Electronics Packaging*, **3**(1), 62 (2010).
6. H. Ishida, T. Ogashiwa, Y. Kanehira, S. Ito, T. Yazaki, J. Mizuno, *Proc. IEEE Electronic Components and Technol. Conf.*, San Diego, California, USA, May 29-June 1, p. 1140 (2012).

Aluminum-Germanium eutectic bonding for MEMS: behaviour and solidification of liquid Al-Ge on different substrates

V. Lumineau[a,b,c], F. Fournel[a,b], B. Imbert[a,b], F. Hodaj[b,c]

a- CEA, LETI, MINATEC Campus, 17 rue des Martyrs, F-38054 Grenoble, France
b- Univ. Grenoble Alpes, F-38000 Grenoble, France
c – CNRS, Grenoble INP, SIMAP, F-38000, Grenoble, France

> In this paper we study the formation and behaviour of liquid Al-Ge alloy, its solidification, the resulting microstructures and the appearance of voids in bonded structure for MEMS packaging. The underneath layer influences the wettability of liquid Al-Ge as well as its solidification which can lead to undesirable Al dendrites. Each underneath layer imposes a critical value for the cooling rate above which dendrites are likely to appear. Blanket and patterned Al/Ge deposited wafers are bonded together with process parameters optimized thanks to previous results. The interfaces present a dendritic-free eutectic structure. Micro-voids are observed at the interfaces between Al and Ge phases. The Kirkendall effect and thermal stresses are two possible reasons for voiding phenomenon.

Introduction

Micro Electro-Mechanicals Systems (MEMS) include moving elements in some cases and they require a controlled atmosphere to operate. Hence they are protected by a cap assembled to the sensor by wafer-level packaging. For specific MEMS packaging, the process should provide strong and hermetic sealing walls of reduced size, less than 100 μm which have also to ensure electrical conductivity.

Eutectic bonding is a special type of metallic bonding. The intermediate layer of this bonding between the two surfaces to be joined is composed of an alloy of elements which lead, under certain conditions of temperature and composition, to a eutectic microstructure. It allows assemblies whose thickness is less than 10 micrometers and ensure electrical conductivity between the two elements. Moreover, thanks to the liquid phase, this type of bonding shows good hermetical performance like for instance the Au-Si eutectic bonding (1). If a high-temperature-resistant sealing (>200 °C) is required and if non CMOS contaminant element are needed, Au-Si eutectic bonding cannot be used anymore and Aluminum-Germanium eutectic bonding could be interesting.

In eutectic bonding, the alloy constituents are deposited on at least one of the two wafers which are brought into contact. The melting and then the solidification of the solder close the bonding interface. The Al-Ge(29.5at%) eutectic alloy has a melting temperature of 424°C (2). This alloy has been shown allowing strong and void-free bonds (3-5).

Nevertheless, during bonding, the eutectic liquid tends to squeeze out of the seal ring which can lead to void formation at the sealing interface and areas with unwanted metal. Moreover, influences of the underlayer on the bonding mechanisms have not been investigated whereas it is known to have an impact on Al/Ge bilayers (6). In this paper

we study the formation and behaviour of liquid Al-Ge alloy, its solidification, the resulting microstructures and the appearance of voids. The main investigated parameters are:
- the underneath layer which is related to the wettability of the liquid Al-Ge eutectic.
- the annealing temperature and cooling rate and their influence on solidified microstructure.
- the seal rings geometry

Materials and Methods

Samples

The 200 mm silicon substrates used in this study are either thermally oxidized or covered by a 50nm-thick titanium nitride layer to prevent metal diffusion into silicon substrate. Aluminum and germanium are deposited by evaporation successively and without air-break to prevent aluminum native oxide formation at the interface. The ratio of thicknesses match the Al-Ge eutectic composition (29.5 at%Ge). The total layers thickness ranges from 0.8 μm to 1.6 μm. As shown by equation 1, the ratio can be calculated taking into account the densities and the atomic weights of Al and Ge:

$$\frac{e_{Al}}{e_{Ge}} = \frac{at\%Al}{at\%Ge} \frac{\rho_{Ge}}{\rho_{Al}} \frac{M_{Al}}{M_{Ge}} = 1.7 \qquad [1]$$

Differential Scanning Calorimetry (DSC)

Differential scanning calorimetry (DSC) is a qualitative measure of the heat exchange between a sample and its environment as a function of temperature. The melting and solidification process of many $1x1$ cm^2 samples from blanket Al/Ge deposited wafers are studied by this technique. Thermal treatments are performed under inert gas (N_2), at atmospheric pressure and with different cooling rates. The microstructures are then analyzed by optical microscopy, Secondary Electron Microscopy SEM and Energy Dispersive X-ray spectrometry (EDX).

Dispensed drop method

In order to evaluate the influence of the underneath layer on the behavior of liquid Al-Ge eutectic alloy during bonding, the wetting of different substrates is investigated by dispensed drop method, under a vacuum of $5x10^{-7}$ mbar. High purity aluminum and germanium (99.999 %) are used to prepare the alloy. It is then place in an alumina crucible ending by a capillary to deposit a droplet on the studied surface (Figure 1). The surfaces are taken from silicon substrates thermally oxidized or covered by a 50 nm-thick TiN layer. During the experiments, temperature ranges from 450 °C to 1200 °C. Observations are made in situ by a camera which is connected to a computer. The results are then analyzed using Drop Shape Analysis software. The size of the droplets and the contact angles are measured once the equilibrium is reached. The contact angle θ is related to the solid-gas energy σ_{sg}, the solid-liquid energy σ_{sl} and the liquid-gas energy σ_{lg} by the Young shown in equation 2:

$$\sigma_{sg} = \sigma_{sl} + \sigma_{lg} \cdot \cos \theta \qquad [2]$$

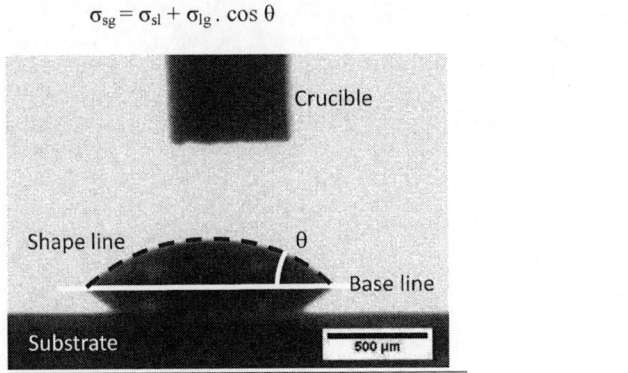

Figure 1 : Contact angle measurement of Ge drop at 1000 °C

Bonding

Blanket and patterned Al/Ge deposited wafers are bonded to silicon wafers either thermally oxidized or covered by a 50nm titanium nitride layer. In the case of a patterned wafer, the seal rings are 5 mm square with widths varying from 50 to 200 μm. Bonding is performed under either vacuum (10^{-3} mbar) or inert gas (N_2) at atmospheric pressure. Experimental temperature is varied from 400°C to 500°C.

During the process, the wafers surfaces should be maintained in intimate contact. However to avoid leakage of the alloy once it is liquid, a low applied force of 5 kN is selected (0.17 MPa over 200 mm diameter area), which is the minimum allowed by the equipment. The temperature is kept constant for 10 minutes until cooling for atmospheric pressure bonding or 1 hour for vacuum bonding.

Squeeze-out and voiding phenomenon are characterized using Scanning Acoustic Microscopy (SAM) and FIB SEM.

Results and Discussion

Liquid Al-Ge eutectic alloy

This part focuses on the behavior of liquid Al-Ge eutectic. Prevent leakage by confining the liquid in the sealing ring is one of the main issues of this process. Studying some of the properties of the liquid should permit to better understand this phenomenon.

First, the melting temperature of many 1×1 cm^2 samples from blanket Al (1000 nm)/Ge (590 nm) deposited wafers is measured by DSC. Both underneath layers SiO_2 and TiN are compared. There is no major differences for the melting temperature (T_m) between the samples and the mean value obtain is $T_m = 426 \pm 0.7$ °C. This value is in good agreement with those found in literature, although it is slightly higher. An excess of one of the two components of the alloy could explain this difference. Indeed, other values of the eutectic composition can be found (7). Even if the composition of the samples is slightly different from the real eutectic one, the melting temperature will increase.

Then, the wettability of pure Al (99.999 %), pure Ge (99.999 %) and Al-Ge eutectic alloy on different surfaces is evaluated by the dispensed drop method. The contact angle formed between the drops and the substrates is measured and the results are given in **TABLE 1**. In the case of a SiO_2 surface, the contact angle of Ge is higher than 90° and it does not wet the substrate unlike Al. On the substrate with a TiN film, the contact angles of both components are smaller than the previous values so the wettability of Al and Ge on TiN is better. These values are in good agreement with those found in literature (8-10). TiN and SiO2 are both wet by the liquid Al-Ge eutectic alloy since the contact angles are less than 90°. The contact angle of the liquid alloy is still smaller on TiN than on SiO2. The TiN, used as an underneath layer allows more intimate contact between the solder and the surfaces to be joined. However as the wetting is better, eutectic squeeze out is more likely to occur.

TABLE 1: Results of contact angle measurement at 1000 °C.

Liquid	Contact angle on SiO_2	Contact angle on TiN
Pure Al	84°	58°
Pure Ge	135°	38°
Al-Ge eutectic alloy	80°	41°

The wettability of the alloy on the substrate is important for two reasons. If the contact angle is greater than 90°, the liquid does not wet the substrate and tends to form drops which could lead to voids in a bonded structure. On the other side, if the contact angle is very small, the liquid can leak out of the seal ring easily. If the angle θ' is greater than the contact angle θ, the liquid wets the vertical walls (Figure 2).

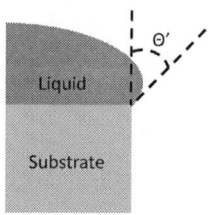

Figure 2 : overflow condition of the liquid outside of the seal ring.

The melting of the eutectic alloy on top of a seal ring coated with TiN is studied. After solidification, the resulting microstructure is analyzed by SEM EDX (Figure 3). There is no Al nor Ge on the sides of the seal rings which means that the liquid was contained in the sealing area. To confirm these results, the wettability of liquid Al-Ge eutectic alloy will be studied in bonded structure. Indeed, in this case, there are two main differences to be taken into account: a free surface is replaced by an interface with another substrate and the liquid is subjected to the stress imposed by the pressure between the two substrates to be bonded.

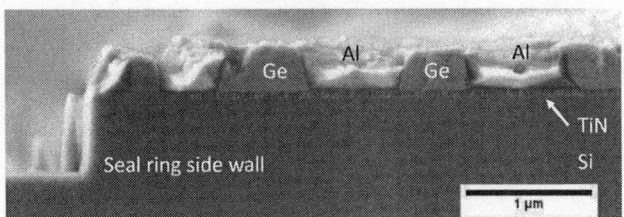

Figure 3: SEM cross section of Si/TiN/Al/Ge seal ring after thermal treatment at 450 °C.

Solidification of Al-Ge eutectic alloy

In this part the solidification process of the liquid Al-Ge eutectic alloy is studied. The final microstructure of the solder is of great importance since it must ensure the mechanical strength and the tightness of the assembly. During the solidification of the alloy, to obtain a homogeneous eutectic microstructure, the appearance of primary dendritic phases has to be avoided. The influence of cooling rate and underneath layer on microstructure is assessed by looking for the presence of dendrites which indicates a metastable solidification.

In a first series of experiments, many $1x1$ cm^2 samples from blanket Al/Ge deposited wafers are melted and then solidified with four different values of cooling rate. The microstructure are then analyzed by SEM EDX (Figure 4).

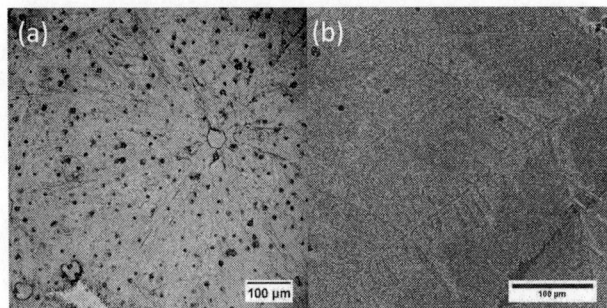

Figure 4 : (a) Optical microscopy of Al-Ge solidified with low cooling rate, radiative eutectic structure with (Ge) phase block at the center, (b) SEM micrographs of Al-Ge solidified with high cooling rate: large primary Al dendrite.

For small values of cooling rates (1°C/min), the alloy shows radiative eutectic structure with small primary (Ge) phase block at the center (11). However, when the cooling rate increases to a critical value, large primary Al dendrites appear. This critical value depends on the underneath layer. SiO_2 promotes the appearance of dendrites and imposes a critical value lower than that of TiN. It is important to note that if the size of these dendrites are in the order of magnitude of the seal ring dimensions, this could be very critical for the reliability of the assembly for two reasons. First, the Al dendrites could be preferential sites to initiate a mechanical fracture of the assembly. Then, they can form "bridges" between the two substrates, which block the diffusion of the liquid and can lead to the presence of voids at the sealing interface (12). To illustrate this point, Al/Ge deposited on a seal ring coated with TiN is melted and then solidified under conditions which permit the appearance of dendrites (Figure 5). These dendrites of Al primary phases are long enough to traverse the entire width of a joint. So, their appearance must be avoided by ensuring a stable cooperative solidification of Al and Ge.

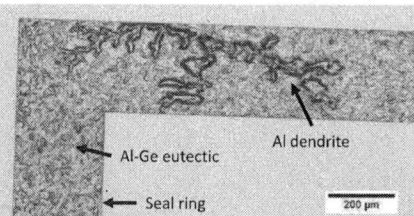

Figure 5 : Optical microscopy of Al/Ge deposited on a Si seal ring coated with TiN after thermal treatment at 450 °C and solidification with high cooling rate.

To better understand this phenomenon, a second series of experiments is conducted. Samples from blanket Al/Ge with different underneath layers are solidified in a DSC equipment. Two values of cooling rates are compared. The results are summarized in **TABLE 2**. The supercooling degree is the temperature difference between the melting temperature and the solidification temperature. No major difference is observed between the samples with different underneath layers. The supercooling degree with SiO_2 underneath layer is even lower than with TiN. This seems to be in contradiction with the fact that SiO2 underneath layer promote the Al dendrites formation. But, the supercooling degree increases much more with the SiO_2 substrate than with the TiN one. The cooling rate in the DSC experiment is quite low (5°C/min) compare to the cooling rate achievable in the bonding tool (20°C/min). This can explain why dendrite appears more quickly with SiO_2 underneath layer than with TiN using the bonding tool. Noteworthy these cooling rate values are lower than those found in the literature when metastable solidification is observed. Indeed, undercooling is generally studied for macroscopic samples but the germination also depends on the size of the system. When the size of the system decrease, it is more difficult for the liquid to germinate which could lead to metastable solidification. Finally, the presence of primary Al dendrites in the microstructure depends on the cooling rate, the underneath layer and the size of the solder joint.

TABLE 2: Results of DSC analysis

Substrate	Supercooling degree with low cooling rate (°C)	Supercooling degree with high cooling rate (°C)
SiO2	2,2	6,7
TiN	5,9	7,1

Bonding

In a first series of experiment, blanket Al/Ge deposited wafers are either bonded together or to silicon wafers coated with a diffusion barrier. The process parameters are chosen to ensure a cooperative solidification of Al and Ge which leads to homogeneous eutectic microstructure. In the first symmetrical bonded structure, blanket Al/Ge deposited wafers containing a SiO2 barrier are assembled. The result is analyzed by SAM (Figure 6). There are two distinct zones for this sealing formed. On the left, there is little porosity (which appear white) and the presence of many metal balls is noted in the border.

The alloy melted and squeezed out of the wafers. However, the right part of the plates has a wide defect and no metal ball is observed. Other tests show that this defect is due to a lack of heating in this area. The temperature was lower than the eutectic point and the alloy does not melted. This defect has been fixed for subsequent experiments.

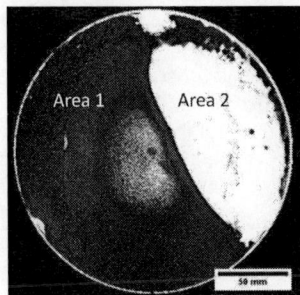

Figure 6 : SAM image of a symmetrical bonded structure with SiO_2 barrier.

After a FIB preparation, the bonding interface is observed by SEM in Area 1 (Figure 7, Figure 8). The morphology presents a succession of Al and Ge phases. The solder completely melted in this zone and solidified in eutectic structure. No Al primary dendrites is observed which is consistent with the low cooling rate imposed. The presence of holes in the Ge layers or at the interface between Ge and Al is noted.

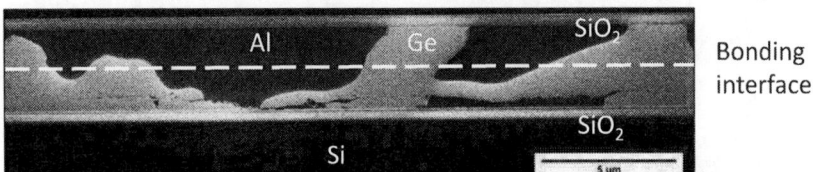

Figure 7 : FIB-SEM cross section of a symmetrical bonded structure with SiO_2 barrier, the interface shows a eutectic microstructure.

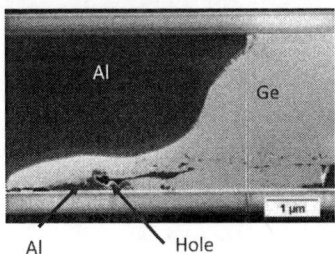

Figure 8 : FIB-SEM cross section, zoom of the Figure 7, holes between Al and Ge phases are noted.

The mechanism responsible for the appearance of these voids is not yet established. The difference between the densities of liquid alloy (Al-Ge) and the solid phases Al and Ge generates thermal stresses during solidification which could be at the origin of these voids. Another reason for this voiding phenomenon is the Kirkendall effect. If one element of the alloy diffuses faster than the other, it creates vacancies which can cluster and form micro-voids.

In a second series of experiment, patterned Al/Ge deposited wafers are bonded to silicon wafers coated with a diffusion barrier. The results are observed by SAM (Figure 9, Figure 10). Both cases present overflows at the borders of the areas containing the seal rings. These overflows are visible on Figure 9 (a), they are uniformly distributed all around the wafers. The applied force, which is uniform around the wafers, is assumed to be one of the causes of this phenomenon. Looking more precisely in the seal rings, holes are seen in both cases. When the SiO_2 is used as the underneath layer, they are mainly located at the corners of the seal rings (Figure 9 (b)). In the assembly using TiN, the holes are bigger and are visible at the edges and the corners of the seal rings (Figure 10). The underneath layer seem to have an influence on the squeeze-out. If the wettability of the liquid Al-Ge eutectic alloy is better, this phenomenon is more likely to occur. FIB-SEM images are ongoing in order to confirm these conclusions.

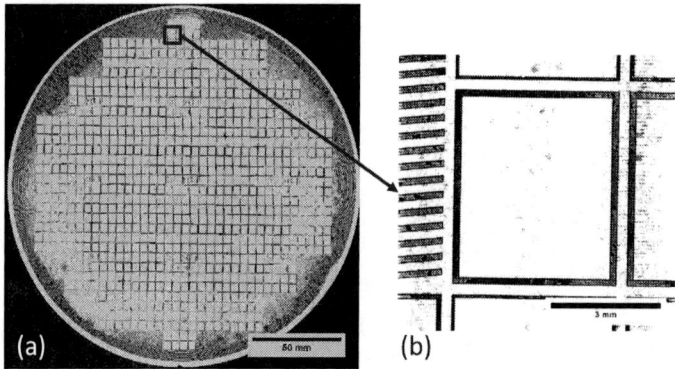

Figure 9: SAM images of a $Si/SiO_2/Al/Ge$ patterned wafer bonded with a blanket SiO_2 wafer (a) full wafer image showing overflow areas, (b) zoom on a seal ring with holes at the corners.

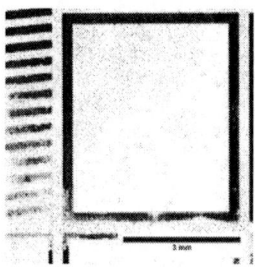

Figure 10 : SAM image of a seal ring from a Si/TiN/Al/Ge patterned wafer bonded with a blanket TiN wafer, holes are visible at the edges and the corners of the seal ring.

Conclusion

In this work, the packaging of MEMS using Al-Ge eutectic bonding was studied. The influence of cooling rate and underneath layer on the formation and behaviour of liquid Al-Ge alloy, its solidification, the resulting microstructures and the appearance of voids was developed. The formation and behavior of liquid Al-Ge alloy were studied through DSC measurement and dispensed drop method. The Al-Ge bilayers used have a melting temperature of 426 °C which is in good agreement with those found in literature. The wettability of liquid Al, Ge and Al-Ge eutectic alloy was assessed by dispensed drop method. The contact angles are smaller on TiN surfaces than on SiO2 surfaces. These results provide information about the ability of liquid to squeeze out of the seal ring regarding the type of underneath layer. Then the influence of cooling rate and underneath layer on solidification and microstructures was evaluated. Care was taken to the appearance of Al dendrites which can be critical in a bonded structure. It was shown that a critical value for cooling rate exists for each underneath layer. Above this value, large Al dendrites appear. Finally, these results served in bonding test. The sealing interface show a eutectic microstructure, without Al dendrites. Micro-voids are observed in the in the Ge layers or at the interface between Ge and Al. The Kirkendall effect and the thermal stresses are two possible mechanisms to explain this phenomenon. The results appear to be dependent on the process conditions. Underneath layer and cooling rate seem of great importance to control the squeeze-out and solidified microstructures in Al-Ge eutectic sealing and to determine the quality of the final assembly. Ongoing work is on studying these phenomena in bonded structure using patterned wafers.

Acknowledgments

The authors would like to thank Constantin Matei for SEM EDX analysis and Oleksii Liashenko for his advice about the Dispensed drop method.

References

1. Y. –C. Lin et al., *Transducers, IEEE*, pp 244-247, 2009.
2. Rodney P. Elliott et al., *Bulletin of Alloy Phase Diagrams*, vol 1, n° 1, pp. 65-68, 1980.
3. B.Vu et al., *J. Vac. Sci. Technol.B*, vol. 14, n° 4, pp. 2588-2594, 1996.
4. F.Crnogorac et al., *J. Vac. Sci. Technol.B*, vol. 30, n° 6, 06FK01, 2012.
5. S.Sood et al., *Advancing Microelectronics*, vol 41, n° 5, pp 30-37, 2014.
6. K. Nakazawa et al., *Jpn. J. Appl. Phys.*, vol 53, n°4, 04EH01, 2014.
7. A.J.McAlister et al., *Bulletin of Alloy Phase Diagrams*, vol. 5, n° 4, pp. 341-347, 1984.
8. V. Laurent et al., *Materials Science and Engineering*, A135, pp. 89-94, 1991.
9. S. K. Rhee, *Journal of The American Ceramics Society*, vol. 53, n° 7, pp. 386-389, 1970.
10. N. Kaiser et al., *Journal of Crystal Growth*, vol 231, pp. 448-457, 2001.
11. N. Yan et al., *Journal of Alloys and Compounds*, vol 607, pp. 258-263, 2014.
A.Garnier et al., *J Mater Sci : Mater Electron*, vol. 24, pp. 5000-5013, 2013.

Chapter 8

Bonding for 3D-Integration

284

Self-Assembly Based Multichip-to-Wafer Bonding Technologies
for 3D/Hetero Integration

T. Fukushima [a, b, c], K.W. Lee [b], T. Tanaka [a, d], and M. Koyanagi [b]

[a] Dept. of Mechanical Systems Engineering, Tohoku Univ., Sendai, 980-8579, Japan
[b] New Industry Creation Hatchery Center, Tohoku Univ., Sendai, 980-8579, Japan
[c] Electrical Engineering Department, UCLA, Los Angeles, CA, 90095, USA
[d] Dept. of Biomedical Engineering, Tohoku Univ., Sendai, 980-8579, Japan

We have proposed and developed 3D integration technologies
based on self-assembly using surface tension of liquid in this
decade. In this paper, microbump bonding and bumpless bonding
in face-up and/or face-down configuration are introduced for fine-
pitch interconnect formation. In addition, "non-transfer stacking",
in other word, flip-chip self-assembly and "transfer stacking"
called reconfigure-wafer-to-wafer using SAE carrier are explained.

Introduction

3D chip stacking approaches with TSV (thorough-silicon vias) are mainly categorized by
chip-to-chip, wafer-to-wafer, and chip-to-wafer 3D integration. The chip-to-chip
technologies are widely used for traditional microelectronic packaging in which KGDs
(known good dies) are assembled using one-by-one pick-and-place techniques. Although
the assembly yield is high due to the use of KGDs, the assembly throughput is not
sufficiently high, compared with wafer-to-wafer approaches. However, wafer-to-wafer
technologies have a serious issue in total production yield that is exponentially decreased
with the number of stacked layers because the failure dies cannot be removed from the
wafers to be stacked. Therefore, chip-to-wafer 3D integration is thought to be a
promising candidate to satisfy both the throughput and yield issues. However, chip-to-
wafer 3D integration potentially has a trade-off between assembly throughput and
alignment accuracies in the conventional robotic pick-and-place techniques. In order to
address the problems, we have proposed and developed multichip-to-wafer 3D
integration using liquid surface tension (1)-(32). From the technical point of view of chip-
to-wafer bonding, we talk about the basic concept and previous studies of the self-
assembly based 3D integration. These approaches using surface tension-driven multichip
assembly are divided into two methodologies: one is "transfer stacking" with carrier
wafers and the other one is "non-transfer stacking" without the handlers. Since no support
wafers are required for the non-transfer stacking based 3D integration, this approach
seems to be cost effective and high throughput, but this is not necessarily advisable.
Target bottom wafers and lower KGDs receive repeated thermomechanical stresses and
damages through the 3D integration processes with thinning and TSV formation, as the
number of stacked layers increases. This means that transfer-stacking based 3D
integration would be useful for multi-layer 3D chip stacking to increase total production
yields. Here, we introduce self-assembly based multichip-to-wafer bonding technologies
for 3D/hetero integration

Capillary Self-Assembly

The first key parameter is "liquid surface tension" as a hosting fluid used in self-assembly processes. Driving forces acting on KGDs to be self-assembled are proportional to liquid surface tension. Water is well known to be a liquid having the highest surface tension at room temperature except for toxic mercury. The value is approximately 72 mN/m at room temperature. Liquid with lower surface tension indicates lower alignment accuracy (16).

The second key parameter is "contact angle differences" so-called wettability contrast between hydrophilic assembly areas and the surrounding hydrophobic areas. If the wettability contrast is low, liquid droplets can easily overflow on the assembly areas to give low alignment accuracy. In contrast, if the contrast is high, liquid droplets can be confined to assembly areas to give high alignment accuracy. When the wettability contrast is 80 degree or more, the alignment accuracy is experimentally found to be within 1-2 μm (5).

The third key parameter is "chip size accuracy" to precisely align KGDs to wafers (7). Chips fabricated by plasma dicing before grinding technique have high-precision chip size accuracy ±1 μm, and thereby, high alignment accuracies of approximately 0.4 μm have been obtained (16). Standard saw dicing can provide high chip size accuracies ±1-2 μm, which have high potential to achieve submicron alignment accuracies (32).

Impact of the other parameters such as wafer tilt (25), liquid volume (1), initial positioning offset (32), and chip edge structure (32) on self-assembly accuracies has been studied so far. The detail cane be seen from the papers listed in the reference section.

Self-Assembly Based Multichip-to-Wafer Bonding Technologies

Self-assembly based multichip-to-wafer 3D integration technologies are classified into two categories: microbump bonding and bumpless bonding. The feature of these bonding technologies with self-assembly is described as follows:

Microbump Bonding.

We have previously employed two microbump materials: one is In/Au microbumps (2)-(6) and the other is Cu/Sn microbumps (7). The former microbumps are mainly composed of In that is capped with Au for preventing the In surface from being oxidized. We have successfully flip-chip self-assembled KGDs on In/Au bumped wafers with an alignment accuracy of approximately 1 μm. The latter microbumps is widely used for advanced microelectronic packaging in industry. However, Sn solder is very sensitive to oxidation, and thus, water droplets used in flip-chip self-assembly would oxidize the surface of the Sn microbumps if the water directly makes contacted with the solder.

In order to address the issues on self-assembly in a face-down embodiment, we have three option: the first one is the use of water-soluble flux. The flux diluted in water can remove the oxidized thin surface while self-assembled. We have demonstrated Cu/Sn microbump bonding with low contact resistances (8), (9). The second one is the use of NCF (non-conductive film) that is typically laminated on bumped wafers prior to saw dicing. The NCF can cover the microbump surface and protect the oxidation. The NCF is a B-staged polymer and includes flux agents. Therefore, the thin surface of oxidized Sn can be removed during thermo-compression flip-chip bonding. In general, NCF is called wafer-level underfill, but the assembly is a one-by-one process. We have combined the NCF technology with self-assembly to realize fully wafer-level underfill in batch processing (10)-(12). The third one is transfer stacking called reconfigured-wafer-to-wafer 3D integration where SAE (self-assembly and electrostatic) carriers can be used as a handler (13), (14). The SAE carries have bipolar electrodes embedded in dielectric layers. The KGDs are self-assembled on the SAE carrier in a face-up embodiment, followed by evaporating the water droplets at room temperature. After DC apply, the KGDs are electrostatically fixed on the carrier. Then, the reconfigured wafer having the KGDs and SAE carrier is aligned and bonded to the target bumped wafers through Cu/Sn microbumps at the wafer level. Since the KGDs are self-assembled right side up on the SAE carrier, the Cu/Sn microbumps are not wetted by the water droplets.

By using twice transfer stacking approaches based on the reconfigured-wafer-to-wafer 3D integration (15)-(17), we have implemented TSV formation using organic (18), (19) or inorganic (20), (21) temporary adhesives. Laser debonding is a key process in both the temporary bonding technologies. Visual laser is employed in the former temporary bonding using the organic adhesive and a release layer (18), (19), whereas near UV laser is employed in the latter one using the inorganic adhesive and another release layer (20), (21). Organic temporary adhesives are widely used in 3D integration. These spin-on thick organic adhesives is cost effective and can fully cover high-topography bumped wafers. However, thermal stability is not so high, and thereby, the temperature of 3D integration processes is limited at below 250℃. On the other hand, the debonding system consisting of SOG (spin-on-glass) as an inorganic adhesive and a-Si:H (hydrogenated amorphous silicon) as a release layer show high heat resistant. Therefore, high-temperature TSV liner deposition techniques can be used to increase the dielectric properties.

Bumpless Bonding.

We have previously integrated two types of 3D chip stacking processes based on bumpless bonding scenario. One is oxide-oxide direct bonding and the other is Cu-Cu hybrid bonding. Underfilling into extremely small gaps between fine-pitch microbumps is very challenging although advanced underfilling using NCF and NCP (non-conductive paste) have been developed. The bumpless bonding requires no underfilling process, which can dramatically reduce the production throughput.

In the first approach of the 3D integration using self-assembly (1), (22)-(27), we have proposed room-temperature load-free direct bonding using diluted hydrogen fluoride. 1% HF solution can precisely and quickly align many KGDs on host wafers at once within 0.1 sec with an alignment accuracy of below 500 nm, and at the same time, the KGDs can be tightly bonded on the corresponding wafers through thermal oxide layers. The HF assisted oxide-oxide direct bonding is applicable to CMP-treated PE-CVD oxide (2), (26)-(27). Nowadays, pure water can be used for the direct oxide-oxide bonding although plasma- and water-assisted thermal compression bonding is required (28), (29). The oxide-oxide bonding needs SiO_2-CMP, but the homogeneous CMP process is not technologically challenging. Sony has employed oxide-oxide direct bonding and fabricated 3D stacked pixel and logic LSI chips in face-to-face configuration with 3-μm-pitch Cu-TSVs through a ultra-thin Si layer for camera modules. On the other hand, Cu-Cu and dielectric-dielectric (polymer/polymer or oxide-oxide) hybrid bonding is a key technology to obtain high-density interconnects. The Cu-Cu hybrid bonding is industrially used for Samsung Galaxy S7 Rear Camera Module as well, where 14-μm-pitch bumpless interconnects are formed in the peripheral regions. However, heterogeneous Cu and dielectric CMP processes dramatically increase the production cost and reduce the yields. Derivation of particle-free extremely high planarity on the CMP surface is still challenging. We have advanced a new hybrid bonding technology using Cu nano-pillars with a diameter of approximately 60 nm (30), (31). This technology has been most recently reported for exascale 2.5D/3D integration toward upcoming IoT (internet of things) and neuromorphic computing systems.

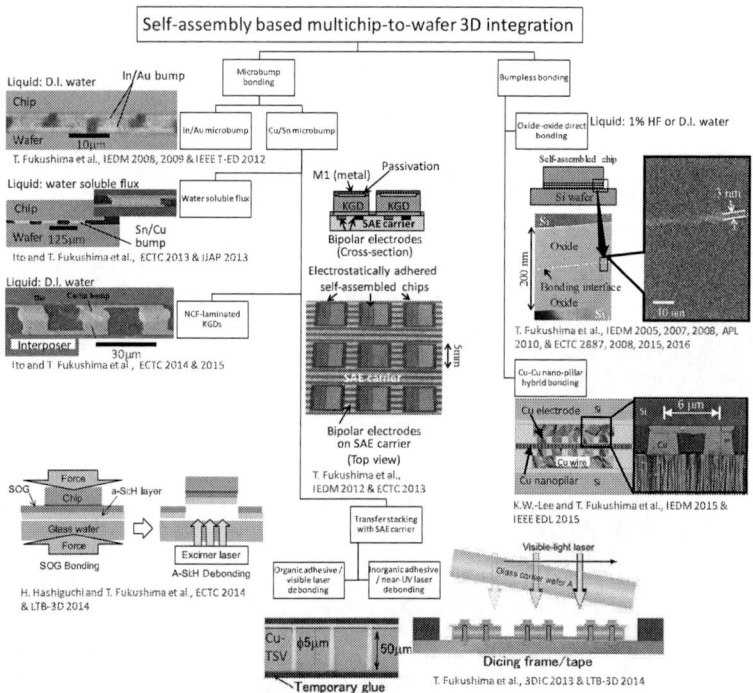

Figure 1. Self-Assembly Based Multichip-to-Wafer Bonding Technologies.

Acknowledgments

This work was performed in the Micro/Nano-machining research and education Center (MNC), Jun-ichi Nishizawa Research Center, and Global INTegration Initiative (GINTI) in Tohoku University.

References

1. T. Fukushima, Y. Yamada, H. Kikuchi and M. Koyanagi, *IEDM Tech. Dig.*, 359 (2005).
2. T. Fukushima, T. Konno, K. Kiyoyama, M. Murugesan, K. Sato, W.-C. Jeong, Y. Ohara, A. Noriki, S. Kanno, Y. Kaiho, H. Kino, K. Makita, R. Kobayashi, C.-K. Yin, K. Inamura, K.-W. Lee, J.-C. Bea, T. Tanaka and M. Koyanagi, *IEDM Tech. Dig.*, 499 (2008).
3. T. Fukushima, E. Iwata, Y. Ohara, A. Noriki, K. Inamura, K.-W. Lee, J. Bea, T. Tanaka, and M. Koyanagi, *IEDM Tech. Dig.*, 349 (2009).
4. T. Fukushima, Y. Ohara, M. Murugesan, J.-C. Bea, K.-W. Lee, T. Tanaka, and M. Koyanagi, *Proc. ECTC*, 2050 (2011).
5. T. Fukushima, E. Iwata, Y. Ohara, M. Murugesan, J. Bea, K. Lee, T. Tanaka and M. Koyanagi, *IEEE Trans. Electron Devices*, **59**, 2956 (2012).
6. Y. Ito, T. Fukushima, H. Kino, K.-W. Lee, K. Choki, T. Tanaka and M. Koyanagi, *Jpn. J. Appl. Phys.*, **54**, 030206 (2015).
7. T. Suzuki, K. Asami, Y. Kitamura, T. Fukushima, C. Nagai, J. Bea, Y. Sato, M. Murugesan, K.-W. Lee, and M. Koyanagi, *Proc. ECTC*, 342 (2015).
8. Y. Ito, T. Fukushima, K.-W. Lee, K. Choki, T. Tanaka, and M. Koyanagi, *Proc. ECTC*, 891 (2013).
9. Y. Ito, T. Fukushima, K.-W. Lee, K. Choki, T. Tanaka and M. Koyanagi, *Jpn. J. Appl. Phys.*, **52**, 04CB09 (2013).
10. T. Fukushima, Y. Ohara, J. Bea, M. Murugesan, K. -W. Lee, T. Tanaka, and M. Koyanagi, *Proc. ECTC*, 393 (2012).
11. Y. Ito, T. Fukushima, K.-W. Lee, K. Choki, T. Tanaka, and M. Koyanagi, *Proc. ECTC*, 856 (2014).
12. Y. Ito, M. Murugesan, H. Kino, T. Fukushima, K.-W. Lee, K. Choki, T. Tanaka, and M. Koyanagi, *Proc. ECTC*, 336 (2015).
13. T. Fukushima, H. Hashiguchi, J. Bea, Y. Ohara, M. Murugesan, K.-W. Lee, T. Tanaka and M. Koyanagi, *IEDM Tech. Dig.*, 789 (2012).
14. T. Fukushima, H. Hashiguchi, J. Bea, M. Murugesan, K.-W. Lee, T. Tanaka, and M. Koyanagi, *Proc. ECTC*, 58 (2013).
15. T. Fukushima, Y. Ohara, M. Murugesan, J.-C. Bea, K.-W. Lee, T. Tanaka, and M. Koyanagi, *Proc. ECTC*, 1050 (2010).
16. T. Fukushima, E. Iwata, Y. Ohara, M. Murugesan, J.-C. Bea, K.-W. Lee, T. Tanaka and M. Koyanagi, *IEEE Trans. Compon., Packag., Manuf. Technol.*, **1**, 1873 (2011).
17. T. Fukushima, J. C. Bea, H. Kino, C. Nagai, M. Murugesan, H. Hashiguchi, K.-W. Lee, T. Tanaka and M. Koyanagi, *IEEE Trans. Electron Devices*, **61**, 533 (2014).
18. T. Fukushima, J. Bea, M. Murugesan, K.-W. Lee, and M. Koyanagi, *Tech. Dig. 3DIC* (2013).
19. T. Fukushima, M. Mariappan, J.-C. Bea, H. Hashimoto, Y. Sato, M. Motoyoshi, K.-W. Lee, and M. Koyanagi, *Proc. LTB-3D*, 69 (2014).
20. H. Hashiguchi, T. Fukushima, A. Noriki, H. Kino, K.-W. Lee, T. Tanaka and M. Koyanagi, *Proc. ECTC*, 856 (2014).
21. H. Hashiguchi, T. Fukushima, M. Murugesan, J.-C. Bea, H. Kino, K.-W. Lee, T. Tanaka, M. Koyanagi, *Proc. LTB-3D*, 71 (2014).

22. T. Fukushima, Y. Yamada, H. Kikuchi, T. Tanaka, and M. Koyanagi, *Proc. ECTC*, 836 (2007).
23. T. Fukushima, H Kikuchi, Y. Yamada, T. Konno, J. Liang, K. Sasaki, K. Inamura, T. Tanaka and M. Koyanagi, *IEDM Tech. Dig.*, 985 (2007).
24. T. Fukushima, E. Iwata, K.-W. Lee, T. Tanaka, and M. Koyanagi, *Proc. ECTC*, 628 (2008).
25. T. Fukushima, E. Iwata, T. Konno, J.-C. Bea, K.-W. Lee, T. Tanaka and M. Koyanagi, *Appl. Phys. Lett.*, **96**, 154105 (2010).
26. K.-W. Lee, A. Noriki, K. Kiyoyama, S. Kanno, R. Kobayashi, W.-C. Jeong, J.-C. Bea, T. Fukushima, T. Tanaka and M. Koyanagi, *IEDM Tech. Dig.*, 531 (2009).
27. K.-W. Lee, S. Kanno, K. Kiyoyama, T. Fukushima, T. Tanaka and M. Koyanagi, *J. Microelectromech. Syst.*, **19**, 284 (2010).
28. H. Hashiguchi, H. Yonekura, T. Fukushima, M. Murugesan, H. Kino, K.-W. Lee, T. Tanaka and M. Koyanagi, *Proc. ECTC*, 1458 (2015).
29. T. Fukushima, H. Hashiguchi, H. Kino, T. Tanaka, M. Murugesan, J. Bea, H. Hashimoto, K.-W. Lee, and M. Koyanagi, *Proc. ECTC*, 289 (2016).
30. K.W. Lee, J. Bea, T. Fukushima, S. Ramalingam, X. Wu, T. Tanaka, and M. Koyanagi, *IEEE Electron Device Lett.*, **37**, 81 (2015).
31. K.-W. Lee, C. Nagai, J.-C. Bea, T. Fukushima, R. Suresh, X. Wu, T. Tanaka, and M. Koyanagi, *IEDM Tech. Dig.* 185 (2015).
32. Y. Ito, T. Fukushima, H. Kino, K.-W. Lee, T. Tanaka and M. Koyanagi, *J. Microelectromech. Syst.*, **25**, 91 (2016).

Wafer-level Vacuum Packaging by Thermocompression Bonding using Silver after Fly-cut Planarization

C. Liu [a,*], H. Hirano [a], J. Froemel [a, b] and S. Tanaka [a]

a. Department of Robotics, Graduate School of Engineering, Tohoku University, Japan
b. WPI-Advanced Institute for Materials Research, Tohoku University, Sendai, Japan

A novel wafer-level vacuum thermocompression bonding using electroplated Ag bonding frames which were planarized by diamond fly cutting was described. At relatively low temperature and press pressure, high shear strength and vacuum sealing were achieved. The diffusion of Ag into a Si substrate was suppressed by the implementation of a TiN barrier layer. Leak rate was measured by zero balance method was at approximately 3.6×10^{-14} Pa·m^3/s. Therefore, the proposed process was useful for both wafer-level heterogeneous integration, such as the integration of MEMS and LSI circuits, and hermetic packaging for various microsystems.

Introduction

With the development of micro-electromechanical systems (MEMS) technology, the integration of MEMS and large scale integrated (LSI) circuits together with hermetic packaging is getting more important for the reduction of device footprints and the improvement of performance.

Among the approaches implemented for the hermetic integration of MEMS and LSI, wafer bonding technology is a good choice. It is a wafer level packaging process, but allows separate design and fabrication of MEMS and LSI until they are integrated, offering an extra degree of flexibility (1-3).

Wafer bonding can be realized through various methods: silicon direct bonding, anodic bonding, liquid phase bonding and solid state bonding. Silicon direct bonding usually requires an additional structure for electrical interconnect (2, 4). Anodic bonding needs electrical field and extra design for electrical interconnect as well. Liquid phase bonding includes eutectic bonding and glass frit bonding. Typically, in order to achieve hermetic sealing, the sealing structures width is hard to control because of the liquid phase and large bonding frame width is adverse to reducing device size (4). Solid state bonding includes thermocompression bonding, thermosonic bonding and surface-activated bonding(5). Thermocompression bonding is based on atomic diffusion and is a rather simple process, which can be realized by applying the pressure and heating without any complex apparatus for example a vacuum chamber merged with a plasma chamber(3, 5, 6). Notably it can inherently realize the electric interconnection as well as hermetic sealing (7). It can also utilize a smaller bonding frame to achieve sufficient bonding strength, which leads to the chip size reduction (4, 7).

The thermocompression bonding based on Au, Al and Cu is well known. Nevertheless the passivation of Cu or Al impairs the performance of bonding, so additional complex pretreatment correlated to the oxide layer is required. Au can be bonded at lower temperature with minimal surface pretreatments. Our previous work used electroplated Au bonding frames after fly-cut planarization (1). The fly-cut planarization reduced the roughness of the electroplated surface to establish enough contact for sufficient atomic diffusion and made it possible to achieve hermetic bonding even on non-flat wafers.

Compared with Au, Ag is not only much cheaper but also of superior thermal and electrical conductivity (8). For the purpose of cost reduction, Ag is proposed using the fly-cut planarization method. However, Ag is known to have relative high diffusivity in Si(9). The diffusion of Ag into Si is a concern for deteriorating device characteristics, especially for the integration with LSI. In addition, the oxidation affinity of Ag is worried to reduce bonding quality because of a surface oxide layer. Consequently we have to solve those problems to develop a new thermocompression method.

In this study, we developed a novel inexpensive hermetic thermocompression bonding method using electroplated Ag bonding frames planarized by diamond fly cutting. A barrier layer was used to block Ag diffusion into a Si substrate during the bonding process. Without any specific pretreatment to remove the oxide layer, high shear strength and vacuum sealing were demonstrated at relatively low bonding temperature and mechanical pressure. The method described here holds the possibility for the wafer-level vacuum packing of MEMS and LSI.

Experimental design and procedures

Barrie layer deposition and wafer process

Minority carrier lifetime in LSI will be severely reduced due to the contamination of metal atoms. To accomplish the integration of MEMS and LSI through thermocompression bonding, the diffusion of Ag atoms to the Si substrate during the heating process must be suppressed. We used a barrier layer to solve this problem. A titanium nitride (TiN) barrier layer was adopted because of its confirmed barrier property in conjunction with good thermal stability and low contact resistance (10, 11). It can meet the requirements of restraining diffusion, simultaneously ensuring the electrical interconnection and the stability of Ag on the surface (12).

The TiN barrier layer can be produced mainly by following techniques: reactive sputtering from a titanium target, reactive chemical vapor deposition, directly sputter deposition from a TiN target and the nitridation of a titanium film via rapid thermal annealing in a suitable atmosphere(10, 13). Stoichiometry control is difficult in reactive sputtering or evaporation, and sub-stoichiometric TiN deteriorates in barrier property (13). Rapid-thermal nitridation usually request high temperature annealing and makes it limited for temperature-sensitive devices (10). Therefore, we chose sputter deposition from a TiN target, by which we could directly get near stoichiometric layer at relatively high deposition rate.

A 20 mm×20 mm Si wafer was used as a sealing wafer, on which the bonding frame would be fabricated. To demonstrate vacuum sealing a 20 mm×20 mm SOI wafer (10/1/400 μm of device layer/box layer/handle layer) was used as cavity wafer. As depicted in Fig.1 (a), on the Si and SOI wafers, a layered structure (20 nm Ti/ 100 nm TiN) was deposited by RF sputtering at room temperature. The Ti layer works as an adhesion layer. TiN was sputtered directly from a TiN target of high purity. Subsequently, the wafers were taken out from the chamber and exposed to air. During the exposure, oxygen diffused in grain boundaries and blocked fast diffusion paths, resulting in better barrier property (11, 14-16). After exposure, 100 nm thickness of Ag as a seed layer for electroplating was sputtering deposited on the TiN layer, as shown in Fig.1 (b).

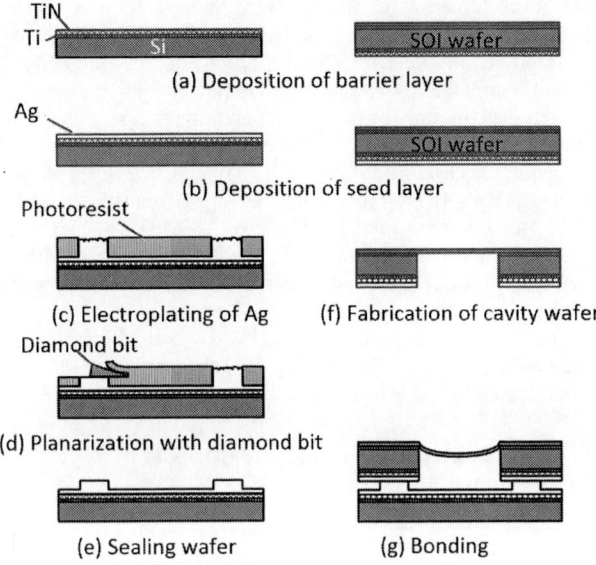

Figure.1 Wafer process and bonding

After that as shown in Fig.1 (c), on the sealing wafer the Ag bonding frames (40 μm width and 10 μm height) were electroplated using photoresist cavities after standard lithographic process. On one wafer, 36 dies were fabricated and each occupied a 2 mm × 2 mm chip region. Next, the bonding wafers were subjected to fly-cut planarization. During this process, the frame was cut to 4 μm height by fly cutting with a diamond bit (Surface Planner DAS8920, DISCO, Japan), as depicted in Fig.1 (d). Finally, after the removal of photo resist, the bonding wafer was ready for the bonding, as shown in Fig.1 (e).

As shown in Fig.1 (f), on the SOI substrate the same stacked layers without electroplated Ag were deposited, and a cavity was fabricated by deep reactive etching (DRIE). The cavity wafer was bonded with the sealing wafer through the bonding process described in the next chapter. At atmosphere, the pressure difference across the diaphragm led to the deflection of the diaphragm (Fig.1 (g)) and the cavity pressure was estimated.

Wafer bonding

The bonding was accomplished with a Süss SB6e wafer bonder. Firstly, both sealing and cavity wafers were exposed to Ar plasma for just 20s, during which wafer surfaces were activated(17). Understandably a certain degree of the oxide was removed as well. Then, under atmosphere and at room temperature, the sealing and cavity wafers were aligned and mounted as a wafer stack.

The wafer stack was transferred into the chamber of the wafer bonder, and then the chamber was vacuumed below 10^{-5} Pa. During vacuuming process the spacers separated the wafers for releasing gas. Thereafter, the spacers were removed and the wafers came into contact with each other. The wafers were loaded with a force of 600 N, corresponding to a pressure of 54 MPa on the bonding frame. Simultaneously the wafers were heated up to 350 °C and held for 40 min. After bonding, the wafer stack cooled down slowly to room temperature in vacuum environment. Finally, the chamber vented and the load force released, putting the end to the bonding process, as shown in Fig.1 (g).

After the bonded wafer was diced into dies, the cross-section was fabricated for observation by scanning electron microscopy (SEM). Die-shear tests were conducted to evaluate the shear strength (PTR-1101, Rhesca, Japan), and fracture surfaces were observed by SEM. Secondary ion mass spectroscopy (SIMS) detection was carried out through stacked layers to ascertain the effect of the barrier layer (SIMS4000, CAMECA, France). The electroplated frame before and after the planarization was also observed by SEM.

The diaphragm was observed at atmosphere by white light interferometer (surface topography measurement system MSA-500, Polytec, Germany). The sealing pressure and leaking rate were evaluated by the zero-balance method(18).

Experimental results and discussion

Planarization and bonding reliability

As mentioned, the planarization was vital for our method. Fig 2 shows the surface of the electroplated Ag bonding frame before and after planarization. From the SEM figures, we can see that the electroplated surface was of high asperity and the bonding frame surface became smooth after planarization. The average surface roughness (Ra) measured by a profiler of the flattened bonding frame was 5.4 nm, which was appropriate to establish enough contact surface for sufficient atomic diffusion. The planarization is a simple method to implement considerably low-cost electroplating for thermocompression bonding.

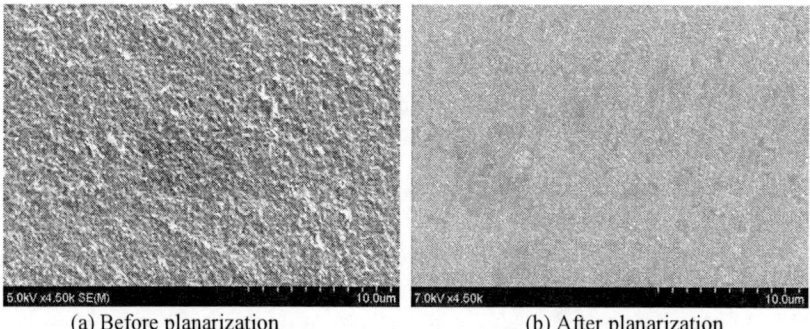

    (a) Before planarization   (b) After planarization

Figure.2 Surface of electroplate Ag bonding frame before and after planarization

Additionally, since the shear strain of the diamond bit, smaller grain size was obtained, which improved grain boundary diffusion. Grain boundary diffusion, rather than solid diffusion, plays a dominant role in solid bonding technique (19). Moreover, during planarization, the electroplated Ag frames were cut from 10 μm height to 4 μm height, meaning that native surface oxide or any other surface contamination was removed.

In the cross-section of the bonded area observed by SEM (Fig.3), there was no visible bonding interface anymore. It could be concluded that grain growth occurred at the bonding interface. The thermocopression bonding consists of three stages: interface formation, crystal misfit accommodation and grain growth (7). With the help of the planarization and mechanical pressure, the electroplated surface and the sputtered surface came into close contact with each other. Thereafter, some extent of crystal orientation alignment occurred within the two metal layers. During the heating process, diffusion advanced further, grain growth was developed and the interface vanished. It was rational to infer high bonding property from the disappearance of the bonding interface.

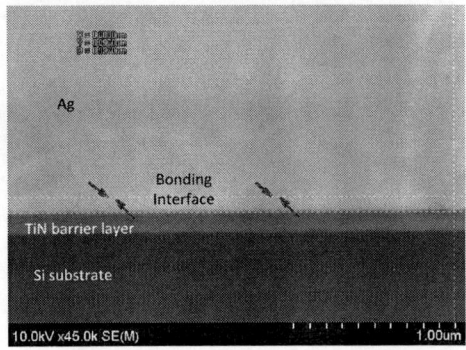

Fig.3 SEM image of bonded interface

To demonstrate it, the shear strength of bonded wafers was measured by a die shear tester. The average strength was up to 200 MPa. Fractures were evaluated by SEM. The dies fractured across the bonding material Ag, and in no case at the bonding interface. The lateral expansion of the sealing frames due to pressure was less than 5%, which made sure to provide an enough cavity space as design.

It was notable that after the Ar plasma activation, although Ag was exposed to air before bonding, the surface oxide of Ag did not have strong negative influence on the bonding property. The newly produced oxide layer after the planarization and 20s Ar plasma exposure was thin because of the slow oxidation rate of Ag. And the thin oxide layer may easily diffuse into the bulk through grain boundaries during bonding process. To confirm such mechanism, detection by transmission electron microscope (TEM) will be done experimentally in the near future.

Barrier layer property

SIMS detections were carried out through Si/Ti/TiN/Ag layers of the die after die shear test and of not annealed die. To avoid the sputtering effects of SIMS, we started the SIMS profile detection from the Si side. For minimizing the etching thickness, Si substrate was thinned mechanically ahead before detection. The SIMS detection outcomes of two kinds of dies were the same and no detectable Ag atom was founded in the Si substrate. In both cases, it seemed that Si diffused through TiN layer. This phenomeon was also found by other researchers and the reason was Si reaction with Ti from the adhesion layer or an unsaturated TiN layer (13, 20). Nonetheless, it did not damage the bonding property and should not have negative impact on the application for the integration of LSI. Thereafter, it is concluded that the barrier layer used in our method was effective to stop the diffusion of Ag during bonding.

Vacuum sealing

After the bonding process of the bonding wafer and cavity wafer, the membrane's clear deflection was observed at atmosphere. The die after 1000 h since bonded was shown in Fig.4 (a), and the deflection was detected by avwhite light interferometer as depicted in Fig.4 (b).

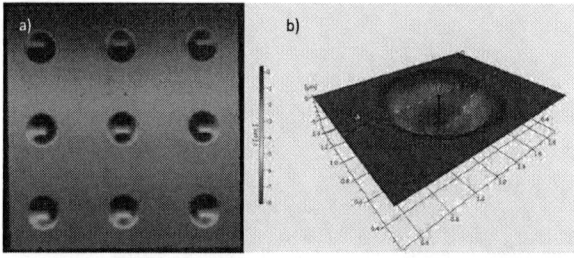

Fig.4 Deflection of Si diaphragms at atmosphere after 1000h since bonding

A small vacuum chamber for the zero-balance method was designed and fabricated. While the diaphragm defections were monitored by white light interferometer, the pressure in the chamber was reduced. Once the diaphragm recovered to flat status, the chamber pressure equaled the cavity pressure. More details about the method can be found in our previous publication(18). The cavity pressure after vacuum packaging was around 6 kPa with a leak rate at approximately 3.6×10^{-14} Pa·m^3/s, where the leak rate was obtained in the detection over 1500 h. To obtain lower cavity pressure, to use a getter layers is an alternative choice (21). The obtained low leak rate indicated high quality of vacuum packaging.

Summary

In conclusion, a novel wafer bonding method was established using planarized electroplated Ag frames. This method achieved vacuum packaging with high shear strength (200 MPa) at relatively low temperature (350 °C). Through the implementation of TiN barrier layer, the diffusion of Ag into a Si substrate was prevented. The leak rate of the bonded cavity was at around 3.6×10^{-14} Pa m^3/s. The proposed process was useful for both wafer-level heterogeneous integration, such as the integration of MEMS and LSI circuits, hermetic packaging and is versatile for various other microsystems.

Acknowledgments

This study was performed in R&D Center of Excellence for Integrated Microsystems, Tohoku University, supported by the Formation of Innovation Center for Fusion of Advanced Technologies, and partially supported by "Nanotechnology Platform" of the Ministry of Education, Culture, Sports, Science, and Technology (MEXT), Japan.

References

1. H. Hirano, K. Hikichi and S. Tanaka, in *18th International Conference on Solid-State Sensors, Actuators and Microsystems, Transducers-2015*, p. 1283 (2015).
2. R. Wolffenbuttel and K. Wise, *Sensors and Actuators A: Physical*, **43**, 223 (1994).
3. C. Oh, S. Nagao, T. Kunimune and K. Suganuma, *Applied Physics Letters*, **104**, 161603 (2014).
4. M. Antelius, G. Stemme and F. Niklaus, *Journal of Micromechanics and Microengineering*, **21**, 085011 (2011).
5. R. Takigawa, E. Higurashi, T. Suga and T. Kawanishi, *IEEE J. Sel. Top. Quantum Electron.* , **17**, 652 (2011).
6. H. Takagi, K. Kikuchi, R. Maeda, T. R. Chung and T. Suga, *Applied Physics Letters*, **68**, 2222 (1996).
7. J. Froemel, M. Baum, M. Wiemer, F. Roscher, M. Haubold, C. Jia and T. Gessner, in *2011 16th International Solid-State Sensors, Actuators and Microsystems Conference* (2011).
8. C.-H. Sha and C. C. Lee, *Microelectronic Engineering*, **99**, 11 (2012).
9. T. C. Nason, G. R. Yang, K. H. Park and T. M. Lu, *Journal of Applied Physics*, **70**, 1392 (1991).
10. T. Brat, *Journal of Vacuum Science & Technology B: Microelectronics and Nanometer Structures*, **5**, 1741 (1987).
11. G. A. Dixit, C. C. Wei, F. T. Liou and H. Zhang, *Applied Physics Letters*, **62**, 357 (1993).
12. T. L. Alford, L. Chen and K. S. Gadre, *Thin Solid Films*, **429**, 248 (2003).
13. H. Joswig and W. Pamler, *Thin Solid Films*, **221**, 228 (1992).
14. W. Sinke, G. P. A. Frijlink and F. W. Saris, *Applied Physics Letters*, **47**, 471 (1985).
15. T. Maeda, T. Nakayama, S. Shima and J. I. Matsunaga, *Electron Devices, IEEE Transactions on*, **34**, 599 (1987).
16. H. G. Tompkins, *Journal of Applied Physics*, **70**, 3876 (1991).
17. G. Kissinger and W. Kissinger, *Sensors and Actuators A: Physical*, **36**, 149 (1993).
18. M. S. Al.Farisi, H. Hirano and S. Tanaka, in *2016 IEEE 11th International Conference on Nano/Micro Engineered Molecular Syst, IEEE NEMS 2016* (2016).
19. T. Shimatsu and M. Uomoto, *Journal of Vacuum Science & Technology B: Microelectronics and Nanometer Structures*, **28**, 706 (2010).
20. A. Kohlhase, M. Mändl and W. Pamler, *Journal of Applied Physics*, **65**, 2464 (1989).
21. R. Knechtel, M. Wiemer and J. Frömel, *Microsystem Technologies*, **12**, 468 (2005).

298

Low Temperature Thermo Compression Bonding with Printed Intermediate Bonding Layers

M. Wiemer[a], F. Roscher[a], T. Seifert[a,b], K. Vogel[a], T. Ogashiwa[c] and T. Gessner[a,b]

[a] Fraunhofer Institute for Electronic Nano Systems, Chemnitz 09126, Germany
[b] Chemnitz University of Technology, Center for Microtechnology 09126 Chemnitz, Germany
[c] Tanaka Kikinzoku Kogyo K.K., 100-6422 Tokyo, Japan

The paper describes capabilities of additive material transfer technologies Aerosol Jet Printing (AJP), Stencil Printing (STP) and Screen Printing (SP) based on nano particle inks with the purpose of intermediate bond layer manufacturing for Chip-to-Chip, Chip-to-Wafer and wafer level bonding. For realization of these applications, the deposition process and the related material morphology, the bonding with nano particle based intermediate bonding layer (IBL) as well as the characterization methods are described. AJP was used to deposit silver (Ag) as IBL to bond copper (Cu) and gold (Au) metallized wafers. Stencil printed Au nano particle IBL layers were used to bond Au metallized wafers. The interface between nano particles and substrate were investigated. For bonding result evaluation, bonding frames and occurring bonding interfaces were analyzed using scanning electron microscopy (SEM), energy dispersive X-ray spectroscopy (EDX) and focused ion beam (FIB). Mechanical properties were qualified using compression shear tests and tensile tests.

Motivation

The manufacturing costs of microelectronic components like Micro Electro Mechanical Systems (MEMS) are strongly correlated to involved packaging technologies like wafer bonding (1). Typically wafer level packaging of MEMS is done by direct bonding with activated surfaces, field assisted bonding or bonding with intermediate bonding layer (IBL) on adhesion promoters if necessary. This IBL can be deposited using printing processes like Screen Printing, Stencil Printing or Aerosol Jet Printing. There are state-of-the-art-approaches on silicon wafers, which contains dry etched rim structures around the active chip structure to precisely define widths of bonding frames manufactured by STP (2). Also lithography patterned, glass based carrier wafers are used as an intermediate step to establish a precise pattern transfer of nano particle based bonding frames printed with STP to a wafer (3). Another approach of printing intermediate bonding layers is using direct writing technology AJP (4). AJP offers high resolution patterns down to 10 μm (5) and makes it possible, to reduce width of bonding frames comparable to lithography processes. By using novel nano particle IBLs not only process-related costs could be saved, but also new possibilities could be realized, like metallization of MEMS at stages of manufacturing, where no wet chemical processes, no electroplating or no lithography steps are suitable anymore. Furthermore, the metallization patterns can be easily generated by CAD data. With this work we present that successful bonds are possible by direct printed Au and Ag based IBLs without any additional intermediate step like described by Ogashiwa et. al. in former publications.

Printing processes for nano particle IBLs

Screen and Stencil Printing

Screen printing and the quite familiar approach of stencil printing (STP) are well investigated mask based additive material transfer technologies in the field of microelectronic packaging (6, 7). For SP a screen is used as a printing template whereby screen open areas represent the printing pattern. A squeegee made from polyurethane, carbon or other materials and adjusted with a considered squeegee angle transfers paste with a defined speed across the stencil. Thereby pressure is applied by the squeegee to the screen what not only brings the screen in contact with the printing substrate but also transfers the paste through the screen open areas onto the printing substrate. At STP a stencil made from metal or polymeric materials is working as printing template whereby apertures – the so called stencil openings – represent the printing pattern. Like in SP also for STP squeegees are used to transfer paste through the apertures onto the substrate. In Figure 1 the working principles for SP and STP are opposed.

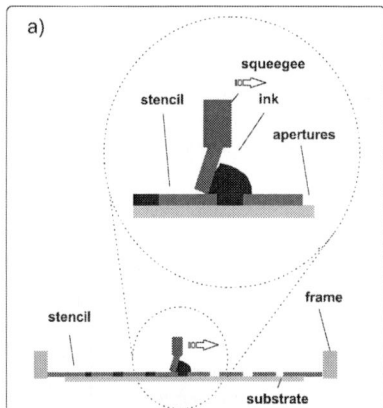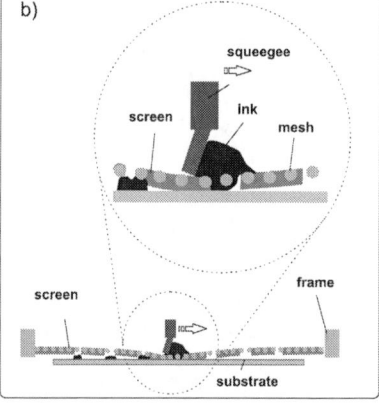

Figure 1: Working principles for Stencil Printing (a), and Screen Printing (b)

The main printing parameters for SP are squeegee angle, squeegee speed, squeegee pressure and snap off distance. The screen itself mainly is defined by wire diameter of the mesh, number of wires per inch, wire distance, mesh thickness, width of mesh openings and thickness of emulsion over mesh. For STP next to the main printing parameters the stencil itself is mainly defined by manufacturing process, stencil thickness, aspect ratio and area ratio of the aperture geometry and the properties of apertures side wall areas. For stencil printing of Tanaka AuRoFuse materials a suspended stencil was used (2) whereby in contrast the wafer was not patterned with rim structures manufactured by deep reactive ion etching (RIE). In Table 1 an overview about general layout based stencil setup is given.

Table 1 Overview about stencil parameters for STP of Tanaka AuRoFuse, Au adhesion layers

Parameter	Values
Outer width of bond frame	4 mm x 4 mm
Frame width	100 µm
Total bonding area	355.7 mm²
Total number of bond frames	228

In Figure 2 the typical printing results gained by STP are shown. In general increasing of lateral resolution to feature sizes small as 100 µm is demanding due to wetting effects of printed material along with widening of printed IBL in comparison to the design. The paste material is dried and pre-sintered before final bonding step in order to remove residuals like binder system or solvents needed to ensure the printability.

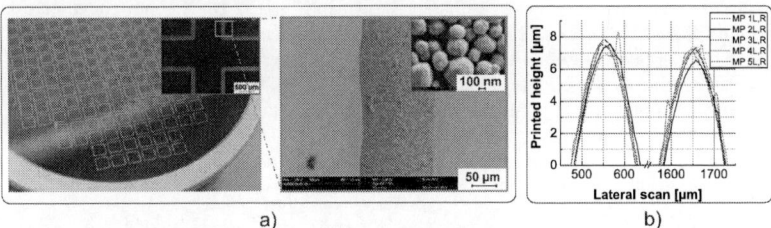

a)   b)

Figure 2 Typical appearance of additive manufactured IBL by STP without deep RIE manufactured rims (a) and cross sectional profiles (b)

In general SP and STP could be performed with systems like the DEK Horizon 03iX shown in Figure 3. The system is equipped with a special, for wafers optimized, platen carrier and includes special software based wafer level alignment solutions.

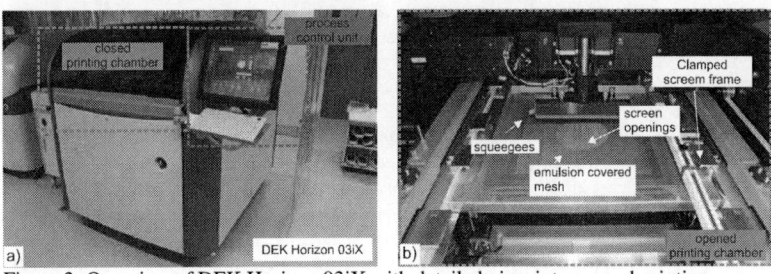

Figure 3: Overview of DEK Horizon 03iX with detailed view into opened printing chamber

The IBLs additive manufactured by STP were investigated regarding morphology after printing, drying and pre-sintering using SEM. In Figure 4 the results are shown.

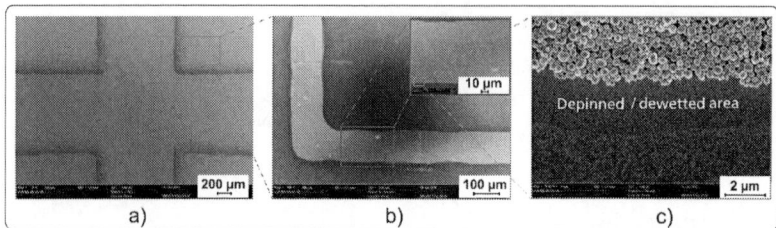

Figure 4: Lateral morphology properties of bonding frames manufactured by STP

It can be seen that printed bonding frames provide good edge sharpness while being wavy. Furthermore residuals of drying and pre-sintering effects were investigated by means of de-pinned or de-wetted areas next to frame borders respectively. The topographic properties were analyzed using Veeco Dektak System (Veeco, New York, USA). In total 20 frames were measured. The results are shown in Figure 2 b). Smooth convex cross sections with an inverse parable like appearance were identified. In maximum a printed frame peak was measured with 8.2 μm whereby the average printing frame peak was calculated with 7.4 ± 0.5 μm. The average frame width was measured with 164.4 ± 5.1 μm what means that printed frame width increased in comparison to original set up frame width of 100 μm by approximately 64 %.

Aerosol Jet Printing

Aerosol-Jet is a registered trademark of Optomec ® (Optomec Design Company, Albuquerque, NM, USA). The Aerosol Jet system is mainly composed of two key components, the atomizer and the deposition head. The raw material to be deposited must be in a liquid form (viscosity between 1 cP and 1000 cP) and is first placed into an either ultrasonic or pneumatic atomizer, which is utilized to generate a dense mist of material droplets between 1 μm to 5 μm in size (10). The working principle of such an Aerosol Jet Printer can be categorized into three parts: (i) atomization of fluids, (ii) guiding the atomized material towards the deposition head and (iii) deposition of atomized material onto a substrate (11). With ultrasonic atomization and pneumatic atomization (Figure 5) the Aerosol Jet Printer 300CE uses two different atomization principles. In this work a pneumatic atomizer was used, realized by a gas flow that is guided to a jet where ink is soaked in from a reservoir due to arising under inflation. Finally, small amounts of ink are disrupted at the disruption zone and are guided through an impactor towards the print head, where a sheath gas aerodynamically focuses the aerosol.

The main parameters of the AJ technology are the gas flow of pneumatic atomizer to generate the aerosol, the sheath gas flow to focus the aerosol beam, the substrate temperature and stand-off between print head and substrate. These parameters have been defined (TABLE 2) for a stable deposition process of a silver nanoparticle ink SW1020. The used ink has 40 wt% silver particle loading (10 nm particle diameter), a sintering temperature starting at 100 °C, a viscosity of ~10 cP and is commercially available from Japanese company Bando Chemical Industries Ltd.

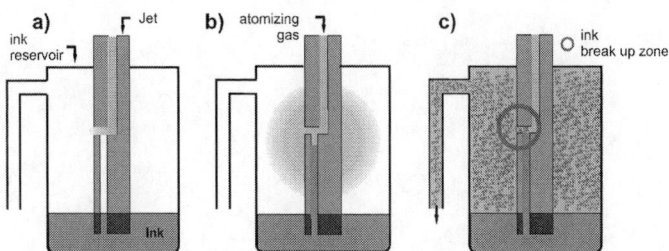

Figure 5 Pneumatic atomizer: feeding atomizer jet with carrier gas of high velocity (a), soaking up ink to ink channel of atomization jet (b), breaking up drops of ink inside the jet and transporting them to impactor exit due to the atomizer gas flow (c)

TABLE 2 Overview about used parameters for AJP Bando SW1020 on Au, Cu and Al adhesion layers

Parameter	Values
Substrate temperature	70 °C
Pre-treatment	Ar plasma (optional)
Atomizer gas flow	500 to 700 [sccm]
sheath gas flow	70 to 90 [sccm]
Stand off	2 to 3 [mm]

Figure 6 shows the typical appearance of deposited Ag tracks and the ability to achieve printed features with less than 30 μm in width (10, 11). Sintering of the ink was observed starting from 100 °C by SEM investigations and electrical measurements.

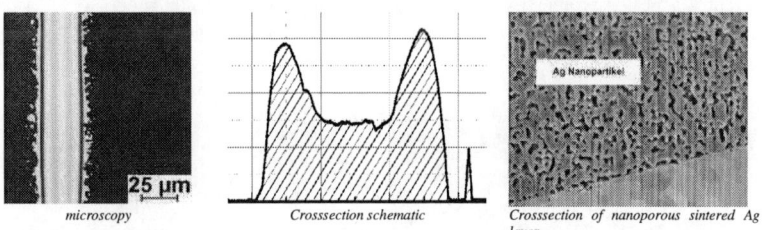

microscopy   Crosssection schematic   Crosssection of nanoporous sintered Ag layer

Figure 6 Typical appearance of a printed Ag track on wafer level

AJP was performed using an Optomec Aerosol Jet 300 CE Printer consisting of a computer including the necessary software (KEWA), a process chamber with print modules, a movable x-y table and atomizer systems as well as a PCM (process control module) to control the gas flows (Figure 7). Within this work the pneumatic atomizer was used to fabricate bond frames in the shape of 4 mm x 4 mm, 100 μm and 115 μm linewidth on wafer level out of the Ag nanoparticle ink for subsequent wafer level bonding. Whereas the output of the AJ300 is a constant aerosol stream the pattern of the desired structures is realized by a mechanical operating shutter arm that opens for deposition of ink onto the substrate and closes to collect the aerosol while printing is not needed.

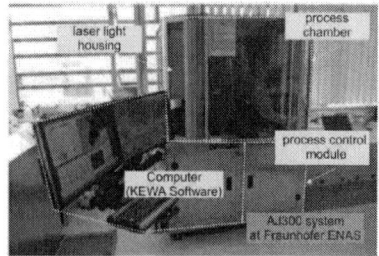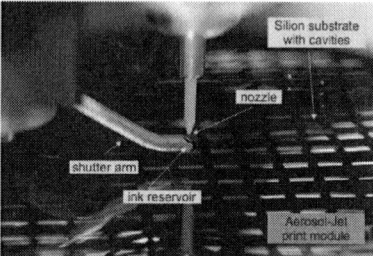

Figure 7 Overview AJ300 system at Fraunhofer ENAS and close-up of the print module

Two types of substrates have been produced for the investigations a) simple 100 mm Si wafers with sputtered Au, Cu or Al adhesion promoters for first feasibility studies and b) 150 mm Si wafers with patterned Au adhesion promoters and wet etched trenches for testing the mechanical stability of the bonding and the hermeticity (see TABLE 3).

TABLE 3 Overview Substrate Materials and adhesion promoters

Substrate	Coatings	Layer Thickness
100 mm Si Wafer	Ta/Cu	20 nm/500 nm
100 mm Si Wafer	Ti/Au	20 nm/500 nm
100 mm Si Wafer	Al	100 nm
150 Si Wafer with cavities	Ti/Au	20 nm/500 nm

As shown in Figure 8 the AJ300 enables a direct deposition of bond frames on non-patterned adhesion promoter coated 100 mm Si wafers as well as on patterned adhesion promoters on 150 mm Si wafers. Standard fiducials etched into the Si wafers using lithography and wet etching have been used for the alignment of the printed patterns. The AJ300 offers an alignment accuracy and repeatability of ± 6 µm.

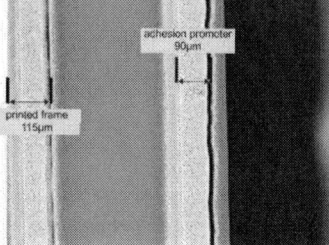

Figure 8 (left) 100mm Si wafer with full area Au adhesion promoter and printed Ag nano particle bond frames, (right) 150 mm Si Wafer with wet etched cavities and Au adhesion promoter (90 µm width) covered by printed Ag bond frames (115 µm width)

Wafer level bonding

Wafer level bonding was performed using a standard substrate bonder equipment from SÜSS MicroTec AG as well as EV Group. Before bonding the substrates are aligned, clamped in a fixture and separated by spacers and loaded to the bonding chamber. Within the chamber, the spacers will be removed and the wafers will get in contact by mechanical load applied by the bond tools. By heating the wafers and applying further bond force F the wafers will get in close contact and the IBL could build an interface or alloy to the adhesion promoters enabled by diffusion between the different materials (Figure 9). For all bonds with Ag based IBL a bond pressure of 80 GPa and a vacuum of 10^{-4} mbar was used while the bonding temperature starting at 350 °C down to 200 °C and bonding time of 30 min and 60 min was varied. For all bonds with Au based IBL the bond force was varied between 6.5 kN and 47 kN and a vacuum of 10^{-4} mbar was used while the bonding temperature was fixed at 200 °C for 10 min.

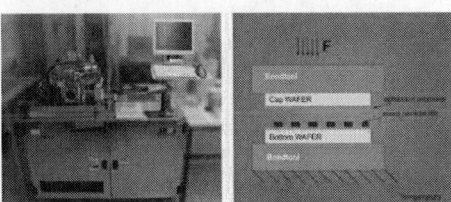

Figure 9 (left) SÜSS SB8 substrate bonder and (right) wafer level bonding schematic process

Results for bonding with Au IBL

The bonded wafer stacks were diced and mechanically characterized regarding tensile strength and shear strength afterwards. After performing the tensile test, the bonding interface was analyzed by light microscopy optically. In Figure 10 the results of chips occurring from cap wafer and printed bottom wafer are opposed.

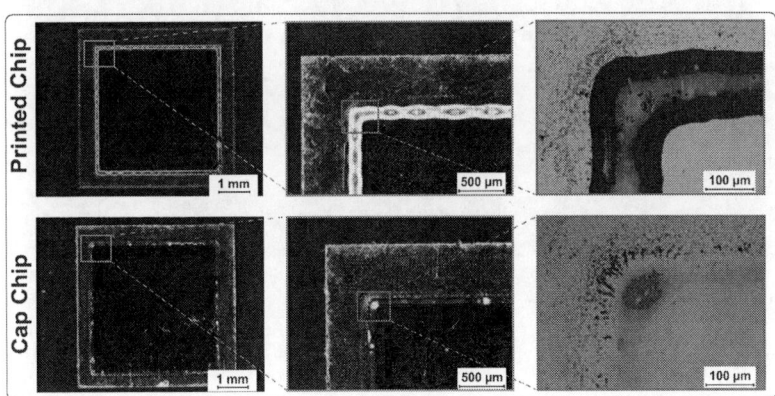

Figure 10 Appearance of IBL on cap and bottom wafer after dicing and tensile testing

It can be seen that in general bonded frames show good resistance against penetration of cooling water used during the dicing process. The difference between level of contamination of chip surface outside the bonding area and within the bonding area is significant. Furthermore the bonding interface is distinguishable depending to the chip side. Whereas the printed chip disposes a large interface comparable to the width of original printed bonding frames, the cap chip provides a significant smaller area of bonding interface. This could indicate limited compressibility of printed IBL at given process setup along with restricted capability of wetting the cap chip. Moreover compressed areas of bonded IBL frames can be identified at printed chip represented by darker colored and smaller areas compared to allover frame area. This is in accordance to occurrence of bonding interface at wetted cap chip (brighter areas). Here color wise the occurring bonding interface can be identified. Failure modes are a combination of adhesive and cohesive failure mechanism. To investigate the bonding mechanism in detail cross sections have been prepared and analyzed by SEM. The results are shown in Figure 11. It can be seen that particles already sintered at chosen bonding temperature of 200 °C along with densification and grain growth. Still, porosity is identifiable all over the bonding interface. The value of porosity is quite higher at the outer areas of the bonding frame cross section. This is also the area where the printed bonding frame did not come into contact with the cap wafer due to restricted compressibility at chosen bonding setup. Non-contact zones are visible which automatically dispose highest grade of porosity due to no mechanically compression of AuRoFuse material within this region. Moreover this effect describes the limit of printed IBL without deep RIE etched RIMS like described by Ogashiwa et al. (2). At given printing setup the bonding pressure of maximum 47 kN is not sufficient to compress 100 % of the IBL due to the achieved topography of the bond frames. An improvement of compressed area could be achieved by adjusting the printing process to reduce the bonding frame width.

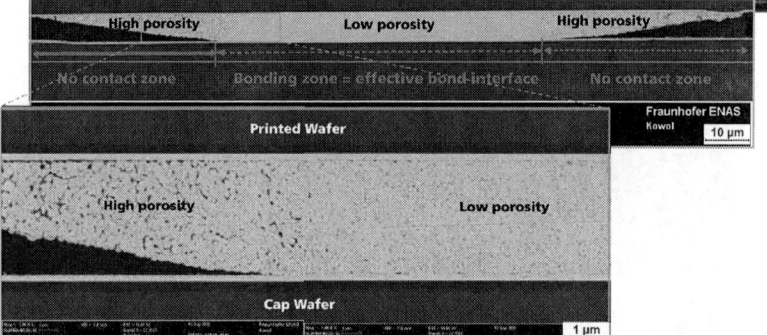

Figure 11 SEM cross section investigations, classification into bonding zone with low porosity and non-contact zones of high porosity

In order to qualify compressibility of AuRoFuse material in correlation to applied bonding force cross sections of occurring interfaces were analyzed using focused ion beam (FIB) technology. In Figure 12 the results are shown.

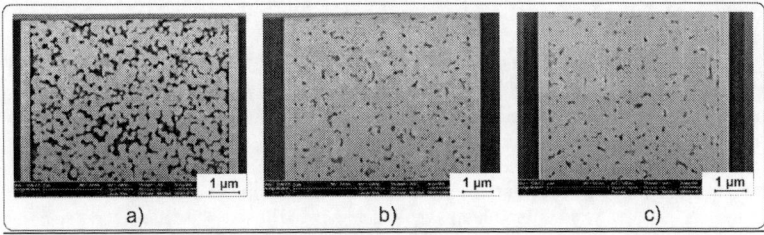

Figure 12 Comparison of bonding interfaces with thicknesses of 7.0 µm and bonding force of 6.5 kN (a), 5.7 µm and bonding force of 27 kN (b) and 5.5 µm and bonding force of 47 kN (c)

It can be seen that with increasing bonding force the progress of grain growth along with densification is much more progressed. Furthermore the interface to adhesion promoting layers of printed wafer and cap wafer are increasing with higher bonding force. Also the thickness of all over bonding interface decreases which is an indicator for higher compressed material along with higher densification. In order to evaluate mechanical properties of gained bonding interfaces, compression shear tests and tensile tests have been performed. The results of tensile strengths and shear strengths are shown in Figure 13.

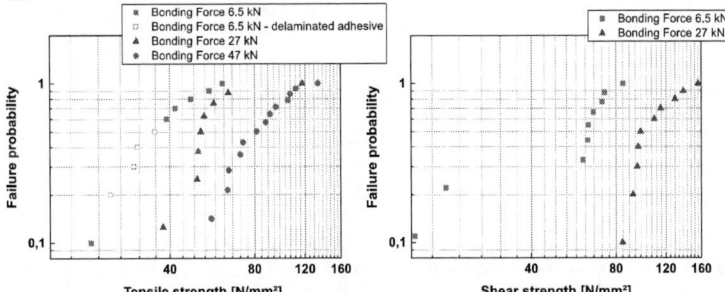

Figure 13 Overview of tensile and shear strength using AuRoFuse IBL for different bonding forces

The difference in the curve gradient for the shear and tensile strength indicated different failure mechanisms. Typical failure mechanisms occurring during both tests were adhesive and/or cohesive failures as well as destruction of specimen during compression shear test could occur. Highest average tensile strength was obtained for bonding force of 47 kN with 83.2 ± 27.1 N/mm² Table 4 the results with varied bond force are summarized.

Table 4 Summary of bonding results in relation different bonding forces

	Bonding Test a)	Bonding Test b)	Bonding Test c)
Bonding Force [kN]	6,5	27	47
Thickness of Interface [µm]	7.0	5.7	5.5
Tensile Strength [N/mm²]	44.3 ± 14.2	60.3 ± 24.4	83.2 ± 27.1
Compression Shear Strength [N/mm²]	57.7 ± 23.7	111.1 ± 23.1	n.a.

Results for bonding with Ag IBL

Using the 100 mm Au, Al and Cu coated Si wafers with 100 μm wide Ag IBLs, wafer level bonding was performed at 300 °C for 30 min in vacuum. A subsequent dicing process (10 mm x 10 mm chip size) shows the feasibility of the different adhesion promoters to build a mechanical stable interface to the Ag nanoparticles. For Al, that spontaneously forms a stable aluminum oxide film in the thickness range of 50 Å, (13) no bonded areas were observed and the yield was 0 % after dicing. For Cu and Au adhesion promoters 100 % yield after dicing were achieved. The cross sections prepared by focused ion beam showed porous Ag IBL with a void free interface to Cu and Au as shown in Figure 14 and reported in (4).

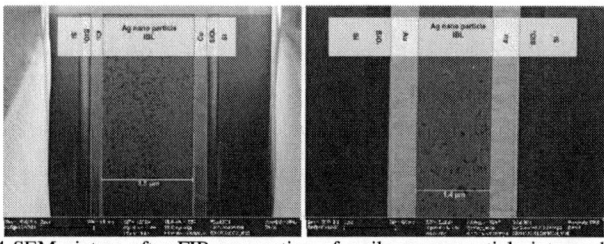

Figure 14 SEM picture after FIB preparation of a silver nanoparticle intermediate layer between adhesion promoter Cu (left) and Au (right) after wafer level bonding at 300 °C for 30 min

The different appearance of the IBL between the Au and Cu adhesion layers as visible in the SEM pictures is likely to be caused by the diffusion between Au and Ag that was observed using EDX inspection and could not be found between Cu and Ag (4). Optimization and adaptation for industrial scale bonding was performed using the 150 mm Si wafers with patterned Au adhesion promoters and deep etched cavities in the silicon. As described the wafers were coated with the Ag ink only on the Au adhesion promoters with slightly wider Ag bond frame width of 115 μm to fully cover the Au adhesion promoter frame. With those wafers WLB was performed at 250 °C and 200 °C with a longer bond time of 60 min to enable the sintering and diffusion of the Ag nanoparticles at those lower temperatures. After dicing (10 mm x 10 mm chip size) with those bonded wafers a yield of ~ 80 % was achieved.

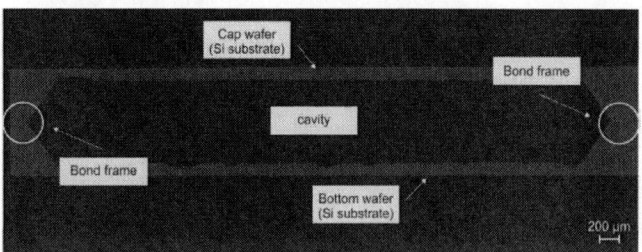

Figure 15 Microscopy of a single chip cross section showing the cavities and bond frames between cap and bottom wafer

The loss was caused by non-homogeneities in the printed patterns where some printed bond frames showed an overlap that hindered a full contact between both substrates in those areas. For interface studies, cross sections have been prepared on bonded single chips after dicing the wafers (Figure 15). Cross section have been used for FIB preparation whereas a sintering of the Ag nanoparticles itself (marked with blue color) and an interface between Ag nanoparticle IBL and Au adhesion layer without voids is visible for both bonding temperatures 250 °C and 200 °C as shown in Figure 16.

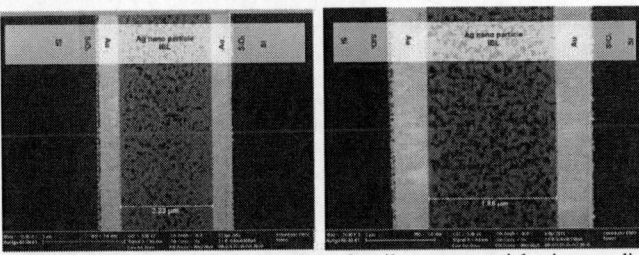

Figure 16 SEM picture after FIB preparation of a silver nanoparticles intermediate layer between adhesion promoter Au after wafer level bonding at 250 °C, 60 min (left) and 200 °C, 60 min (right)

A clear difference in the distribution and amount of the pores is visible when comparing the IBL bonded at 300 °C, 250 °C and 200 °C. Compressive shear test was performed to evaluate the average shear strength for all Au coated Si wafers at 350 °C, 300 °C, 250 °C and 200 °C (Figure 17). Whereas the shear strength lowered when going down from 350 °C to 300 °C, the shear strength could be increased for 250 °C and 200 °C by increasing the bonding time from 30 min to 60 min as shown in Figure 17. Hermeticity was evaluated by using the thin 100 µm thick Si membranes that should show a deflection after bonding in vacuum and exposing those wafers to normal atmosphere. Whereas the deflection was visible for all bonding temperature directly after bonding, it was gone latest after 24 h. As the cross sections indicate and the membrane deflection tests proofed there was no real hermeticity achievable with Ag nano particle IBLs so far. The maximum achieved shear strength of ~ 27 N/mm^2 is still lower in comparison to widely used glass frit bonding, whereas the results indicate that by adjusting the bonding time the stability of the bond could be increased.

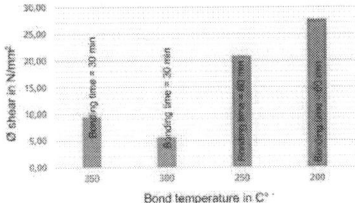

Figure 17 Shear strength of Ag nano particle IBL for different bonding temperatures and bonding times

Conclusion

Wafer level bonding using additive AJP and screen as well as stencil printing for deposition of Ag and Au nano particle IBLs was investigated on 100 mm and 150 mm Si wafers coated with Au, Cu and Al adhesion promoters. For Ag nanoparticles deposited with AJP the results show that there is a high potential to enable a mechanical stable bond of the IBL towards Au and Cu adhesion promoters whereas Al (due to its native and very stable oxide) could not be used for this approach.

Compared to glass frit bonding the Au nano particle IBL enables three times higher bond strength at lower bonding temperatures of 200 °C. Increasing tool pressure while bonding leads to increasing tensile strength, shear strength, grain growth and densification. The achievable results are strongly related to the printed pattern morphology. The cross sections of an Au nanoparticle bond frame after bonding shows 2 typical zones (i) an inner contact zone responsible for the joint with almost no porosity and (ii) the outer contact zones produced by the printing process but not in contact with the counter substrate. The process is capable to be adapted to chip level or can be expanded to 3D integration applications.

Acknowledgments

The work related to wafer level packaging using Ag nanoparticle inks was supported by the German Federation of Industrial Research Associations (AiF) (contract 16990 BR).

References

1. M. Esashi, J. Micromech. Microeng, vol. 18, no. 7, p. 073001, 2008.
2. T. Ogashiwa et al, International Symposium on Microelectronics, vol. 2015, no. 1, pp. 000073–000078, 2015
3. H. Ishida et al, Sixty Second Electronic Components & Technology Conference: The Printing House, Inc, 2012, pp. 1140–1145
4. F. Roscher, T. Seifert, K. Vogel, M. Wiemer, and T. Gessner, GMM-Fb. 81: Mikro-Nano-Integration, 2014
5. T. Seifert et al, Ind. Eng. Chem. Res, vol. 54, no. 2, pp. 769–779, 2015
6. R. Kay and M. Desmulliez, Soldering & Surface Mount Technology, vol. 24, no. 1, pp. 38–50, 2012.
7. G. Wu, D. Xu, X. Sun, B. Xiong, and Y. Wang, IEEE Trans. Compon, Packag. Manufact. Technol, vol. 3, no. 10, pp. 1640–1646, 2013
8. F. Roscher, T. Seifert, M. Wiemer and T. Gessner, CSTIC 2015 - Semicon China, March 2015
9. F. Roscher, T. Seifert, M. Baum, M. Wiemer and T. Gessner, Conference on Wafer Bonding for Microsystems 3D- and Wafer Level Integration, Braunschweig, Germany, 2015
10. M. J. Renn, Patent: US20030020768
11. M. J. Renn, Patent: US6636676B1
12. J. Evertsson, Applied Surface Science 349, 2015

Plastic Deformation of Thin Si Membranes in Si-Si Direct Bonding

E. Poppe, G. U. Jensen, S. T. Moe, D. Wang

SINTEF, Department of Microsystems and Nanotechnology, Oslo, Norway

The effect of bond anneal in Si-Si direct bonding of laminates with thin membranes suspending closed cavities is studied. For membranes of a certain size and thickness, it is found that the under-pressure in the cavity during bond anneal leads to plastic deformation of the membrane. By controlling the cavity pressure it is found that the Si crystal of the membrane can be kept intact during bond anneal.

Introduction

High temperature annealed silicon direct - or fusion - bonding is a widely used process, enabling the fabrication of robust, long-term stable devices with 3D structures [1]. To achieve good bond strength a temperature above at least 900 °C is required [2]. For device fabrication a wider thermal range is required, with oxidation and diffusion processes up to 1150 °C. During these high-temperature steps, thin membranes suspending sealed cavities may become permanently deformed. We have investigated the extent and cause of such deformations and we have established how these deformations can be avoided by adjusting the cavity pressure in the bonded device prior to the high temperature anneal step.

Experimental

Fusion bonding employs pre-bonding in controlled atmosphere at typically 50 °C. The bond anneal furnace works at atmospheric pressure, and a desired maximum temperature of 1050 °C was used in this study. Assuming that the gas inside the closed cavity follows the ideal gas law, a pressure of 247 mbar at 50 °C corresponds to 1 atm at 1050 °C. In this study, pre-bonding pressures of $207 - 247$ mbar N_2 were chosen to achieve a sufficiently small deflection of the membranes at the 1050 °C anneal step. In addition, one wafer pair was pre-bonded in vacuum as a reference.

Test structures were fabricated on eight 100 mm wafer pairs. SOI wafers with 43 μm device layer, 500 nm Buried Oxide layer (BOX) and 380 μm handle wafer were used, together with 400 μm thick bulk wafers. Both wafer types were p-type with (100) orientation. The bulk wafers were oxidized, and the oxide was patterned to allow for etching of an inlet hole after bond anneal.

Membranes were fabricated by Deep Reactive Ion Etching (DRIE) of the SOI wafers through the handle wafer, using the BOX as an etch stop. Circular membranes with a diameter range of $840 - 3720$ μm and square membranes with side edges $1105 - 3270$ μm were defined, as summarized in Table 1. The same lithography mask was used on the front side of the wafers to etch the oxide on top of the membranes, followed by

TABLE I. Summary of test structure dimensions, calculated maximum stress at 1 atm pressure difference, and crystal property results for pre-bonding in vacuum

Cavity name	Membrane type and size	Calculated max stress @ 1 atm [MPa]	Bonding results in vacuum
S1	Square 1105*1105 µm	25	Intact crystal
S2	Square 1885*1885 µm	74	Slip lines
S3	Square 2465*2465 µm	126	Slip lines
S4	Square 3265*3265 µm	221	Slip lines
C1	Circle dia 840 µm	7.2	Intact crystal
C2	Circle dia 1880 µm	36	Slip lines
C3	Circle dia 2640 µm	72	Slip lines
C4	Circle dia 3720 µm	142	Slip lines

thermal oxide growth. This makes a step in the Si on the front side, making it easier to identify the membrane position after bonding. To ensure symmetric membranes all masking oxide and the exposed BOX were stripped, and a new 190 nm thick thermal oxide was grown.

Wafer pairs were cleaned in piranha, rendered hydrophilic in RCA-1, aligned, and loaded into an EVG510 wafer bonder, separated by 50 µm thick spacers. The bond chucks were kept at 50 °C. The chamber was brought to vacuum ($<1*10^{-3}$ mbar), and purged with N_2 to reach the desired pressure level. Wafers were brought to contact, and a bond force of 1000 N was applied for 2 min. A cross section of the test structures is shown in Figure 1. A Zygo New View white light interferometer (WLI) was used to measure membrane deflection. For each of the pre-bonded laminates, some reference membranes were measured before the laminates were annealed at 1050°C for 2 hours. After bond anneal, wafers were measured again, and an inlet to each cavity was etched by DRIE. A final deflection measurement was then performed. The membranes were also inspected by an optical microscope with a halogen lamp and a Hamamatsu C8800 IR sensitive camera.

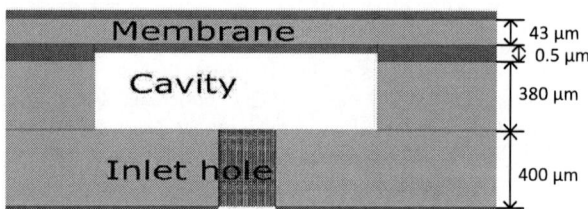

Figure 1. Sketch of test structure cross-section (not to scale). The inlet hole is etched by DRIE after bond anneal.

The maximum membrane stress was calculated at 1 atm differential pressure, using the standard analytical expressions [3] for flat plates in the linear regime. At the relatively small resulting deflections, this approach is sufficiently accurate, as evidenced by comparison with the FEA derived values by Ren et al. [4]. The corresponding calculated maximum deflections range from 0.04 µm to 13.5 µm for the circular membranes, and 0.15 – 11.4 µm for the square ones.

Results

The WLI measurements before bond anneal of the S4 type membranes showed deflections of around 12 μm for the pressure compensated membranes, and around 15μm for the membranes of the vacuum reference wafer, in good correspondence with calculated values.

After bond anneal, WLI measurements of the pressure compensated membranes showed essentially the same deflection as before for most devices, with exception for the lowest part of the compensating pressure range combined with the largest membranes. The vacuum cavities, however, showed severe increase in deflection. Profile plot from WLI measurements of a vacuum cavity C3 type membrane after bond anneal is shown in Figure 2, together with fringe pattern picture from the WLI microscope. The C3 and S3 membranes had a deflection of around 5 μm before bond anneal, which changed to 50-56 μm after bond anneal. The C4 and S4 vacuum membranes bend in a way that gave problems with range limitations in the WLI measurements, but a deflection of 108 μm was estimated on one S4 membrane.

Inspection of the membranes in an IR capable microscope revealed a clear difference between the vacuum and the pressure compensated membranes. Figure 3 shows IR images of two S4 membranes after bond annealing: one pre-bonded in vacuum, and one pre-bonded in a 207 mbar N_2 ambient. What we interpret as slip lines were clearly seen in all the vacuum pre-bonded square and circular membranes except the smallest of each type, as concluded in Table 1.

An IR microscope image of a C2 type membrane pre-bonded in vacuum is shown in Figure 4. Slip lines are visible all over the membrane, and not only in the regions that are expected to see the highest stress.

Ventilation of the membranes by opening the inlet holes provided little change in deflection of the vacuum pre-bonded reference cavities. The pressure compensated

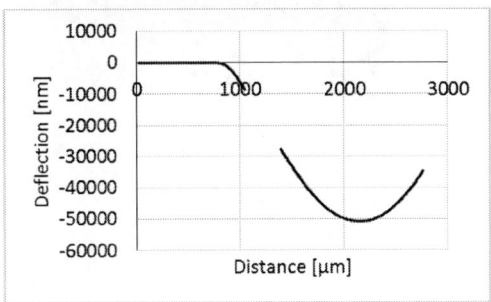 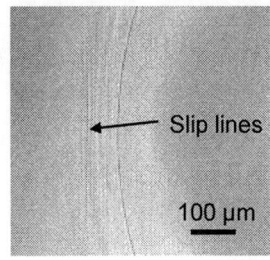

Figure 2. WLI measurement of a C3 type membrane pre-bonded in vacuum, after bond anneal (left). The break in the otherwise solid line appears because the wide angle of reflected light cannot be picked up by the WLI microscope. The right-hand picture shows the fringe mode picture from the WLI, with visible slip lines. Bar in lower right corner indicates 100 μm length.

membranes however essentially turned flat (+/-0.1 μm), except for many C4 and S4 membranes pre-bonded at 207 mbar. Deflection measurements for these two membrane types vs pre-bonding pressure are shown in Table II. Figure 5 shows WLI measurements of different C4 membranes before and after INLET etch. For membranes from 207 mbar cavities, both types of "after INLET"-deflections shown in Figure 5 were seen, and deflections less than 100nm are excluded from the 207 mbar averages of Table II. In Figure 5, please note that the visible surface step does not coincide with the actual membrane edge. This step was only intended as a crude means of making the location of the membrane visible from the front-side; thus, the two involved lithography masks were not precisely aligned. Furthermore, the DRIE that defines the membrane produces a negative etch angle and a slightly larger membrane. The deflection of the non-deformed membranes is not symmetric relative to this step, but it is not clear if this is due to misalignment alone or to the asymmetric stress imposed by the misaligned step. The most important information from this figure lies in the maximum deflection values.

TABLE II. Average deflection and maximum calculated stress at bond anneal conditions for C4 and S4 after INLET etch vs pre-bonding pressure, for pressure compensated membranes. 1 atm pressure is assumed on the outside. Deflections less than 0.1 μm are excluded from average for 207 mbar pre-bond pressure, as these devices fell into two different categories; one with and one without permanent deflection.

Pre-bond pressure [mbar]	Average deflection, C4 [μm]	Average deflection, S4 [μm]	Max stress @1050 °C, C4 [MPa]	Max stress @1050 °C, S4 [MPa]
207	1.42	1.28	23	36
232	0.08	0.09	9	14
247	0.08	0.09	0	0

By visual inspection before bond anneal it was noted that the membranes of pressure compensated cavities fell into two categories on each laminate: "large" and "small"

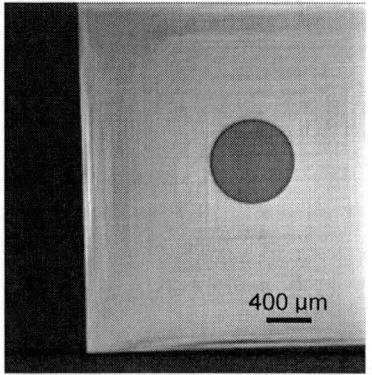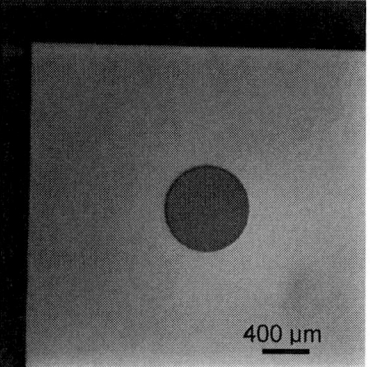

Figure 3. IR microscope picture of S4-type membranes pre-bonded in vacuum (left) and at 207 mbar (right). The light gray area is the membrane. The straight lines seen in the left hand membrane are slip-lines in the Si crystal. The dark circle is the oxide mask opening for inlet DRIE. Bar indicates 400 μm length.

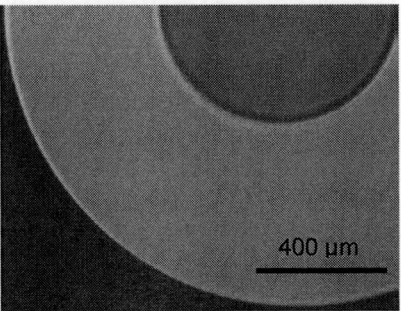

Figure 4. IR microscope picture of parts of a C2 membrane pre-bonded in vacuum. It is worth noting that there are slip-lines all over the membrane. Bar indicates 400 μm length.

deflection. The wafer positions of some of the membranes with "large" deflection before bond anneal were noted, to check if membranes at these specific locations behaved differently in bond anneal. It was found that after bond anneal these membranes also had slip lines and deflections of the same level as the corresponding membranes on the vacuum reference wafer. An example of this is shown in Figure 6, where the permanently deflecting membranes are easily distinguishable.

Discussion

The maximum stress for a 1 atm pressure difference over the membranes is calculated by analytical expressions assuming deflections in the linear regime, as shown in Table 1. Our calculations show that a 1 atm pressure difference over the S1 and C1 membranes corresponds to a stress in the membranes of 25 MPa and 7.2 MPa, respectively. Both

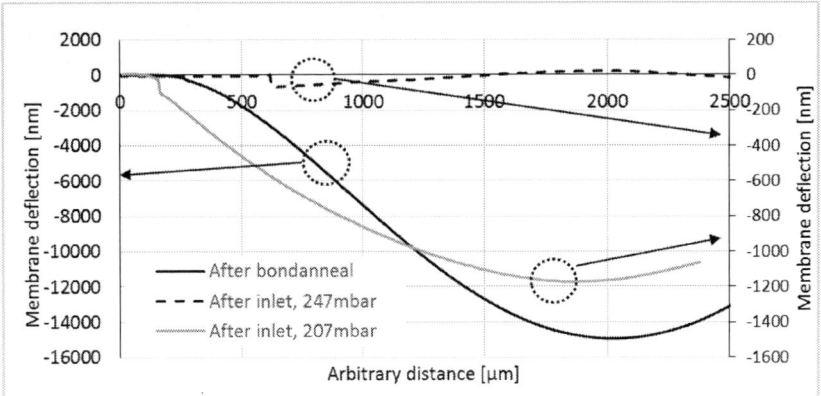

Figure 5. Profile plots from WLI measurements of C4-type membranes just after bond anneal and after INLET etch for 207 and 247 mbar. Please note the difference in scale. After INLET etch the membrane location has been made visible by the step created in the Si before bonding, but note that this step does not coincide precisely with the membrane edge.

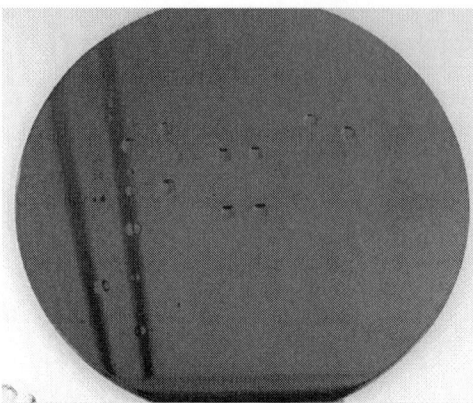

Figure 6. Picture of a 100 mm diameter wafer pre-bonded at 207mbar, after bond anneal. The membranes with visually clear deflection were found to show slip-lines similar to those seen on the vacuum reference wafer.

these smaller membrane types were found to have intact Si crystal after pre-bonding in vacuum. That C1 behaves this way corresponds well with the results in the recent work by Ren et al. [4], who report on similar experiments accompanied by FEA simulations. The fact that S1 with 25 MPa stress seems to be below the yield strength warrants further discussion, see below. All the other - hence larger - membranes were found to have slip lines when pre-bonded in vacuum, a result that also corresponds well with Ren et al. [4], as our calculations show a membrane stress of $\geq$36 MPa for these geometries.

Ren et al. used a Si yield strength value at 1000 °C of around 20 MPa [4]. They referred to Patel et al.'s experimental work [5], from which they extracted yield strength values at various temperatures; 700, 800, 900, and 1000 °C. For this extraction process, they chose a specific strain rate from Patel's three rates. Since the strain rate is a critical parameter in discussions regarding the dynamic process of silicon yield, we include some clarification and discussion on this theme here. The experimental temperature dependence of the Si yield strength was summarized in Figure 14 in Patel's work, for the case of initially dislocation-free Si crystals. (The data are for the maximum stress σ_M of the stress-strain curve, which is also referred to as the "upper yield point".) The experiments were performed by pulling a 2.5 cm long silicon specimen in a fixture, with constant pull rate (and thus, constant strain rate) at varying temperatures. Patel et al. provide these data at three different "crosshead velocities", i.e. pull rates. These rates must be divided by the 2.5 cm total length in order to obtain the strain rate. Thus, to the best of our understanding, ref. [4] chose Patel's data for the lowest strain rate out of the three, $3.3*10^{-5}$ s^{-1}. This is not an obvious choice *per se*, but the corresponding maximum stress values are indeed those in [5] that best fit Ren's experimental results.

In our work a temperature of 1050 °C is used. If we extrapolate the above-mentioned curve from [5] to 1050 °C, we obtain roughly 12 MPa. This is clearly lower than the calculated 25 MPa that our S1 devices experienced without noticeable plastic deformation (although in accord with the C1 situation). The 25 MPa value is in better agreement with Patel's data for their *medium* strain rate, as extrapolation of the medium strain rate curve to 1050 °C gives a maximum stress of about 27 MPa. This strain rate of

$3.3*10^{-4}$ s^{-1} is ten times larger than the rate used in [4]. This difference demonstrates that caution is required when establishing design and process guidelines for membrane yield. Improved understanding of involved effects, of correlations between modelling and experiments, and of correlations between different experimental situations, is needed.

The large dependence of yield behavior on strain rate was further investigated by Sumino [6], as well as the influence of initial dislocation density and of oxygen contents. Sumino as well as Patel furthermore stress the fact that yield is a highly dynamic process, and Sumino carefully discusses the generation, motion, and multiplication of dislocations due to strain. It may be fruitful to try to reason around the question of how strain rates and other parameters in such experiments as Patel's and Sumino's are relevant in our experimental situation. Such experiments as theirs are characterized by *steadily increasing strain* on a bulk sample under tension with *no bending*, at a *constant temperature*. In contrast, our situation is one of i) a near constant strain (except during a very limited period of gas expansion in the case of gas filled cavities), ii) a strongly bending membrane (which accentuates the role of microscopic details in the silicon surface rather than the bulk situation of the pull experiments), and iii) a temperature that is in fact steadily increased until the final 1050 °C is attained. These are far from two identical cases. The most obvious difference between the two experimental situations is the increasing strain as opposed to the constant strain, but the other factors can certainly not be disregarded.

In [4] it is reported that the larger membranes continue to evolve toward somewhat larger deflection between 1 hr and 16 hrs at the anneal temperature. (In fact, they also report negative creep in some cases.) In our work, we have used an anneal time of 2 hrs at 1050 °C, a time that falls between Ren's values. This should be noted when comparing results. In addition, we speculate that our anneal temperature ramp protocol may also have implications. This could be a matter of interest for further studies.

Another remarkable observation in [5] and [6] is how much the upper yield stress increases with diminishing initial dislocation density. Although modern silicon wafers typically are delivered with densities that are much lower than those of the highest range presented in these works, this is another reason for caution when comparing different works. In fact, Sumino presents data that do not go below $2*10^4$ cm^{-2}, which may still be a somewhat higher density than in modern silicon wafers such as those used in our study. Arguably, most silicon wafers today can probably be assumed dislocation-free in this regard.

For those of our wafers that were pre-bonded with a cavity pressure of 247 mbar, the estimated membrane stress is close to zero at bond anneal conditions. All membrane types that were pre-bonded between 233 and 247 mbar were found to be slip-line free in IR microscope pictures, and they turned essentially flat after INLET etch. This proves that this accuracy in the compensating pressure range is sufficient to avoid permanently distorted membranes, even for 43 μm thick membranes as large as at least 3265*3265 μm square.

Wafers pre-bonded at 207 mbar cavity pressure also had flat membranes except for some of the C4 and S4 types. 207 mbar at pre-bonding corresponds to 847 mbar at bond anneal, which gives a calculated maximum stress of 23 MPa and 36 MPa, respectively, for the C4 and S4 membranes. As can be seen from Figure 5, even the C4 types can show some permanent distortion, but no slip-lines are visible in IR microscope. (Future studies could incorporate pit etching to better reveal whether dislocations exist.) As the wafers pre-bonded at 207 mbar show a combination of undistorted and distorted C4 and S4 membranes, the stresses of these membranes are probably close to the onset of plastic deformation for the bond anneal process used in this study. Uneven gas filling cavity-to-cavity and/or pre-bending of the wafers might give just the amount of stress variation needed for plastic deformation in some wafer positions and no deformation in others.

The highest stress occurs close to the membrane edge. Hence, one would expect that in membranes that are just above the border for plastic deformation, slip-lines would be observed only close to the membrane edge, while for membranes far above this border slip lines could be expected for the entire membrane. The latter is indeed the case for e.g., S4 in Figure 3 (left), pre-bonded in vacuum. For the C2 membranes pre-bonded in vacuum, that with their calculated stress of 36 MPa arguably are around the border values, the slip lines also appear all over the membrane, as in Figure 4. On S4 and C4 membranes pre-bonded in 207 mbar, slip lines are not visible at all, despite the remaining deflection after INLET etch, as seen in Figure 5. The calculated maximum stresses for "C2 in vacuum" and "S4 at 207 mbar" are nearly equal (both around 36 MPa), and they both experience plastic deformation. Yet, the slip-line observations give very different results. It seems that crystal dislocations are too small to be revealed by IR microscopy in the C4 and S4 membranes pre-bonded at 207 mbar, but they are clearly visible on C2 membranes pre-bonded in vacuum. These observations are also candidates for future studies.

When the wafers have been loaded into the bond chamber they are kept apart with 50 µm thick spacers only. Thus, the clamping force from the bond chuck, possibly combined with wafer bow, can easily make the two wafers touch each other in some areas. During the pump stage of the pre-bonding, remaining gas in the cavities will have a chance to escape even where wafers touch; but during the chamber filling stage the under-pressure of the cavities will suck the two wafers together, thus hindering gas filling of cavities. This is most likely the case for the wafer pictured in Figure 6. Thicker spacers and slower gas filling were in a later study found to increase the gas filling yield.

Conclusion

In conclusion, our work shows that flexible silicon structures such as membranes can be permanently deformed during bond anneal if a stress exceeding the yield strength of silicon is present during anneal. This deformation can be avoided by applying a compensating N_2 gas pressure during pre-bonding, thus actively reducing the differential pressure across the membranes during bond anneal. Crystal slip lines can quickly be detected by IR-sensitive microscope inspection for severe crystal slip, but can be difficult to identify when the dislocations cover a limited area of the crystal. Our results provide guidelines for proper design. We underline, however, the need for carefulness, as devices and experimental conditions are not necessarily directly comparable across technologies and set-ups, and dislocation generation, motion, and multiplication are complex and dynamic processes that still require further studies in order to be properly understood and quantified in the context of silicon device design and implementation.

Acknowledgments

Financial support from The Research Council of Norway through the project NBRIX (Contract No. 247781/O30) is gratefully acknowledged. The authors wish to thank Kari Schjølberg-Henriksen and Trond A. Hansen from SINTEF ICT for fruitful discussions.

References

1. E.g., Chapter 5 in P. Ramm, J. J.-Q. Lu, M. M. V. Taklo (Eds.), *Handbook of Wafer Bonding*, Wiley, 2012
2. Q.-T. Young and U. Gösele, *Semiconductor Wafer Bonding: Science and Technology*, Wiley, 1999
3. W. C. Young and R. G. Budynas, *Roark's Formulas for Stress and Strain*, 7th ed, McGraw Hill, 2002, Tables 11.2 and 11.4
4. J. Ren, M. Ward, P. Kinnell, R. Craddock, and X. Wei, *Sensors* 2016, *16*, 204 (2016).
5. J. R. Patel and A. R. Chaudhuri, *J. Appl. Phys 34 (9)*, 2788 (1963)
6. K. Sumino, *Metallurgical and Materials Trans. A*, Vol 30A, 1465 (1999)

Surface Protection for Semiconductor Direct Bonding

Roy Knechtel, Holger Klingner

X-FAB MEMS Foundry GmbH, D-99097 Erfurt, Haarbergstrasse 67, Germany

Semiconductor direct wafer bonding is a widely used process for fabricating 3-dimensional structures, especially engineered substrates such as SOI wafers or cavity wafers with and without insulating layers at the bonding interface. The investigations described here concern cavity SOI wafers without insulating layers, as used for discrete and integrated pressure sensors. Semiconductor direct bonding is the initial bonding by mechanical contacting of high-quality surfaces activated by wet chemical and/or plasma treatments, followed by thermal annealing at temperatures and times related to the chosen activation. This ultimately results in a very strong mechanical bond, which can survive all of the subsequent process steps, such as those needed to process MEMS pressure sensors. The requirement for the direct bonding is that the wafers to be bonded have a very high quality; this means very low roughness, low waviness and no surface residues or contamination. In this paper, the focus is on roughness in relation to the bonding process and process integration.

Introduction

It is well known that prime wafers with vendor polishing as the finishing step have a very good mirror polished surface, and can be bonded very easily – therefore they are often used for basic wafer bonding investigations, such as the effectiveness of surface activation. On the other hand, fresh polishing as the final step before wafer bonding is often used to reconfigure the surface, to allow a very good wafer bonding. In practice, especially in industrial processes, the problem is that wafers need to be processed before bonding to generate the required functional structures (the reference pressure cavity in case of the pressure sensors and alignment marks on the wafer backside to retrieve the sealed cavities after bonding). On the other hand, fresh polishing directly before bonding is not possible, either due to availability or because it cannot be used for the already-structured wafers (either the structure on the wafers or the polishing pads would be damaged). Here, one option is to protect the bonding surfaces with a hard coating, such as by growing a thermal oxide on a silicon wafer. In the case of the pressure sensors, this layer can fulfil a double function – protecting the bonding surface and acting as a hard mask for the cavity etching. Hence, the flow of the cavity wafer can be summarized as follows: thermal oxidation, back side lithography of the alignment marks, plasma etching of oxide back side (hard mask structuring), resist removal, front-to-back side lithography, plasma etching of the cavity hard mask on the front side, resist removal, KOH etching of silicon in an etch bath (simultaneously etching of cavities on front side and alignment marks on wafer back side), hard mask stripping using oxide etching baths – this oxide stripping defines the wafer bonding surface and by this the bonding behaviour and is therefore investigated and discussed in this publication, using the example of the production process for MEMS absolute pressure sensors [1], where a very high bonding quality especially a void free bond is extremely important.

Process for MEMS Absolute Pressure Sensors

There are a large number of applications for absolute pressure sensors. Because the sensor chip contains a hermetically sealed vacuum cavity as reference pressure, the absolute pressure can be measured without the requirement of a special package or an external reference pressure. This type of sensor is used for a wide range of automotive (tire pressure, side air bag sensors), medical and other applications.

The most widely used types of pressure sensors are piezoresistive pressure sensors. The transduction of piezoresistive pressure sensors [2] is based on the change of the electrical resistance of doped silicon resistors by mechanical stress. The relation between mechanical stress and the resistance change is given by the piezoresistive coefficients (π_i). In this way, the deflection of the pressure sensor membrane is transferred to a resistance change (Figure 1). Because the piezoresistive coefficients depend on the silicon crystal orientation, resistors with negative and positive resistivity changes can be realised. These can be connected together in a Wheatstone bridge, which yields an output voltage that is proportional to the applied pressure. However, since the piezoresistive effect is strongly temperature dependent, piezoresistive pressure sensors require external temperature compensation.

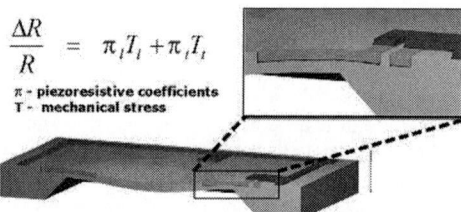

Figure 1. Functional principle of a piezoresistive MEMS pressure sensor

In order to realise absolute pressure sensors a reference vacuum has to be sealed in a cavity at one side of the sensor membrane. The most efficient technique to create such a sealed cavity on wafer level is wafer bonding. If two silicon wafers are bonded together, a single material system is created. This eliminates the thermo-mechanical stresses, which are always present in heterogeneous material systems.

For absolute pressure sensors the following technological process flow is used by pre-processing of special substrates (Figure 2):

Base wafers with shallow cavities are pre-bonded after RCA and O2-Plasma activation to a non-structured cap wafer. The bond is annealed at 1050°C. Afterwards the cap wafers are grinded and polished back to leave a defined membrane over the cavities. By using either a complete or a condensed CMOS process, integrated or discrete pressure sensors can be implemented (Fig. 2a). This technology provides the advantage that all pressure sensor specific micromechanical process steps are carried out before the piezoresistive elements are created. After the substrate pre-processing the wafer can be handled like a standard wafer in a CMOS-like process to realise piezoresistive pressure sensors. An advantage of the technology is the overload protection that is provided by the shallow cavity. If a high pressure is applied, the membrane will touch the bottom of the cavity, which decreases the maximum stress within the membrane and avoids fracture. The drawback of this technology is the fact that the membrane is created by a mechanical process. The membrane wafer is thinned by grinding and polishing to the target thickness. The tolerances for these processes are relatively high. Only the use of handle wafers with low total thickness variation (TTV) and the optimisation of the grinding and polishing

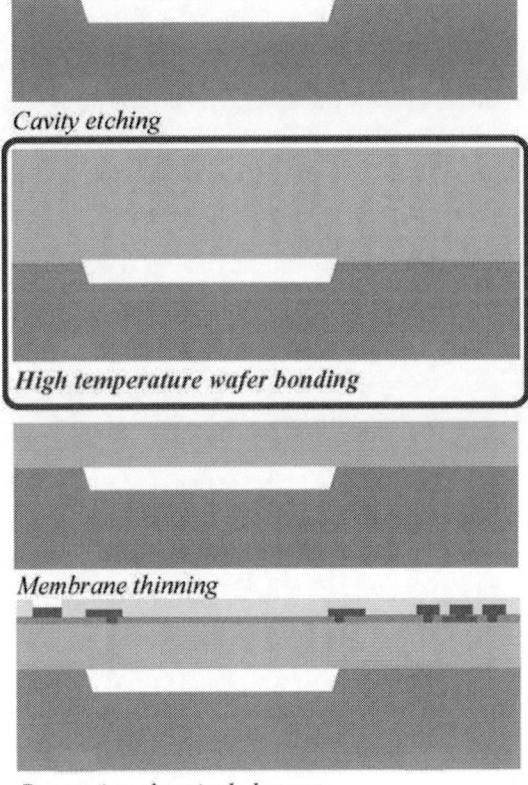

Figure 2. Process flow for MEMS absolute pressure sensors utilizing semiconductor high temperature direct bonding

Surface Characterisation for and by Direct Wafer Bonding

In semiconductor direct wafer bonding processes, the wafers are typically cleaned (wet processes) and their surfaces are activated either by this cleaning itself or by plasma treatment, often followed by a DI water rinse. The purpose is to hydrophilize the wafer surfaces to allow good hydrophilic bonding. This is followed by mechanical contacting of the wafers (pre-bonding) and then thermal annealing to transform the OH bonds into oxygen bonds with much higher strength [3]. In the pre-bonding step, a so-called bonding front travels across the wafers after they are initially contacted at just one point. This bonding front can be easily observed by infrared (IR) light transmission. The bonding front is the actual position where the pre-bond is formed between the wafers, and through the surface forces of the bonding process, it travels by self-propagation across the wafer, to finally form a full-area wafer bond. The behaviour of the bonding front can be evaluated in different ways in order to describe the bonding process and its influencing factors. If the activation process is kept constant, differences in the front speed and its behaviour at obstacles, such as particles or scratches at the bond interface, can be used to evaluate the bonding behaviour of the surfaces to be bonded. While the bonding front speed and direction can be easily observed by an infrared camera during the bonding of standard silicon wafers, the obstacles which are a measure of the actual surface energy in the bond interface need to be well defined. For the method described here in this paper, two blades are placed between the wafers before they are brought into contact (Figure 1). The bonding front stops before reaching the blades, since this is a surface separation, and the open area between the bonding front and blade is measured as an indication of the surface energy in the pre-bonding process. This is ultimately a modification of Maszara's blade test [4] for pre-bonding evaluation. Using an infrared video camera system and a reference ruler placed on the bonded wafer pair it was possible to evaluate both the bonding front speed and the open area at the blades, in order to evaluate the pre-bonding behaviour, and by this, the quality of the surface layers to be bonded [5].

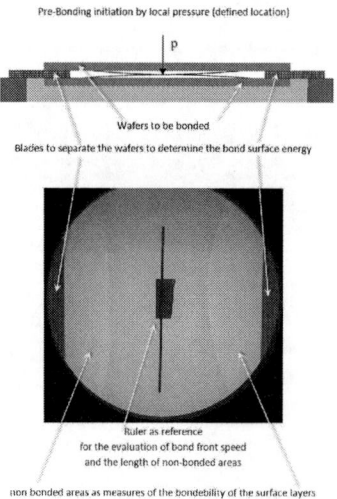

Figure 3.   Configuration and example of the bonding test step with pre-inserted blades

Figure 3 shows the initiation and observation of the traveling bond front interaction with the pre-inserted blades.

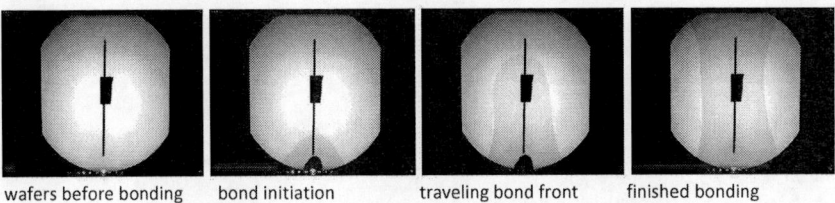

wafers before bonding   bond initiation   traveling bond front   finished bonding

Figure 4. Example of traveling bond front when using pre-inserted blades

To investigate surface roughness differences and their relation to the cavity wafer bonding process, which is the topic of this paper, the bonding front travel speed, which is actually the bonding front traveling time (from bonding initiating point at float, the wafer edge opposite flat) over the 6" wafer, is the interesting aspect. Therefore the in this case the pre-inserted blades were not used. Figure 5 illustrates the setup and the bonding time measurement.

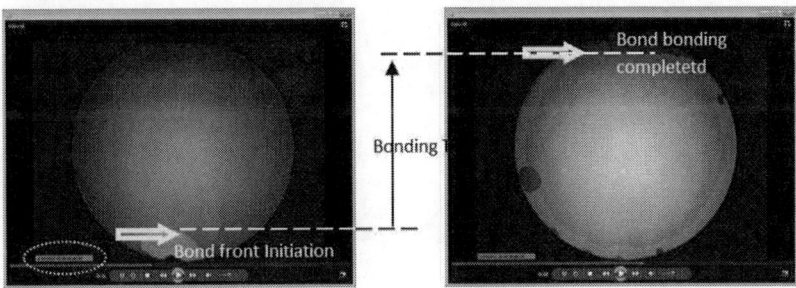

Figure 5. Bond front traveling time measurement in an IR-inspection setup

Evaluation of the Hard Mask Stripping using Oxide etching Baths

As already mentioned in the introduction and the process flow, the cavities need to be etched into the handle or cavity wafer. For this etching a KOH wet process is used. This is a well-known, widely used and economically attractive technology, since at one time a batch of up to 25 wafers can be etched. For the etching hot and concentrated KOH has to be used, this is resulting in quite harsh etching condition, so even for a short etching used for the shallow cavities, enough for the pressure sensors, a hard mask is need. Due to the mandatory good adhesion of that mask to the silicon, thermal oxide is here used. Since the surface of the oxide is as well attacked by the KOH, it has to be removed before the direct wafer bonding. Dry etching of that oxide would also leave some surface roughness as well as often hardly to remove polymers. An oxide wet etching process with very high selectivity to silicon is therefore here the best choice. In our case, due to availability, Buffered Oxide Etchant (BOE) has to be used for stripping the hard mask from the wafer, and it was found that this etchant can slightly attack the silicon surface, making it a little rougher, which makes the bonding more difficult. Pre-investigation by etching blank silicon for different times in BOE followed by AFM measurements confirmed this hypothesis as shown in table 1.

TABLE I. Experimental investigation surface roughening of silicon by BEO etching

Etching time BOE	0s - no etching	60s etching	180s etching
Roughness Rq [nm] - RMS	0.065	0.107	0.232
Profile Picture scales x 0.5 µm/div y 0.5 µm/div z 3.0 nm/div			

TABLE II. Experimental data direct bonding of wafers with different BOE etching times

	Standard BOE Etching Time			**Reduced BOE Etching Time**		
	Time Initiating Bonding Front	Time Bond Front at Opposite Wafer Edge	**Traveling Time Bonding Front over 6" Wafer**	Time Initiating Bonding Front	Time Bond Front at Opposite Wafer Edge	**Traveling Time Bonding Front over 6" Wafer**
Process Lot	[s]	[s]	[s]	[s]	[s]	[s]
	Wafer 2			**Wafer 8**		
E28903	6,29	8,70	**2,41**	4,00	5,93	**1,93**
	Wafer 22			**Wafer 10**		
E28980	13,26	15,86	**2,60**	7,03	9,04	**2,01**
E28981	7,17	9,98	**2,81**	5,28	7,57	**2,29**
E28982	5,67	8,29	**2,62**	6,76	8,70	**1,94**
E28983	4,46	6,95	**2,49**	4,61	6,70	**2,09**
	Wafer 4			**Wafer 16**		
E29335	7,14	9,68	**2,54**	5,26	7,42	**2,16**
E29337	13,57	16,46	**2,89**	7,28	9,51	**2,23**
E29338	4,95	7,56	**2,61**	5,23	7,23	**2,00**
Average Traveling Time Bonding Front			**2,62**			**2,08**

Based on the etching and atomic force (AFM) investigations a comprehensive study of etching conditions and bonding behaviour was done. The reverence was set by the standard etching setup of the bonded cavity wafers – the BOE etching time here is set for etching process safety quite long, especially for the overetch part, as usually done for non-critical wet etch processes. To learn about the behaviour with reduced overetch time and less surface roughening, the etching time was reduced by about 50% to a theoretical minimal overetch, still long enough to cover all process variations – tolerances in thermal oxide thickness and etching rates – to ensure in all the cases a complete oxide removal. To get a first level of statically confidence wafers from 8 lots were compared – the investigation plans as well as the bond front traveling times are shown in table II.

The results of the investigations are shown in graphical manor in figure 6. The bonding front traveling time is plotted for the 8 wafer pairs. It can be concluded that for every pair of wafer the bonding front travels faster at the wafer with the shorter BOE etching time. This confirms a lower surface roughness and by this a better, more safe direct wafer bonding process. While the absolute times show some variation, the difference in the bond front traveling time are very constant, about 0.5s. The differences in the absolute times indicates some other influencing factors, which are not yet investigated, but the constancy in the difference confirms how significant the reduced etching time improves the direct bonding behaviour. Figure 7 shows that with the recued overetch time very good bonding yield – a void freed direct bond can be achieved.

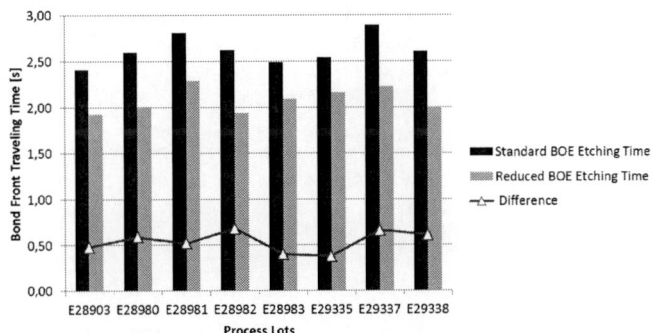

Figure 6. Comparison BOE etching conditions regarding bond front traveling time

Figure 6. IR inspection results of wafer bonded using the reduced BOE time (part of the wafer shown to have the right magnification to see the cavities – dark squares)

Conclusions

Based on the results shown in this paper we can conclude the following for the oxide stripping of the cavity hard mask before direct silicon wafer bonding: With respect to a safe process, the etching time has to be long enough to remove the oxide in all cases, (e.g. grown oxide thickness in upper spec range, BOE oxide etching range in lower spec range), to reduce all the oxide everywhere on the wafer, but it must be short enough to prevent the roughening of the surface. Therefore, the extra etching time the wafer is left in the bath after the oxide has been removed, which is needed to cover natural process variations, must be as short as possible. This was evaluated experimentally. As a measure of the surface roughness, the bonding process itself was used, in particular, the travelling speed of the bonding front across the wafer – since the higher this is, the better the surface and the safer the overall process. As result, it was found that reducing the over-etch time by about 50%, reduced the travelling speed of the direct wafer bonding front by 20%, which is very significant. From this, it can be concluded, that the bonding behaviour of the silicon is strongly influenced by the time the free silicon surface remains in the BOE – the longer this time, the rougher the silicon, and the more difficult the bonding. This is ultimately very important for the overall process integration and to achieve a safe bonding process for almost void-free engineered substrates such as cavity wafers.

References

1. R. Knechtel et. al: Low- and High-Temperature Silicon Wafer Direct Bonding For Micromachined Absolute Pressure Sensors; 8th International Symposium on Semiconductor Wafer Bonding: Science, Technology, and Applications 205 Spring Electrochemical Society Meeting Quebec Canada

2. M.H. Bao: Micro-Mechanical Transducers, Pressure Sensors, Accelerometers and Gyroscopes, Elsevier, Amsterdam, 2000

3. Q.-Y. Tong, U. Gösele: Semiconductor Wafer Bonding Science and Technology, John Wiley &Sons, Inc. 1999

4. W. P. Maszara et. al.: Bonding of Silicon Wafers for Silicon-on-Insulator, J. Appl. Phys. 64, 4943 (1988)

5. R. Knechtel et. al: Surface Protection for Semiconductor Direct Bonding, 13th International Symposium on Semiconductor Wafer Bonding: Science, Technology, and Applications 2014 Fall Electrochemical Society Meeting Cancun Mexico, - ECS Transactions, 2014 64(5): 133-140, © The Electrochemical Society

Chapter 9

Equipment & Applications

Direct Bonding of Multiple Curved, Wedged and Structured Silicon Wafers as X-Ray Mirrors

B. Landgraf[a], R. Günther[a], G. Vacanti[a], N. Barrière[a], M. Vervest[a], D. A. Girou[a,] A. Yanson[a], M. J. Collon[a]

[a]cosine measurement systems, Oosteinde 36, Warmond 2361 HE, The Netherlands

> In this paper, we present the technological basis for the production of structured silicon X-ray mirrors using direct wafer bonding. We dice 12" silicon wafers into rectangular plates, which are then structured to have a thin membrane and several ribs. As a next step we wedge the plates on both sides. Using in-house developed stacking robots, we then elastically deform the plates into conical shapes and bond the plates into a stack, which finally consists of several tens of such plates in a stiff self-supporting structure. Throughout the production of the structured plates and their assembly into stacks, we focus on maintaining pristinely clean surfaces as this is a prerequisite for direct wafer bonding. Several metrology tools have been developed, operating in-line, to ensure precise plate to plate alignment, high cleanliness and an accurate figure. The process is adaptable to work with plates of various dimensions, with metal coatings and with different bend radii.

Introduction

We have, together with the European Space Agency, developed a novel X-ray optics technology for space applications, the so-called Silicon Pore Optics (SPO) [1, 2, 3, 4, 5, 6]. The SPO technology is based on high-quality monocrystalline silicon wafers which are getting processed based on existing and modified methods and tools available in the semiconductor industry. Figure 1(a) illustrates our SPO production sequence. After the dicing, ribbing and wedging process steps, we utilize the principle of hydrophilic silicon wafer bonding during stacking in order to obtain a monolithic self-supporting matrix-like X-ray mirror structure. In most applications of silicon fusion bonding, commercially available tools are used to bond Si/Si, Si/SiO_2 and SiO_2/SiO_2 wafer surfaces. We have further expanded the application of silicon bonding by developing new tools and metrology that now allow bonding tens of curved silicon substrates [7, 8]. The image in Figure 1(b) shows an example of such a SPO stack of 35 structured silicon plates.

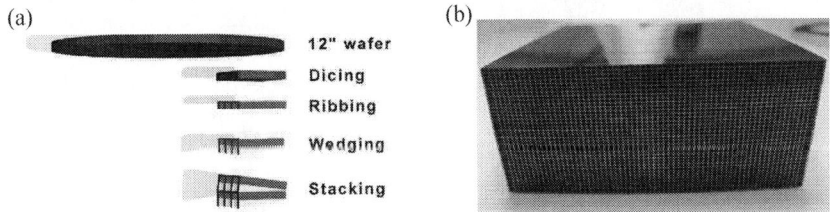

Figure 1. The left image illustrates the production of SPO stacks. It starts with a commercial high-quality double-side polished 12" silicon wafer which is thermally

oxidized and is then diced into rectangles using a dicing machine. At the same time, we also groove the rectangles to obtain ribs and a thin membrane. Subsequently, the oxide of the silicon plates gets wedged by etching. The stacking is performed automatically using an in-house developed stacking robot. The right image shows a stack of 35 ribbed and wedged silicon plates, with a radius of curvature of 737 mm. The ribs are 0.17 mm wide with a spacing of 1 mm and the membrane is 0.17 mm thick, too. The Newton fringes visible on the top plate indicates an oxide thickness gradient from the wedging process.

Plate manufacturing

The production of structured silicon plates for SPO starts with standard thermally oxidized 300 mm monocrystalline silicon wafers, which are double-sided polished and about 0.8 mm thick. A surface roughness better than 0.1 nm and a Total Thickness Variation (TTV) of less than 0.34 µm meets the requirements to form high-quality X-ray optics. A standard semiconductor dicing saw is used to cut the wafers into smaller individual rectangular shaped plates. During the subsequent ribbing process, a number of long grooves get cut into the silicon plate leaving walls (called ribs) and a thin bottom (called membrane). Parameters such as rib width, pitch, number of ribs, membrane thickness, plate width and its length can be adjusted to optimize the optical performance and mechanical properties of specific applications. Typically, the ribs have a width of 0.17 mm with a pitch of 1 mm and a membrane thickness of 0.17 mm. The plate ribbing process results in a well-defined pore geometry; but a known side effect of dicing is the generation of micro-cracks in the brittle silicon [7]. An anisotropic and selective damage is being applied in order to reduce the length and number of residual micro-cracks. In subsequent assembly, the plates can then bend elastically and with a significant curvature without breakage. In view of focusing X-ray optics, we require a taper over the whole plate. The wedged shape is formed by growing oxide layers on both sides of the silicon substrate and then gradually immersing the plate into an isotropic oxide etching bath using custom-built equipment. Typically, the wedge angle is a few arcseconds, resulting on plate lengths of 66 mm in a wedge layer with a maximum thickness of a few 100 nm and an accuracy of 3 % of the nominal wedge angles.

Stacking process

We have developed a number of fully automated assembly robots [4, 5, 9, 10] to produce X-ray mirrors of different radii. A complete system has a footprint of a few square meters only, is installed in a class 100 clean environment and can be operated remotely from outside of the cleanroom for maintaining the highest grade of cleanliness required for the direct bonding process. Inside of the clean room tent, we are monitoring the cleanliness using particle sensors in order to assure high-quality bonding interfaces of our SPO products. The image in Figure 2 illustrates our current production systems for SPO with radii of 0.74 m and 0.25 m. As parts of our systems, linear stages, hexapods and a robotic arm control alignments of high-precision during the entire stacking process. The alignments are monitored and ensured by using high-resolution cameras and image recognition software as well as laser sensors. After the stacking and bonding process step of individual mirror plates, a surface metrology based on phase measuring deflectometry is used to quantify small scale defects due to trapped particles in between bondable areas and large scale figure errors and thus assure the quality of the optics.

Figure 2. The assembly system, developed by cosine executes the stacking process automatically at radii of 0.74 m and 0.25 m. It is set up inside of a class 100 clean environment. The system is installed on a vibration isolated table, is fully automated and designed to build stacks up to 35 plates. The plates can be positioned with micrometer accuracy and automatically bent into required shapes.

Prior to the stacking process of individual silicon plates, the plates need to be cleaned and the surface to be activated for bonding. Regarding direct wafer bonding, particulate contamination is a known problem. We are using a state-of-the-art cleaning wet bench including a standard semiconductor plate drying process specialized to dry silicon plates with high aspect ratio structures and without mechanical force to remove the moist. In particular, it is suitable for the grooved and rectangular shaped silicon mirror plates with numerous sharp edges. During the wet chemical cleaning process, plates undergo a standard SC-1 clean in order to activate and make their surfaces hydrophilic [11]. Normally, this is followed by a SC-2 clean to remove metallic contamination. In view of SPO, it is not important because the metal contamination does not influence the bonding process.

After loading the plates into the stacking system, the robot picks up a plate and inspects it to localize any residual particles and drying stains on the surface. Standard wafer inspection systems are only applicable to smooth surfaces. In our case, light is scattered from the sharp rib edges giving erroneous particle measurements. Therefore, we have developed a machine vision system with image analysis software to resolve scratches and particles on the reflective mirror and ribbed side of the plate. By scanning the silicon plate's surface with a line scan camera and in combination with a high-intensity light source at an oblique angle (dark-field), the system detects the light scattering of particles. Sub-micro-meter particles cannot be directly imaged with our system due to the 7 μm CCD camera pixel size. But, the intensity of the scattered light does give a measure for the particle size (Rayleigh scattering: $I \propto d^6/\lambda^4$) and provides enough information to localize particles. In addition, particle inspection of plates provides valuable information about the quality of the cleaning process. After inspection the robot takes, based on the level of contamination, a decision whether to proceed with the plate or to reject it. As an example, Figure 3 shows a processed particle map of a plate's non-ribbed side. The plate got rejected by the robotic system due to particle contamination at bondable areas.

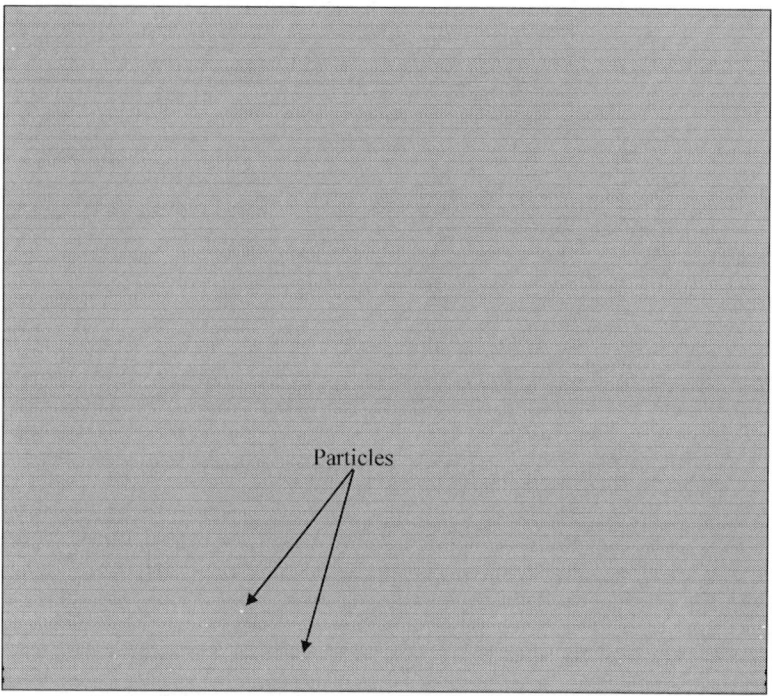

Figure 3. The image shows a particle map of a plate's non-ribbed side. The dark horizontal stripes display where ribs are located on the ribbed side of the plate and the brighter stripes the location of the membranes. The white dots highlight detected particles.

Subsequently, the robot brings the plate to the actual stacking tool, which elastically bends it into a cylindrical or conical shape. Furthermore, it ensures a uniform bond due to a thin compressible layer. This tool, called a die, is then lowered onto a mandrel, where it places the plate, or stacks it, onto already existing ones. In close contact of the plate's surfaces, hydrophilic bonds form at the interface based on intermolecular interactions [11] so that each stacked plate perfectly copies its predecessor. The die is capable of adjusting its radius of curvature with a peak to valley figure error smaller than 2 μm in order to match the change in figure of the top plate. With each bonded plate, the radius of curvature of the next plate is diminished by the thickness of the plate. Repeating the stacking process creates a stiff, monolithic and lightweight silicon block, called a stack, with hundreds of pores and an open area ratio larger than 65 %. The final optic is an arrangement of the individual modular blocks that is within itself a X-ray optic.

During the stacking step, the plates get precisely aligned to less than 5 µm of the target alignment position to match the rib locations on each plate. This alignment gets monitored and controlled with a high-resolution camera system and image recognition software as well as laser sensors. In addition to controlling the stacking force, we can measure in-situ the pressure distribution with spatial resolution in the millimeter range when bonding two plates together. Figure 4 shows an example of a plate's pressure distribution during the bonding process.

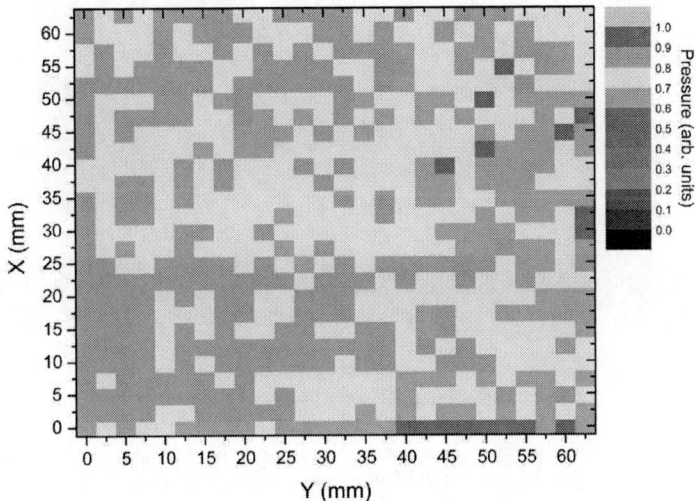

Figure 4. Pressure distribution map when bonding two plates together.

The stacking is performed from outer radii inwards, thus always exposing the last integrated mirror surface to the surface metrology system. We have developed a method based on a fringe reflection technique (FRT) [12] that measures highly-reflective free form surfaces with an accuracy better than 1 ". Thereby, local surface deviations are resolvable to within ± 100 nm and global radii to within ± 250 µm of their specified values. Figure 5 depicts a sketch of the FRT system.

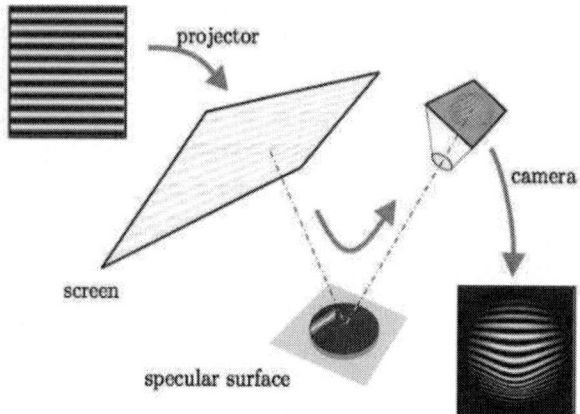

Figure 5. The FRT system is composed of a TFT screen on which fringes are displayed and a camera which records distortions of the fringes by the object. An algorithm calculates the figure of the object from the fringe distortions.

With this surface metrology, it is possible to control and monitor deviation from the desired global figure of the stack but also particular contamination. Particles trapped in between the bonded plates are detected as residual figure errors as shown in the left residual height map in Figure 3. These particles deform silicon plates locally and transfer the distortion to the next plate. Smoothening can be observed with higher stacked plates if these distortions are in the middle of the plate. But, they tend to grow in size if they are located along the edges of the plate. The right image in Figure 6 shows stack with a lot less particular contamination. In both images, the elevated parts along the edges of the plates in y-direction disclose not perfectly bonded areas of the plate. The faintly visible vertical lines are caused by the 1 mm-wide membranes sagging between the ribs. Thereby, the shape of the membranes slightly deviates from the ideal shape.

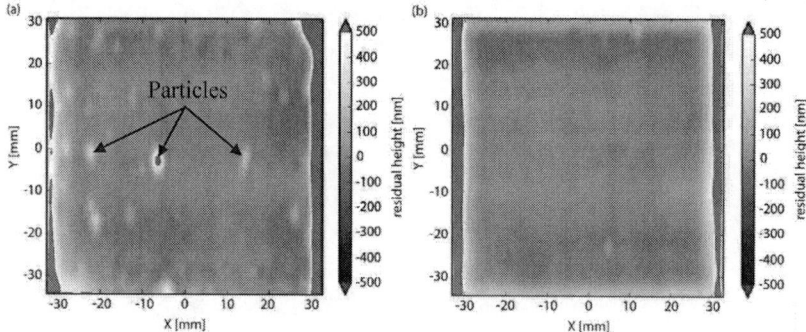

Figure 6. The images present residual height maps calculated from FRT data for stacked and bonded silicon plates with sizes of 65.7 x 65.7 mm² and bending radii in x-direction

of 0.74 m. (a) shows a height map with particles trapped at the bondable areas between two silicon plates protruding through the top silicon plate. (b) depicts a stack with little particular contamination.

After assembly, we characterize the X-ray reflectivity of SPO with an X-ray beam at a synchrotron facility. By scanning the plate with a pencil beam, aligned to be at grazing incidence angle with respect to the surface along the pores and recording the position of the reflected spot, a complete map of local slope errors on the stack is created. Currently, we are developing a method to directly relate the local slope error map measured with X-rays to the height profiles measured with the fringe reflection technique.

Conclusion

In this paper, we presented and discussed the status of the technology development for the production of Silicon Pore Optics. Regarding the rectangular structured silicon plates, we had to overcome many technological challenges compared to using round wafers. We have successfully developed a process to stack and directly bond structured silicon plates in curved configuration with a high-bond yield. The developed tools and processes demonstrate our capability to provide unique solutions to demanding challenges. In view of other possible and attractive applications ranging from material science to medical diagnostics and security equipment, the presented technology can be transferred to any silicon based wafer bonding technology, e.g. MEMS and microfluidic devices.

References

1. M. Bavdaz, et al., *Progress at ESA on high energy optics technologies*, Optics for EUV, X-Ray, and Gamma-Ray Astronomy, vol. 5168 of Proceedings of SPIE, pp. 136–147, San Diego, Calif, USA, (2003).
2. M. Beijersbergen, et al., *Development of X-ray pore optics: novel high-resolution silicon Millipore optics for XEUS and ultra-low mass glass micropore optics for imaging and timing*, Design and Microfabrication of Novel X-Ray Optics II, vol. 5539 of Proceedings of SPIE, pp. 104–115, Denver, Colo, USA, (2004).
3. M. Beijersbergen, et al., *Silicon pore optics: novel lightweight high-resolution X-ray optics developed for XEUS*, Proc. SPIE 5488, 868-874, (2004).
4. R. Günther, et al., *Production of silicon pore optics*, Proc. SPIE 6266, 626619, (2006).
5. M. J. Collon, et al., *Silicon pore optics for astrophysical X-ray missions*, Proc. SPIE 6688, 668813, (2007).
6. M. J. Collon, et al., *Silicon pore optics for astrophysical X-ray missions*, Optics for EUV, X-Ray, and Gamma-Ray Astronomy III, vol. 6688 of Proceedings of SPIE, San Diego, Calif, USA, (2007).
7. M. Bavdaz, et al., *X-Ray Pore Optics Technologies and Their Application in Space Telescopes*, X-Ray Optics and Instrumentation, (2010).
8. M. J. Collon, et al., *Stacking of Silicon Pore Optics for IXO*, Proc. SPIE 7732, 77321F (2010).
9. R. Gunther, et al., *Production of silicon pore optics*, Space Telescopes and Instrumentation II: Ultraviolet to Gamma Ray, vol. 6266 of Proceedings of SPIE, Orlando, Fla, USA, (2006).

10. M. J. Collon, et al., *Design, fabrication, and characterization of silicon pore optics for ATHENA/IXO*, Proc. SPIE 8147, 81470D (2011).
11. Q.-Y. Tong, et al., *Semiconductor Wafer Bonding: Science and Technology*, John Wiley & Sons, New York, NY, USA, (1999).
12. W. Juptner, et al., *Sub-nanometer resolution for the inspection of reflective surfaces using white light*, Proc. SPIE, 7405 (2009).

High Efficiency Cleaning Processes for Direct Wafer Bonding

D. Dussault[a], J. Rothballer[b], F.Kurz[b], M. Reichardt[b], and V. Dragoi[b]

[a] ProSys Inc., 1745 Dell Av., Campbell, California, 95008, USA
[b] EV Group, DI E. Thallner 1, 4782 – St. Florian am Inn, Austria

Substrate cleaning is a very important process step in direct wafer bonding. Current work describes development, testing and verification of a single wafer megasonic cleaning method utilizing a transducer design that meets the extreme particle neutrality, Particle Removal Efficiency (PRE), and repeatability requirements of production scale wafer bonding and other applications requiring extremely low particle levels. The results were obtained using 300 mm diameter Si wafer which were processed as received, without any wet bench cleaning process. These experiments simulated real case production scenario in which the particle counts on incoming wafers are typically 0.1 LPD/cm^2 and lower.

Introduction

Direct or fusion wafer bonding has been adopted by the semiconductor industry since the early 1990s'. Used initially for Silicon-on-Insulator (SOI) manufacturing (1), direct bonding emerged as a powerful technology in various wafer-level 3D integration applications as Backside Illuminated CMOS Image Sensors (BSI CIS) (2).

The future of direct bonding looks very promising as processes based on this technology are foreseen to bring unique benefits for fabrication of complex architectures (e.g. for DRAM applications) either as low temperature direct bonding (3) using Via-last process flows or used for metal/dielectric hybrid wafer bonding process (4).

The direct wafer bonding process is extremely sensitive to particle contamination so cleaning is a very important process element. Besides the standard wet bench cleaning procedures typically available in semiconductor fabs, a single wafer cleaning step is usually introduced in the direct bonding process flow prior to bringing the two substrates in contact. This single wafer step is meant to remove any airborne and other residual particles which may be deposited on the substrate surface during handling from the cassette to the bonding process station. This cleaning step typically utilizes deionized water (DIW) as the cleaning medium, the use of other chemicals is restricted in most cases due to possible reaction with the substrate surface and interference with the bonding mechanism (direct wafer bonding is based on molecular bonds formation).

Different methods and mechanisms can be employed for such a cleaning step: brush scrubbers, high pressure and binary nozzles, megasonic nozzles and transducers among others. As the goal is to perform a final cleaning on substrates exhibiting very low particle count when introduced to the bonding equipment, the quantification of the cleaning result cannot be evaluated using the standard particle removal efficiency (PRE) calculation and comparison model A typical specification for incoming wafers is to show less than 0.1 LPD/cm^2, with minimum particle size of 90 nm and often the cleaning results are specified in terms of "no added particles".

In prior work the use of a large area megasonic transducer (the MegPie® from Prosys Inc.) was successfully demonstrated as being compatible with this type of process by documented high levels of PRE and particle neutrality processing intentionally contaminated test substrates (contamination with silicon nitride particles) (5).

In this current work paper we demonstrate the use of the MegPie® transducer for pre-bond cleaning of wafers with low levels of particles, simulating real Mass Volume Production MVP process conditions in an Automated MVP Production tool. Experimental data will show the compliance of the method with the demanding specifications of fusion wafer bonding.

Experimental

Si wafers, 300 mm diameter and (100) orientation were used as test materials. The wafers were used "as-received" from vendor, without using any wet bench cleaning. The wafers were measured for particles before and after cleaning using an SP1 particle counter from KLA Tencor. Some of the wafers were plasma activated: after plasma activation the surface becomes more reactive and the probability of trapping airborne particles is higher. Wafer processing and handling were performed in an EVG Gemini®FB fully automated wafer bonding system equipped with class 1 mini-environment (fan-filter unit ensuring class 1 environment inside the equipment).

Two types of MegPie® transducers setups were tested: a simple setup using single point fluid dispense, and an enhanced setup using an additional manifold allowing for better fluid distribution and increased rotation speed during cleaning.

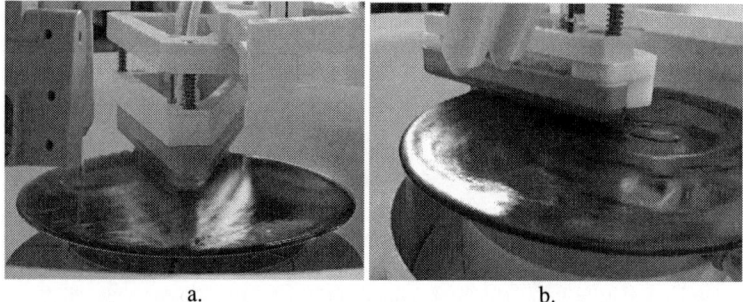

a.   b.

Figure 1. a. Standard MegPie® setup using single point fluid dispense, and b. enhanced MegPie® setup using a manifold for fluid dispense (5).

The enhanced manifold version of the MegPie® utilizes the identical sapphire transducer and control electronics, but adds a PTFE fluid manifold to the leading edge of the transducer. The manifold dispenses the process fluid directly into the gap between wafer and transducer allowing more uniform fluid distribution in the active area as well as allowing higher speed rotation of the substrate and requiring less fluid than the single point of dispense. The higher RPM provides for more centrifugal force during the process enabling a more positive, accelerated particle removal path away from the substrate, therefore a shorter cleaning time (6).

In establishing the Design of Experiments (DOE) for this production simulation test series, we chose a three tier approach. The first portion of the DOE dealt with establishing a particle baseline for this production sequence, including wafer transport, DIW and wafer conditions as well as measurement repeatability and accuracy. The stability of particle performance was established and found to be acceptable and stable. We proceeded to the second tier of our DOE, the addition of the MegPie® system to the cleaning process, first through Best Know Method (BKM) recipe and then with planned modifications to BKM parameters in order to improve performance and widen the process window without extending the budgeted overall process time for this step. The parameter modifications included RF power, spin speed, fluid dispense sequences and timing as well as switching to manifold dispense option from the single point dispense.

The final tier included the completion of a plasma activation series under different conditions and testing to see the possible impact on cleaning performance: for this test the wafers were plasma activated and handled manually before being placed in the fully automated equipment for cleaning. The meaning of this test is to prove the cleaning efficiency under the circumstances of a modified surface condition (higher reactivity due to plasma activation may change the nature/strength of the forces responsible for holding the particles on wafer surface) which may impact on the cleaning process by the amount of particles trapped on the surface during the manual handling. The experimental results obtained are presented in next section.

Results and Discussion

At the start of tier one of this DOE, the particle stability of the complete HVM production system was tested. Pre-Test A was designed as a handling baseline test: the wafer is scanned, then loaded to pre-aligner, then transferred to cleaning station spinner (no water, no rotation), then unloaded, and transferred back to measurement. Pre-Test B is primarily a DIW neutrality test: after initial measurement the test wafer is loaded to pre-aligner, then to cleaning station spinner (15s center water dispense, no rotation), then unloaded and re-measured.

TABLE I shows the particle levels for the handling of the wafer through the system (no cleaning process – Pre-Test A), and deionized water source particle neutrality (just water dispense, no cleaning – Pre-Test B).

TABLE I. Particle Tests Performed for the Handling Process and for the Deionized Water Line.

Test	Wafer Run	Pre- Particle Count >90 nm	Post - Particle Count >90 nm	Particle Difference (before-after) >90 nm
Pre-Test A	1	12	20	8
	2	12	23	11
	3	18	16	-2
	4	10	11	-1
Pre-Test B	15	24	17	-7
	16	16	12	-4
	17	16	14	-2
	18	22	16	-6
	19	15	12	-3

A system baseline for particle neutrality was established in the Tier 1 experiments (results in Table 1) and was found to be stable and repeatable with DIW. It was then decided to proceed to tier two cleaning experiments.

TABLE II shows a summary of cleaning results by varying MegPie® RF power density and fluid introduction method. As an important remark, in the Particle Difference column a negative value represents the amount of particles removed, while a positive value is considered to represent adders.

TABLE II. Overview of the Particle Removal Performance of Megpie for Various Power Densities.

Test	Wafer Run	Pre- Particle Count >90 nm	Post - Particle Count >90 nm	Particle Difference (before-after) >90 nm	RF Power Density W/cm²
MegPie®1	1	26	16	-10	0.38
	2	14	13	-1	0.38
	3	27	28	-1	0.38
MegPie®2	4	11	10	-1	0.54
	5	9	12	-3	0.54
	6	15	17	-2	0.54
MegPie®3	7	10	7	-3	1
	8	30	27	-3	1
	9	10	10	0	1
MegPie®4	10	41	30	-11	1.46
	11	26	20	-6	1.46
	12	49	34	-15	1.46
DI rinse1	13	14	2	-12	0
DI rinse2	14	32	18	-14	0
Handling check	15	18	14	-4	0
MegPie®6	16	21	20	-1	1
	17	19	14	-5	1
	18	18	8	-10	1
	19	24	11	-13	1
	20	129	28	-101	1
	21	107	40	-67	1
MegPie®7	22	20	18	-2	1
	23	14	14	0	1
	24	8	7	-1	1
	25	11	13	2	1
	26	28	11	-17	1
	27	40	11	-39	1
DI rinse/ manifold	28	28	21	-7	0
	29	22	19	-3	0

In the tests MegPie® 1 – 4, the BKM recipe parameters had already been time optimized, and only the transducer power density was varied. For these four tests the deionized water was dispensed in a single point. The analysis of SP1 measurement results indicated that 1.0 W/cm² RF power setting provided the best particle performance in the broadest range of particle sizes. This power setting was used for the balance of the experiment. The tests DI rinse 1 and 2 as well as Handling check used the same parameters as tier one testing described above, and provided a successful cross check of baseline results achieved in the first test series.

The tests MegPie® 6 and 7 involved the processing and reprocessing of the same 5 wafers. In both tests the process recipe parameters were identical to those of the optimized BKM with the only exception of rotation speed and fluid dispense procedure. For this test series the cleaning spin speed was 100RPM and the fluid dispense was through the manifold. The particle removal data indicates a clear trend in particle removal and particle neutrality within baseline. The DI Rinse/Manifold tests were added to verify the neutrality of the alternate fluid dispense method.

As a final test the cleaning process was applied for wafers simulating real plasma activated bonding process. In this process the plasma activation produces a surface change allowing for a specific bonding mechanism at low temperature. As a drawback, the increased surface reactivity after plasma activation may increase the risk of trapping airborne particles on the wafers' surfaces in case the environment is not very clean. Fully automated bonding equipment is preventing such situation as it includes a fan-filter unit which ensures class 1 environment inside the equipment. However, in order to check the impact of the plasma activation on cleaning efficiency the cassettes with activated wafers were removed from the automated equipment, the wafers were manually handled out of the cassette, then placed back into their cassettes and further processed in the automated equipment for cleaning: with this it was intended to increase the risk of particle contamination by manual handling in class 100 environment, which is definitely a "worst case scenario" compared to the clean class 1 environment inside the Gemini®FB equipment. The plasma process of choice was maintained constant in terms of RF conditions but the time was increased to double than usual (30 seconds as baseline, 60 seconds for cleaning test purpose).

TABLE III shows an overview of the cleaning results obtained for four different plasma activation times, while maintaining the MegPie power density to 1 W/cm^2 at 100 RPM and identical enhanced BKM from tier two testing.

TABLE III. Overview of the Particle Removal Performance of Megpie for Various Plasma Process Times

Test	Wafer Run n	Pre-Particle Count >90 nm	Post-Particle Count >90 nm	Particle Difference (before-after) >90 nm	RF Power Density W/cm^2	Plasma Activ. Time s
MegPie®1	1	10	2	-8	1	30
	2	12	14	2	1	30
	3	17	14	-3	1	30
MegPie®2	4	17	13	-4	1	60
	5	13	8	-5	1	60
	6	29	16	-13	1	60
MegPie®3	7	37	22	-15	1	30
	8	18	14	-4	1	30
	9	33	29	-4	1	30
MegPie®4	13	2	6	4	1	30
	14	14	11	-3	1	30
	15	14	11	-3	1	30

The cleaning process results in TABLE III indicate particle neutrality within baseline parameters and a clear indication of particle removal in "worst case scenario" process conditions.

Conclusion

The single wafer pre-bond cleaning step in the direct wafer bonding process flow is primarily a safety step that should be capable of removing particulate of unknown type or source, if removable, and accomplish this using only DIW and not add particles during this sequence. We have demonstrated that these requirements can be met through use of the sapphire MegPie® and manifold in real life MVP conditions.

References

1. M. Bruel, *Nucl. Instr.Methods Phys. Res.B*, **108**, 313 (1996).
2. V. Dragoi, G. Mittendorfer, A. Filbert and M. Wimplinger, in *Advanced Interconnects and CMP for Micro- and Nanoelectronics/2010*, C. Bonafos, Y. Fujisaki, P. Dimitrakis and E. Tokumitsu, Editors, PV 1249, p. 1249-F08-06, MRS Proceedngs Series (2010).
3. T. Plach, K. Hingerl, S. Tollabimazraehno, G. Hesser, V. Dragoi and M. Wimplinger, *J. Appl. Phys.*, **113**, 094905 (2013).
4. P. Enquist, *Sens. and Mat.*, **17**(6), 307 (2005).
5. D. Dussault, F. Fournel and V. Dragoi, *Solid State Phenom.*, **187**, 269 (2012).
6. D. Dussault, F. Fournel and V. Dragoi, *ECS Tans.*, **50**(7), 41 (2012).

High Precision Low Temperature Direct Wafer Bonding Technology
for Wafer-Level 3D ICs Manufacturing

F. Kurz, T. Plach, J. Süss, T. Wagenleitner, D. Zinner, B. Rebhan and V. Dragoi

EV Group, DI E. Thallner 1, 4782 – St. Florian am Inn, Austria

The development of low temperature direct wafer bonding
processes paved the way for new categories of applications based
on semiconductor devices. Precise optical alignment of wafers
prior wafer bonding plays a key role in manufacturing of current
and future applications based on wafers stacking. In order to
address the continuous feature size shrinking and increasing
integration levels the need for high alignment accuracy imposed
significant hardware and process improvements. The future
microelectronics applications are foreseen to require wafer-to-
wafer alignment accuracy as low as ±100 nm and better.
This work reviews the main contributors to the misalignment
budget and presents experimental alignment results for alignment
accuracy in the range of 50 – 100 nm.

Introduction

Wafer bonding techniques faced a significant development over the past decade, mainly
driven by applications used in consumer electronics. Used for fabrication of various
sensors, memories, miniature cameras, wafer bonding is nowadays a mature technology
offering unique solutions in terms of materials/device integration.

Wafer bonding technology was first used in semiconductor industry for Silicon-on-
Insulator (SOI) substrates manufacturing (1). This process flow was based on direct
(fusion) wafer bonding of two blank substrates and the main requirement for the wafer
bonding process was related to contamination levels (particles and particularly metal
traces), which had to be compatible with the allowed contamination levels in CMOS
technology.

In the recent years wafer bonding applications increased significantly in complexity
and transitioned from bonding two or more non-patterned substrates to bonding multiple
fully-patterned wafers in complex optically-aligned wafer bonding process flows.

The development of low temperature direct bonding processes in early 2000's
enlarged significantly the applications range for wafer bonding as it allowed for bonding
of patterned wafers (e.g. fabricated in CMOS technology) and for building complex
microelectronic structures (2). The optical alignment technology followed rapidly the
new developments as it allowed for multiple wafers stacking and fabrication of real 3D
structures.

The features sizes on the two substrates to be bonded essentially determine the
needed bond alignment accuracy (fig.1). It can be observed that for the three
interconnects dimensions used as reference the required alignment accuracy for a pads
overlap of >80% has to be in the range of ±0.2 µm and better.

Thus, the accuracy of the wafer-to-wafer optical alignment is a key enabling factor allowing for high density of interconnects. The state-of-the-art equipment currently available provides alignment accuracy significantly better than ±0.5 µm. Additional improvements had to be implemented in order to further increase alignment precision and to reach values lower than ±0.2 µm: wafers quality (significant decrease of all patterning tolerances), aligner hardware and alignment processes had to be reviewed based on new criteria, which for alignment accuracy in the range of ±1 to ±5 µm were negligible, but for sub-micrometer range became critical. Such alignment performance is used in conjunction with low temperature plasma activated bonding (3) or with dielectric-metal hybrid wafer bonding (4) for wafer-level 3D interconnects. Fusion wafer bonding is a two-steps-process where wafers are firstly pre-bonded at room-temperature and then further processed in a subsequent thermal annealing step. Using this process, high precision alignment at room-temperature is fixed and wafers are bonded. While pre-bonding is based on reversible, hydrogen bridge bonds, thermal annealing results in the formation of permanent, covalent bonds.

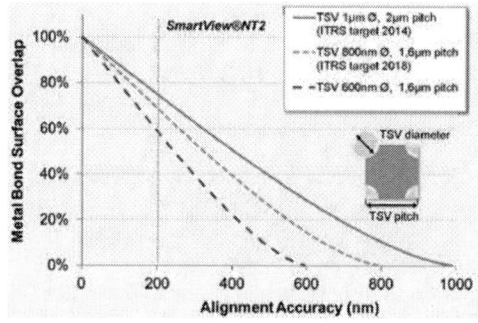

Figure 1. Calculation of the metal pads overlapping area as a function of alignment accuracy for different ITRS roadmap-relevant size/pitch of the electrical interconnects (5).

All other wafer bonding processes are typically adding to the optical aligner accuracy a misalignment budget of 1 – 5 µm, depending of the type of process (impact factors on alignment: compression force, temperature value and heating/cooling ramps) or the type of interface (rigid or fluid/compressible).

This paper aims to review the main factors influencing the optical alignment accuracy for the sub-micrometer accuracy range and is presenting experimental results showing the current performance of optical alignment in terms of accuracy and repeatability.

Experimental

Basics

Optical alignment of the two wafers is performed right prior to placing the wafers in contact. At this moment the two wafers are already cleaned and their surfaces are plasma activated. For the highest alignment accuracy the wafers are pre-bonded directly inside the optical aligner.

Generally there are two types of optical alignment processes: direct alignment (if at least one wafer is transparent to light - visible or IR - the alignment keys from both wafers are observed in live mode) and indirect alignment methods (wafers are not transparent and one wafer is aligned using as reference the digitized image of second wafer).

The most used wafer-to-wafer optical alignment methods and their main features are summarized in Table I.

TABLE I. Optical wafer-to-wafer alignment methods

Item	Direct Alignment		Indirect Alignment	
	Visible	Reflective/ Transmission IR	Backside	SmartView® (Face-to-Face)
Live image	Yes	Yes	No	No
Accuracy	±0.5 µm	±0.5 µm	±1 µm	< ±0.2 µm
Substrate	1 wf. transp.	1 wf. IR transp.	Any	Any
Alignment keys	Wf. 1 Front + Wf. 2 Front	Wf.1 Front + Wf. 2 Front	Wf. 1 Front + Wf. 2 Back	Wf. 1 Front + Wf. 2 Front
Limitations (general)	Minimum 1 glass wafer	- Low doping - Backside polished - No metal close to keys - Image quality decreases for multiple wafers stacks	Misalignment budget front-to-back keys	None

The test materials used for this work were patterned 300 mm diameter Si wafers, (100) orientation. The experimental work described in this paper was performed using an EVG SmartView®NT2 optical aligner (6, 7) integrated in a fully automated production bonder EVG Gemini®FB. The process flow in this equipment is the following:

- The two wafers are first plasma activated (3);
- After surface activation both wafers are cleaned with deionized water using megasonic-enhanced process (8)
- The two wafers are loaded face-to-face into the SmartView® aligner. After the alignment performed using the pre-defined alignment keys the two wafers are placed into contact (pre-bonded).
- After pre-bonding the alignment performance is evaluated inside the Gemini® FB by an alignment verification module (AVM) and the misalignment-data is sent in a feedback loop to the SmartView® aligner to be used for further adjusting the alignment settings for the next alignment processes.
- The pre-bonded wafers pairs are unloaded from the alignment verification process station and transferred to an oven for a low temperature thermal annealing (typically ~300°C for 1-2 hours).

- In case the alignment does not fulfil the specified alignment accuracy values the pre-bonded pairs can be transferred in another optional process station where they are separated and reworked (cleaned, aligned and pre-bonded, no need for additional plasma activation).

Misalignment Sources

Starting from the premises that the equipment is fully functional according to its specifications, three types of misalignment errors can be mentioned (fig. 2): i) translation (misalignment in x-, y-, or x- and y-direction), ii) rotation (misalignment due to rotation of the patterns at wafer level)), and iii) run-out misalignment (mismatch of global alignment due to scaling factors).

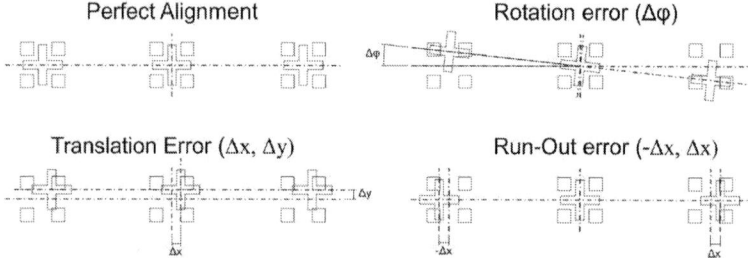

Figure 2. The main optical alignment errors (misalignment types).

In fig. 2 can be observed that translation and rotational errors can be easily compensated by the equipment by adding into the settings an offset to compensate for the misalignment. However, when it comes to run-out errors (fig. 3), they cannot be compensated by conservative aligners and require special measures: one has to differentiate between the different types of run-out errors based on their root causes and implement various methods of compensation.

A run-out error is shown as an increasing misalignment between the alignment targets from the two wafers from center towards the edge when the centers of the two wafers are perfectly aligned.

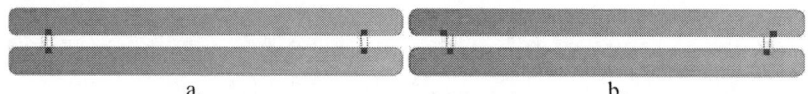

Figure 3. a. Perfect alignment with no misalignment error, and b. wafers showing run-out misalignment error.

The different root causes for run-out errors are the following:
- Incoming wafers patterning: due to the complex patterning processes a certain intrinsic run-out of the patterns can be induced (fig. 4a).
- The bow/warp of one or both wafers may result in run-out misalignment (fig. 4b).

- The different temperature of two identical wafers showing no intrinsic run-out may result in a run-out misalignment due to the absolute thermal expansion of the two wafers at the moment when they are placed into contact (fig. 4c).
- The bonding process condition may induce a run-out error due to the deformation of the wafers during bond propagation process (fig. 4d).

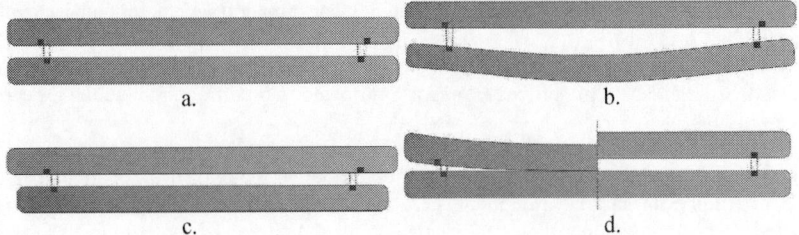

Figure 4. Schematic illustration of the run-out errors based on their root cause: a. intrinsic run-out (produced by patterning processes), b. bow/warp-generated run-out, c. thermal mismatch-induced run-out, and d. bonding process-induced run-out.

From the run-out root causes listed above can be observed that they are very different in nature so a complex solution has to be adopted to compensate for all these factors.

Results and Discussion

Experiments were performed using the Gemini®FB equipment with the target to demonstrate alignment accuracy lower than ±100 nm and the reproducibility of the results.

First, the optical aligner hardware in contact with the wafers was redesigned according to the targets. The experiments were performed using 300 mm diameter Si CMOS wafers processed with the process flow shown schematically in fig. 5.

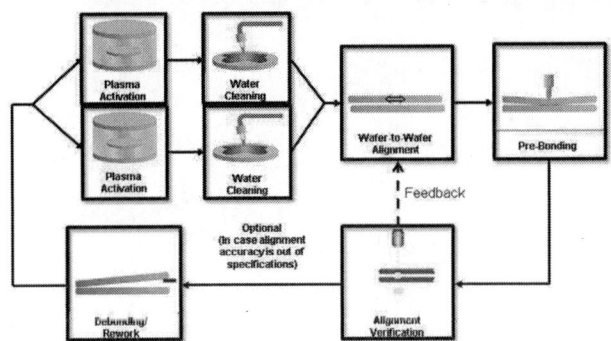

Figure 5. Schematic process flow of the aligned wafer bonding experiments.

Pre-bonding process, together with hardware optimization, were the measures considered for minimization of the run-out misalignment error in this work.

With respect to alignment accuracy evaluation, one has to distinguish between two different ways to measure/quantify the wafer-to-wafer alignment accuracy: measurement in a limited number of locations (typically close to wafer edges in two points at the locations of the two alignment keys used for the alignment process) and the alignment performance across the entire wafer. The wafer-level evaluation method is considering and compensating all disturbing factors (e.g. run-out errors) and is the most relevant measurement of the alignment results as it has a clear impact on yield and subsequently on device performance.

Besides the hardware optimization, the process sequence was fine-tuned and particularly the pre-bonding process step was optimized in order to minimize the process-generated run-out.

The process-generated run-out is mainly rooted in the deformation of the wafers during the pre-bonding process (fig. 6). In direct bonding, the two surfaces are placed in close proximity during the alignment step. After alignment completion, the bond is initiated at a single point, typically located at wafer center or close to wafer edge, depending on the particular bonding scenario. During the bond initiation the van der Waals forces are responsible for the establishment of the bond between the two surfaces at the initiation point: from here the bond is self-propagating until filling entirely the wafers' surfaces. The speed of the bond front is usually given by the surface quality (e.g. a very low microroughness would allow for high speed) and, of course, by the nature of the bonds established between the two surfaces. For example, a silicon oxide surface (native, thermally-grown or CVD-deposited oxide) showing hydrophilic behavior can form different types of bonds when placed in contact with a similar surface: due to the surface nanometer-scale topography there are silanol – to – silanol groups bonds but also bonds between silanol groups and water molecules adsorbed on the surface. A high degree of moisture on the bonding surfaces (e.g. few water monolayers) would result in an increased number of weaker bonds and subsequently in a slower bond front propagation and lower pre-bonding bond strength. Also gas molecules trapped on the two surfaces have to be removed prior the van der Waals forces can force the localized contact of the two surfaces (9). When the same hydrophilic surfaces are placed in contact under vacuum environment a certain surface dehydration/other gas molecules removal occurs and this results in an increase of the number of stronger bonds and subsequently in a faster bond front propagation and higher pre-bonding bond strength (9, 10).

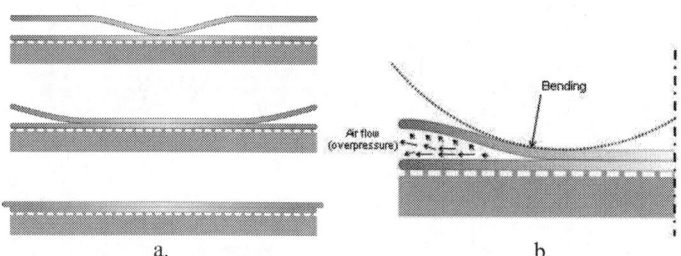

a.   b.

Figure 6. Schematic illustration of the process-generated run-out alignment error: a. run-out generated due to stretching of the top wafer with bottom wafer fixed on the chuck, and b. close-up image showing the deformation of the top wafer under the action of the van der Waals forces responsible for the bond propagation across the wafer after the initiation in a single point.

During the bond front propagation the wafer/wafers suffer a certain deformation. The process sequence was optimized in accordance with the used hardware for an optimum control of wafers' shape during pre-bonding: the bond front speed is not directly responsible for the run-out generation but the wafers deformation was found to be a significant factor. The gap between the wafers at the pre-bonding initiation moment is responsible for the initial deformation of the wafers and thus becomes an important factor in generating misalignment and , more specific, run-out errors.

In a first step of development the target was to achieve an overlay alignment within ±200 nm: after achieving this intermediate development stage the technology was further optimized in order to reach the final target of ±100 nm. Figure 7 shows results on run-out optimization process development.

The measured data show good results with the ultimate development step reaching the targeted specification of ±100 nm. In fig. 7 can be observed that the alignment accuracy is within the targeted specifications in all measured points.

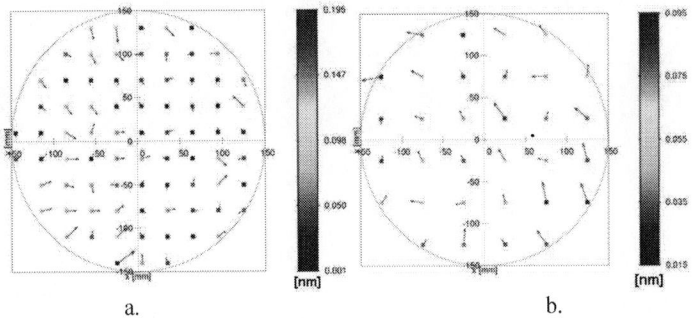

a.   b.

Figure 7. Overlay alignment data after run-out compensation: a. alignment accuracy within ±200 nm, and b. alignment accuracy within ±100 nm.

Figure 8 shows the overlay alignment values measured after run-out compensation for a batch of 25 bonds running in automated mode on the equipment.

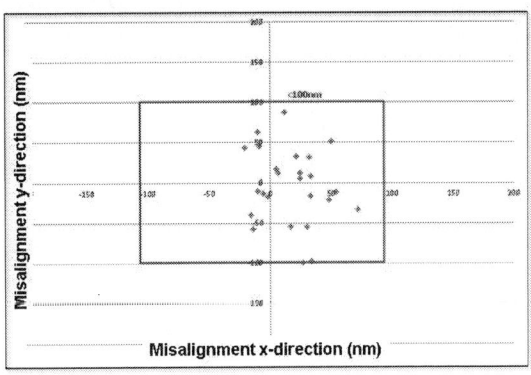

Figure 8. Alignment data measured on two points for a batch of 25 bonded wafers pairs.

It can be observed that all results are within the ±100 nm specified alignment accuracy, proving that the equipment is able to run the specified process.

As an additional technology qualification topic, process reproducibility was tested as this is an important parameter for volume manufacturing. For reproducibility evaluation test batches of typically 10 bonds were running periodically and measured for process reproducibility and stability. Figure 9 is showing an example of process reproducibility tested on a batch of 10 bonds after 2000 consecutive alignment processes: it can be observed that the misalignment for all 10 bonds is within the specification of ±100 nm alignment accuracy.

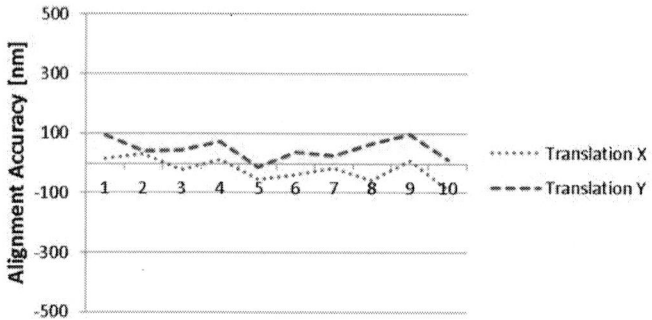

Figure 9. Alignment data reproducibility demonstrated on a batch of 10 bonded wafers pairs: misalignment values on x- and y-direction are within ±100 nm.

Conclusion

The continuous feature size shrinking demands very high alignment accuracy across the entire bonded area, as good as ±100 nm and better.

In order to reach such values, the alignment concept had to be completely revised by considering a number of factors contributing to misalignment, which were ignored for the past application generations, for which the required alignment accuracy was typically ranging from ±1 µm to ±5 µm. The new alignment concept requires significant hardware changes and process optimization in order to deal particularly with the run-out error produced by various factors.

The entire development process started with materials (wafers) optimization, equipment improvement and process optimization. Using a state-of-the-art commercially available fully automated Gemini®FB bonding equipment alignment performance with an accuracy within ±100 nm could be demonstrated, with a reproducibility within the standard specifications of semiconductor manufacturing environment. Such extremely high alignment accuracy can be used for wafer-level 3D applications using TSV processes namely low temperature fusion bonding or metal/isolator hybrid wafer bonding.

References

1. M. Bruel, *Nucl. Instr.Methods Phys. Res.B*, **108**, 313 (1996).
2. V. Dragoi, E. Pabo, J. Burggraf and G. Mittendorfer, *Microsys. Techn.*, 18, 1065 (2012).
3. T. Plach, K. Hingerl, S. Tollabimazraehno, G. Hesser, V. Dragoi and M. Wimplinger, *J. Appl. Phys.*, **113**, 094905 (2013).
4. P. Enquist, *Sens. and Mat.*, **17**(6), 307 (2005).
5. International Technology Roadmap for Semiconductors (ITRS), 2011edition.
6. V. Dragoi, P. Lindner, M. Tischler, and C. Schaefer, *Mat. Sci. in Semicond. Proc.*, **5**(4), 425 (2003).
7. L. Di Cioccio, *ECS Trans.*, **3**(6), 19 (2006).
8. F. Fournel, L. Bally, D. Dussault and V. Dragoi, *ECS Trans.*, **33**(4), 495 (2010).
9. U. Gösele, S. Hopfe, S. Li, S. Mack, T. Martini, M. Reiche, E. Schmidt, H. Stenzel and Q.-Y. Tong, *Appl.Phys.Lett.*, **67** (6), 863 (1995).
10. W.B. Yu, C.M. Tan, J. Wei, S.S. Deng and G.Y. Huang, *J. Micromech.Microeng.*, **15**, 1001 (2005).

354

Bonding of SiO$_2$ and SiO$_2$ at Room Temperature Using Si Ultrathin Film

J. Utsumi[a], K. Ide[a], and Y. Ichiyanagi[b]

[a] Advanced Manufacturing System Development Center, Mitsubishi Heavy Industries
Machine Tool Co., Ltd., Ritto, Shiga 520-3080, Japan
[b] Department of Physics, Graduate School of Engineering, Yokohama National
University, Yokohama 240-0067, Japan

> The bonding of metal electrode and insulator hybrid interfaces is
> very important in three-dimensional integration technology.
> Surface activated bonding (SAB) is expected to be a suitable
> method for three-dimensional integration technology, as the
> bonding method is carried out at room or low temperatures.
> Though metal materials such as Cu or Al are easy to directly bond
> using the SAB method, insulator materials such as SiO$_2$ and SiN
> are more difficult. In this study, we examined the bonding
> technique at room temperature using only Si ultrathin films for the
> insulator material, and investigated the relationship between the
> SiO$_2$/SiO$_2$ bonding strength and the thickness of the Si ultrathin
> film. We confirmed that the surface energy was about 1 J/m^2 for a
> Si film thickness of more than about 3 nm.

Introduction

Three-dimensional integration technology is expected to be a possible solution in the post-Moore era (1,2). This technology is based on the stacking of homogeneous or heterogeneous devices and connecting devices with fine through-silicon vias (TSVs). Thus, the bonding of metal electrode and insulator hybrid interfaces is a critical technique. The hybrid interface serves as both an electrical connection and a mechanical bond. However, the bonding of these hybrid interfaces is a challenging issue. Since the conventional bonding process requires high-temperature annealing to achieve strong bonds (3,4), it is associated with various problems such as thermal damage, low throughput, and low alignment accuracy. As surface activated bonding (SAB) is a bonding method carried out at room or low temperatures (5), it is expected to solve these problems. Using the SAB method, metal materials such as Cu or Al can be directly bonded easily. However, it is very difficult to directly bond insulator materials such as SiO$_2$ and SiN. The bonding of Si/Si using an amorphous Si layer as an adhesion layer was reported (6), and the bonding of Si/Si, SiO$_2$/SiO$_2$, and SiO$_2$/SiN using an Fe nano-adhesion layer was also investigated (7–9).

We have reported on the bonding technique at room temperature using only Si ultrathin films for the insulator materials, and we have shown that a high bonding strength is achieved (10). As the surface of the electrode is also covered by Si film during this bonding method, the Si thin film should be as thin as possible. Thus, it is very important to reveal the influence of the thickness of this Si film on the SiO$_2$/SiO$_2$ bonding. In this study, we have investigated the relationship between the SiO$_2$/SiO$_2$ bonding strength and the thickness of the Si ultrathin film.

Experimental

Silicon blanket wafers (diameter of 8 in. and thickness of 725 μm) with a thermal oxide (thickness of about 300 nm) were prepared for this experiment. Figure 1 shows a schematic illustration of the bonding apparatus (Mitsubishi Heavy Industries, Ltd., MWB-08AX) used in this experiment. Surface activation is carried out by an Ar fast atom beam (FAB). The FAB source generates a neutralized Ar atom beam with a voltage of 1.8 kV and a current of 100 mA. In the normal SAB process, the upper and lower wafers are irradiated with the Ar-FAB at the same time. The procedure for SiO_2/SiO_2 bonding was as follows. The upper wafer was held by the electric static chuck (ESC), and the Si blanket wafer was placed on the lower side as the sputtering target. First, by irradiating only the Si target wafer with FAB1 (first irradiation), a Si thin film was deposited on the upper wafer surface at a rate of about 1.2 nm/min. The background vacuum pressure was about 2×10^{-6} Pa before the first irradiation. The blanket Si wafer was then exchanged with another 8-in. Si blanket wafer with a thermal oxide. Only the upper wafer was irradiated with FAB2 (second irradiation). The Si film surface on the upper wafer was etched by FAB2 at a rate of about 2 nm/min, and then a Si thin film was deposited on the lower wafer surface at a rate of about 0.4 nm/min. The upper and lower wafers were then brought into contact.

The surface energy of the bonded wafer was evaluated by the crack-opening method (11). The bonding strength was also measured by a tensile test. The bonding interface was investigated by transmission electron microscopy (TEM). The elemental composition across the bonding interface has been analyzed by energy-dispersive X-ray spectroscopy (EDS).

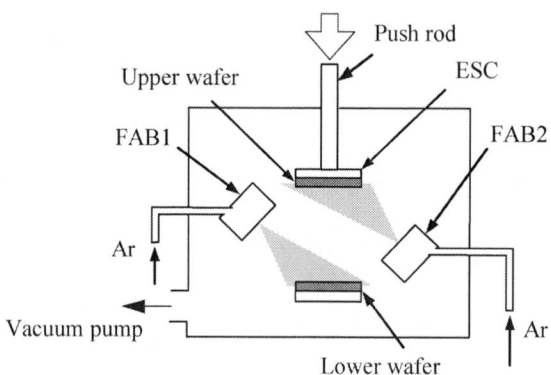

Figure 1. Schematic illustration of bonding apparatus.

Results and Discussion

Bonding Results

Figure 2 shows the surface energy as a function of the Si thin-film thickness. The thickness of the Si film was estimated as the total thickness. The surface energy is about 1 J/m² for total Si film thicknesses of more than about 3 nm. Figure 3 shows an infrared (IR) transmission image of bonded SiO_2/SiO_2 wafers obtained using a first irradiation time (T_1) of 5 min and a second irradiation time (T_2) of 1 min. It seems that the total thickness of the Si film is about 3 nm according to the bonding procedure used in this experiment. No voids were visible in the bonded wafers as shown in Fig. 2.

The bonding strength was evaluated by a tensile test. The specimens for the tensile test were prepared by dicing the bonded wafers into 12×12 mm² sections. Figure 4 shows a top view of the bonded SiO_2/SiO_2 wafers ($T_1 = 5$ min, $T_2 = 1$ min; total Si film thickness of about 3 nm) after the sawing process. All of the pieces withstood the external stress during the wafer sawing process. The bonding over the whole area of the 8-in wafer remained tight. The specimens were glued to metal attachments and mounted on the tensile testing machine. The bonding strength was found to be about 15 MPa, but the specimen fractured at the glue interface and not at the bonded interface (Fig. 5). The tensile tests were also carried out for other specimens that bonded with a Si film thickness of more than about 3 nm. All specimens fractured at the glue interface and not at the bonded interface. This means that the bonding strength of the SiO_2/SiO_2 interface is higher than the measured value.

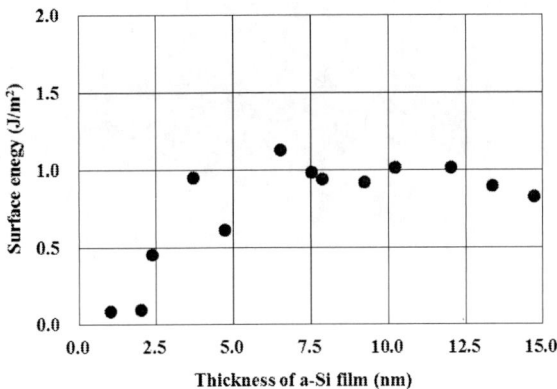

Figure 2. Relationship between the surface energy of the SiO_2/SiO_2 bond and the total thickness of Si intermediate layer.

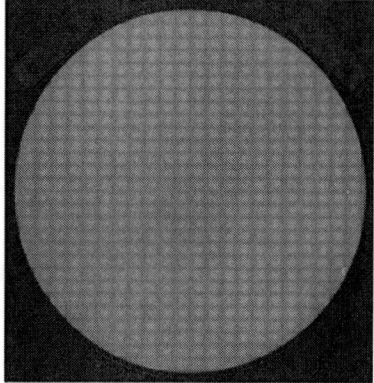

Figure 3. IR image of the bonded 8-in. SiO_2/SiO_2 wafers using an amorphous Si layer. The first irradiation time was 5 min, and the second irradiation time was 1 min. No voids are visible.

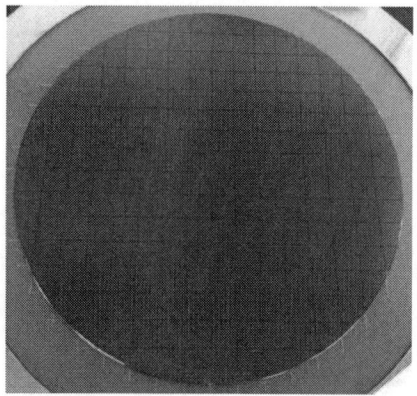

Figure 4. Dicing of the bonded 8-in. SiO_2/SiO_2 wafers shown in Fig. 3 into 12×12 mm^2 sections. No pieces debonded during the wafer sawing process.

TEM Observation and EDS Analysis of SiO_2/SiO_2 Bonding Interface

Figures 6a and 6b show cross-sectional TEM images of the SiO_2/SiO_2 bonding interface in wafers bonded for $T_1 = 5$ min and $T_2 = 1$ min. No microvoids or gaps are observed at the bonding interface, but the intermediate layer is visible. This layer had a thickness of about 3 nm and was assumed to be the Si layer that formed during the bonding process. This thickness value is comparable to the estimated thickness.

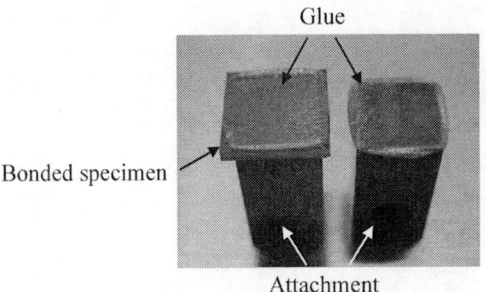

Figure 5. Image of SiO_2/SiO_2 bonding specimen fractured in the tensile test. Fracture occurred from the glue interface.

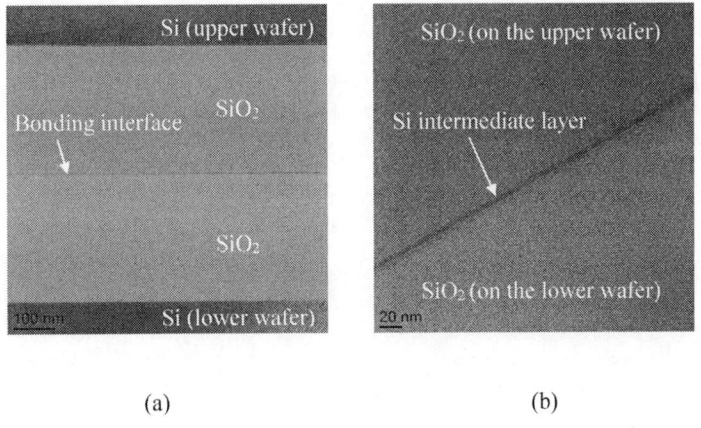

(a)   (b)

Figure 6. TEM cross-sectional images of the bonded SiO_2/SiO_2 interface: (a) low magnification and (b) high magnification.

The elemental composition across the SiO_2/SiO_2 bonding interface was measured by EDS. As the specimens with a Si film thickness of less than 5 nm were damaged during the EDS measurement, the EDS analysis was only carried out for specimens that had a thickness of more than 10 nm across the scanning line shown in Fig. 8a. Figure 8b shows the Si and O concentrations across the interface. Fe and other metal materials were not confirmed either in the intermediate layer or at the interface between the SiO_2 and the layer. We have succeeded in achieving strong SiO_2/SiO_2 bonding using only Si thin layers.

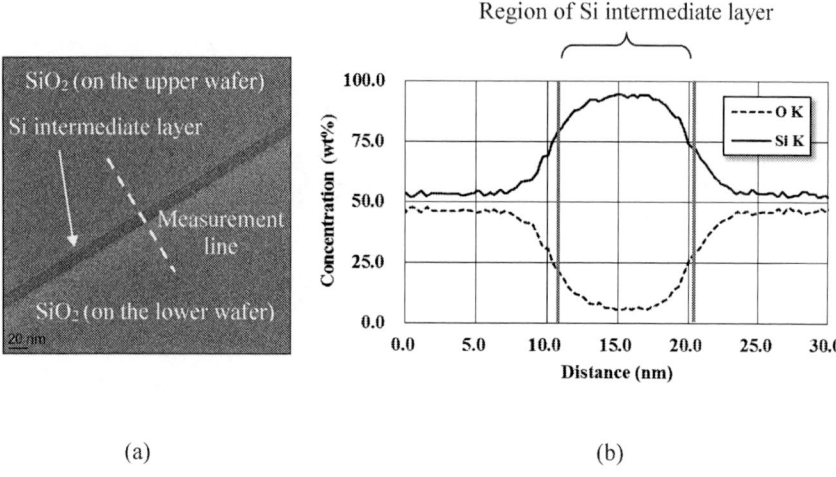

Figure 8. (a) TEM cross-sectional image of the bonded SiO_2/SiO_2 interface for EDS analysis. EDS line analysis was carried along the dashed line. (b) EDS analysis across the SiO_2/SiO_2 bonding interface. The solid and dashed lines are for Si and O, respectively. The red lines indicate the SiO_2–Si interfaces. It was confirmed that there were no metal materials present.

Summary

We have investigated the relationship between the bonding strength and the thickness of the Si intermediate layer in the bonding of SiO_2/SiO_2, which is difficult to directly bond using SAB. It was confirmed by using the crack-opening method that for a Si film thickness of more than about 3 nm the surface energy was about 1 J/m^2. The specimens for the tensile test were prepared by dicing the bonded wafers into 12×12 mm^2 pieces. All pieces withstood the external stress during the wafer sawing process. The whole bonding area of the 8-in. wafer was sufficiently tight. Tensile testing revealed that the bonding strength was higher than 15 MPa. No metal materials were found by EDS analysis either in the layer or at the interface between the SiO_2 and the layer. We showed that a high bonding strength was achieved in SiO_2/SiO_2 bonding using only Si ultrathin films.

References

1. N. Sillon, A. Astier, H. Boutry, L. Di Cioccio, D. Henry, and P. Leduc, *IEDM Tech. Dig.*, 1 (2008).

2. L. Di Cioccio, P. Gueguen, F. Grossi, P. Leduc, B. Charlet, M. Assous, A. Mathewson, J. Brun, D. Henry, P. Batude, P. Coudrain, N. Sillon, L. Clavelier, G. Poupon, and M. Scannell, *Proc. IMAPS*, 477 (2008).
3. A. Fan, A. Rahman, and R. Reif, *Electrochem. Solid-State Lett.*, **2**, 534 (1999).
4. P. Gueguen, L. Di Cioccio, P. Gergaud, M. Rivoire, D. Scevola, M. Zussy, A. M. Charvet, L. Bally, D. Lafond, and L. Clavelier, *J. Electrochem. Soc.*, **156**, H772 (2009).
5. H. Takagi, K. Kikuchi, R. Maeda, T. R. Chung, and T. Suga, *Appl. Phys. Lett.*, **68**, 2222 (1996).
6. T. Shimatsu and M. Uomoto, *Materia Japan*, **49**, 521 (2010) [in Japanese].
7. R. Kondou and T. Suga, *Proc. 61st IEEE Electronic Components and Technology Conf.*, 2165 (2011).
8. R. Kondou, C. Wang, A. Shigetou, and T. Suga, *Microelectron. Reliab.*, **52**, 342 (2012).
9. R. Kondou and T. Suga, *Scr. Mater.*, **65**, 320 (2011).
10. J. Utsumi, K. Ide, and Y. Ichiyanagi, *Jpn. J. Appl. Phys.*, **55**, 026503 (2016).
11. W. P. Maszara, G. Goetz, A. Caviglia, and J. B. McKitterick, *J. Appl. Phys.*, **64**, 4943 (1988).

Author Index

Abadie, K.	203	Fujino, M.	33, 77, 117, 185
Arai, M.	221	Fujioka, A.	53
		Fujita, T.	229
Barriere, N.	331	Fukushima, T.	285
Baumgart, H.	191	Furuna, K.	25
Beaudoin, G.	169		
Besson, P.	247	Gervasoni, M.	241
Bowers, J. E.	179	Gessner, T.	299
Bridoux, C.	129, 145	Girou, D.	331
		Goorsky, M. S.	39, 241
Chan, W.	87	Goto, M.	103
Chang, L.	179	Gunther, R. Sr.	331
Choowitsakunlert, S.	215	Gutierrez-Aitken, A.	87
Chua, S. L.	109		
Collon, M. Sr.	331	Hagiwara, K.	103
		He, R.	117
Davenport, M. L.	179	Heller, M.	229
De Nigris Brandolisi, I.	247	Higurashi, E.	97, 103
Dempwolf, S.	255	Hingerl, K.	15
Dhamrin, M.	25	Hinterreiter, A.	15
Dillmann, H.	137	Hiramoto, T.	103
Dragoi, V.	15, 45, 339, 345	Hirano, H.	291
Dussault, D.	339	Hobart, K. D.	191
		Hodaj, F.	273
Eichler, M.	137, 229	Honda, Y.	103
Elmustafa, A. A.	191	Hoshi, T.	97
Endo, S.	53	Huang, D.	179
Enot, T.	247		
Enyedi, G.	247	Ichikawa, M.	53
		Ichiyanagi, Y.	355
Ferizovic, D.	87	Ide, K.	355
Flötgen, C.	45	Iguchi, K.	77
Fournel, F.	129, 145, 163, 203, 247, 273	Iguchi, Y.	103
		Imbert, B.	273
Froemel, J.	291	Ishida, H.	265

Jensen, G. U.	311	Mohammed, Y.	191
		Monier, C.	87
Kachtouli, R.	247	Montméat, P.	247
Kagiwada, R.	87	Morales, C.	129, 145, 203
Kaneshiro, E.	87	Moriceau, H.	129, 145, 163,
Kashio, N.	97		203
Klages, C. P.	137	Mu, F.	77
Klingner, H.	255, 321	Mukai, T.	53
Knechtel, R.	255, 321		
Kobayashi, K.	215	Nagel, K.	137
Kobayashi, M.	103	Nakamura, E.	87
Koehler, A. D.	191	Nakano, Y.	33
Kononchuk, O.	163	Nakazawa, H.	77
Kosugi, T.	53	Nanba, M.	103
Koyanagi, M.	285	Nishio, Y.	25
Kurashima, Y.	3		
Kurishima, K.	97	Ogashiwa, T.	265, 299
Kurz, F.	339, 345	Oki, A.	87
Landgraf, B. Sr.	331	Pantzas, K.	169
Landru, D.	163	Patriarche, G.	169
Larrey, V.	129, 145, 163	Plach, T.	345
Le Bourhis, E.	169	Poppe, E.	311
Lee, K.	285	Poust, B.	87
Liang, J.	25, 221		
Liao, M.	39	Radisson, D.	145
Lin, N.	87	Razek, N.	45
Liu, C.	291	Rebhan, B.	15, 345
Liu, Y.	153	Reichardt, M.	339
Lumineau, V.	273	Rieutord, F.	129, 145, 163,
			247
Machness, A.	241	Roscher, F.	299
Malik, N.	15	Rothballer, J.	339
Matsubara, M.	25		
Matsumae, T.	191, 197	Sagawa, H.	53
Matsuzaki, H.	97	Saraya, T.	103
Mauguen, G.	145	Sasaki, T.	9
Moe, S. T.	311	Sato, K.	87

Schjølberg-Henriksen, K.	15	Wang, C.	153
Scott, D.	87	Wang, D.	311
Seal, M.	39	Watabe, T.	103
Seifert, T.	299	Watanabe, K.	33
Shigekawa, N.	25, 221	Wiemer, M.	299
Shimatsu, T.	53, 67	Wimplinger, M.	45, 203
Shimizu, S.	221		
Shiratori, Y.	97	Yamada, I.	9
Smorchkova, I.	87	Yamashita, D.	33
Suga, T.	3, 9, 33, 77, 117,	Yamauchi, A.	117
	153, 185, 191,	Yanson, A.	331
	197	Yee, M.	39
Sugiyama, M.	33	Yokoi, H.	215
Süss, J.	345		
		Zeng, X.	87
Takagi, H.	3	Zinner, D.	345
Takagiwa, K.	215	Zoberbier, M.	229
Takahashi, Y.	77		
Takeuchi, K.	185		
Talneau, A.	169		
Tan, C. S.	109		
Tanaka, S.	291		
Tanaka, T.	285		
Tardif, S.	163, 247		
Tedjini, M.	129, 163		
Thai, K.	87		
Toshiyoshi, H.	103		
Toyoda, N.	9		
Uomoto, M.	53, 67		
Utsumi, J.	355		
Vacanti, G.	331		
Vervest, M.	331		
Vogel, K.	299		
Volet, N.	179		
Wagenleitner, T.	345		